MEMCOMPUTING

MemComputing

Fundamentals and Applications

Massimiliano Di Ventra

University of California, San Diego

OXFORD

UNIVERSITY PRESS

OXFORD
UNIVERSITY PRESS

Great Clarendon Street, Oxford, OX2 6DP,
United Kingdom

Oxford University Press is a department of the University of Oxford.
It furthers the University's objective of excellence in research, scholarship,
and education by publishing worldwide. Oxford is a registered trade mark of
Oxford University Press in the UK and in certain other countries

Published in the United States of America by Oxford University Press
198 Madison Avenue, New York, NY 10016, United States of America

British Library Cataloguing in Publication Data

Data available

Library of Congress Control Number: 2021948719

ISBN 978-0-19-284532-0

DOI: 10.1093/oso/9780192845320.001.0001

Printed and bound by
CPI Group (UK) Ltd, Croydon, CR0 4YY

To
Elena, Matteo, and Francesca
undeserved, yet the most beautiful gift of my life

Preface

In questions of Science, the authority of a thousand is not worth the humble reasoning of a single individual.
Galileo Galilei (1564–1642)

MemComputing is a portmanteau word I introduced in a Nature Physics article of 2013[1] to represent '*any* physical system that computes *in* and *with* memory'. In other words, it is a type of computation performed by physical units able to concomitantly *process* and *store* information in the *same* region of space.[2]

However, although the prefix 'mem' stands precisely for *mem*-ory, I do not necessarily mean 'long-term memory' or 'storage'. Rather, I have in mind the following equivalence:

> **Memory is *equivalent* to time non-locality.**

A MemComputing machine then exploits *time non-locality* (the ability to *remember* its past dynamics) to perform the necessary tasks involved, whether this memory is short- or long-term.[3]

As such, the MemComputing paradigm is a *radical* departure from the accepted Turing paradigm of computation, and from its traditional implementation in a von Neumann architecture, where a central processing unit (CPU) is physically distinct from its storage unit.[4] It is also *fundamentally* different from another computing paradigm of current interest: Quantum Computing.

It is loosely inspired by the operation of the brain, although this analogy, at this time, is quite superficial and, as I will discuss in this book, only limited to a few features.

The rationale behind MemComputing is simple, and yet very powerful. Unlike the traditional algorithmic/combinatorial way in which problems are solved nowadays, MemComputing represents them first as physical systems, including those combinatorial and optimization problems that seem difficult to imagine in physical terms. In fact, as I will explain at length, this physical representation can preserve the *digital* structure of the input and output, so that the digital version of MemComputing machines (*digital MemComputing machines* or DMMs), like our modern

[1] M. Di Ventra and Y.V. Pershin, *Nature Physics* **9**, 200 (2013).

[2] MemComputing is *not* synonymous of 'computing with memristive elements' as some authors narrowly imply (see Section 4.5). Of course, as I will show in this book, resistive memories may be of use. However, the notion of MemComputing encompasses much more than resistive memories. In fact, it can be realized with many different physical systems and devices, and requires *active* elements as well.

[3] Time non-locality could be due to any type of physical property of the system, e.g., its spin or charge polarization, atomic configuration, geometric shape, etc. (Pershin and Di Ventra, 2011*a*).

[4] Let me also point out that MemComputing is *not* just 'in-memory computing' (or 'computing in memory'). In fact, the latter typically refers to storing of data in random access memory (RAM), often in a distributed way, across different RAMs of a parallel computer or server. This speeds up the processing of information considerably, compared to processing that requires access of data from a hard disk. However, it still suffers from the limitations of the von Neumann architecture and *does not* fundamentally change the computing paradigm employed. See Section 4.8 for the consequences of taking away the 'with memory' part from MemComputing.

[5]Even though quantum features of the underlying devices used in MemComputing may be of help, or even necessary for their operation, the physical systems I will consider in this book exhibit *non-quantum* dynamics. This means that they do not require the full description of Quantum Mechanics, based on the time evolution of a state vector in a Hilbert space. Rather, the dynamical state of the non-quantum systems representing DMMs resides in a *phase space*, which has a mathematical structure quite distinct from a Hilbert space. Therefore, MemComputing is *fundamentally* different from Quantum Computing both in its *physical* realization and its *mathematical* formulation.

[6]This is somewhat similar to the challenges one faces when explaining Quantum Computing, which is also a physical approach to computation.

Shaded boxed Sections
The reader will notice that I have put a few Sections of the book in shaded boxes. These are Sections that cover quite advanced material, requiring substantial background in specialized fields. As such, I could only provide a cursory introduction to such subjects. However, these Sections can be skipped at first read without (I hope) much detriment to the understanding of the following material. The interested reader can then return later to those Sections, and use them as a starting point to learn more about these advanced topics with the aid of the suggested literature.

---

[7]Unlike quantum computers that *cannot* be simulated efficiently on our modern computers, and thus require a *hardware* implementation to show any advantage.

computers, are *scalable*.

The solutions of the original problem (if they exist) are the (long-time) equilibrium points of an appropriately designed *non-quantum* dynamical system.[5] These dynamical systems can be engineered so that they do not manifest (quasi-)periodic orbits and chaos, and offer several advantages compared to *quantum* systems.

At this point, it shouldn't come as a surprise that I face a considerable challenge in explaining such a topic.

Since efficient computation is our ultimate goal, we want to address a Computer Science audience. However, MemComputing is first and foremost a *physical approach*, hence some background in Physics and Engineering is necessary.[6] To complicate matters further, some Physics aspects that help us understand the operation of these machines—such as (supersymmetric) Topological Field Theory—are not even common knowledge to all physicists, let alone scientists in other fields of research.

Finally, determining whether a MemComputing machine actually computes efficiently requires demonstrating a lot of statements, with specific knowledge of diverse subjects such as Dynamical Systems Theory, Functional Analysis, Algebraic Topology, and even some Differential Geometry. The range of topics needed is thus quite broad. Therefore, to make MemComputing accessible to a wide audience, a compromise is in order.

This book is my attempt to present this new computing paradigm in relatively simple terms, by keeping the Mathematics to the necessary minimum. The reader who wants to delve deeper into the Physics and Mathematics (often quite heavy) behind these machines will be urged to look into published papers, and references therein. However, I hope that the exposition in this book will be much easier to follow than in the technical papers, and thus provide a much-needed introduction to MemComputing accessible to a wide range of scientists, in a single, comprehensive volume.

Since, for practical purposes, we are only interested in scalable machines, most of the book will be devoted to DMMs that map a *finite* string of symbols (such as 0s and 1s) into a *finite* string of symbols, so that we can write the input and read the output with *finite precision*. I will discuss that when implemented in *hardware* these machines can solve complex (non-convex) problems by employing time, energy, and space resources that scale only *polynomially* with the size of the problem.

Quite unexpectedly, however, since these same machines are represented by *non-quantum* dynamical systems, their *ordinary differential equations* of motion can also be efficiently simulated on our modern computers.[7] By 'efficient' simulations I mean those that require time and memory resources of the computer processor employed, that scale only polynomially with input size. I will indeed present in this book several examples of hard problems solved with polynomial resources by simply simulating DMMs on standard processors that we find even in our laptops. In general, these simulations outperform, by orders of magnitude, state-of-the-art algorithms specifically designed to solve those problems. Therefore, these results show that a Physics-inspired paradigm of computation, such as MemComputing, offers advantages that are not easily achieved by traditional algorithmic means.

This, then, begs the following question I often get asked. Since our modern computers are the closest physical realization of a deterministic Turing machine, and since we have solved some instances of hard problems in polynomial time on these computers, does this mean we have answered the famous question of whether NP = P or not?

The answer is a resounding *no*: one cannot prove a mathematical conjecture using numerical simulations! In addition, the demonstrations we have carried out, and I will point out in this book, pertain to the *continuous-time* dynamics, not the *discrete* version one necessarily employs in numerical simulations.[8] Therefore, the resolution of such a question would require more rigorous mathematical work, and has not been settled at the time of the writing of this book.

Of course, the answer to that question is somewhat irrelevant to the *practical* application of MemComputing to hard problems of interest to both academia and industry. What matters in that case is whether MemComputing offers any particular advantage compared to other computational methods. As I will amply discuss in this book, this is indeed the case and has already been demonstrated in several published works.[9]

As a final note, I would like to offer a disclaimer.[10] It is naïve to think that a technical book, like the present one, would be completely free of errors, despite the best efforts of the author. Therefore, if I uncover any errors or confusing statements in the book after its publication, I will post corrections on a link to my website:

> https://diventra.physics.ucsd.edu/

Irrespective, I hope this book will be a useful starting point to think more deeply about the relation between *physical systems* and *computation*, with MemComputing an interesting representative example.

> —*per intellectum et voluntatem*—
> Massimiliano Di Ventra
> Carlsbad, San Diego
> Some time in 2021

Important remarks
The book contains several 'important remarks', highlighted in shaded boxes. Unlike the shaded boxed Sections, I *strongly* suggest the reader to go through these remarks, as they are intended to clarify and stress, in a succinct way, very important points.

Margin notes are not marginal!
I have taken advantage of 'margin notes' (like this one) to define and clarify concepts I use in the main text. This way, the reader (hopefully) should be able to readily understand most of the jargon before an in-depth study of the relevant literature. These notes are typically within ($\pm$) one page away from the place where the concept they expand on is first introduced.

[8]Let us also recall that the famous NP vs. P question has a meaning *only* for Turing machines. MemComputing machines are *not* Turing machines.

[9]As well as in numerous ones performed at the company MemComputing, Inc. (www.memcpu.com).

[10]Not intended though as a justification for my mistakes.

Acknowledgements

Showing gratitude is one of the simplest yet most powerful things humans can do for each other.
Randolph Frederick Pausch (1960–2008)

The work on MemComputing could not have been done without the invaluable collaboration of a few people, and to them goes my heartfelt gratitude.

First of all, I am grateful to Yuriy Pershin with whom I started exploring the very concept of MemComputing at an early stage, and its application to a variety of optimization problems. I am also greatly indebted to Fabio Traversa, with whom I proposed the digital version of these machines, and its practical realization using self-organizing gates. Igor Ovchinnikov has introduced me to the supersymmetric Topological Field Theory of non-quantum dynamical systems. This powerful theoretical approach has provided us with a better physical understanding of these machines. Some of my students, present and past—Sean Bearden, Haik Manukian, Forrest Sheldon, Yan Ru Pei, and Yuanhang Zhang—have done a lot of research on several aspects of MemComputing, adding considerably to its development.

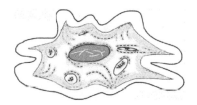

Special thanks to all the amoebas of the world. They have no brain, but their memory has been my original inspiration for MemComputing.

I also want to thank Eric Fullerton, who, through the Center for Memory and Recording Research at the University of California, San Diego, has provided much-needed funding for my research in this field, when I could not get any from either government agencies or private foundations.[11] John Beane has supported our efforts in the field with contagious enthusiasm and continuous encouragement, as well as provided seed funding for the company MemComputing, Inc., we have co-founded (together with Fabio Traversa) and which aims to develop this computing paradigm for industrial applications.

Finally, no words can fully describe the support and encouragement my family has given me during the most difficult times. I therefore express my deepest gratitude to my wife, Elena, and my children, Matteo and Francesca, for their love, support, and patience toward an often absent husband and father.

[11] However, in the past couple of years, I have been very fortunate to find open-minded program directors at the National Science Foundation, the Department of Energy, and the Defense Advanced Research Projects Agency, who have funded different research directions in MemComputing. Of course, these program directors and funding agencies are not responsible for the ideas expressed in this monograph.

Contents

Part I

Preliminary concepts

We are all agreed that your theory is crazy. The question that divides us is whether it is crazy enough to have a chance of being correct.
Niels Bohr (1885–1962) to Wolfgang Pauli (1900–1958)

'Tain't What You Do (It's the Way That You Do It)
'Sy' Oliver (1910–1988) and 'Trummy' Young (1912–1984)

For it is owing to their wonder that men now both begin and at first began to philosophize; ...

In these preliminary Chapters I will try to convey the notion of 'computing' as it is intended in Theoretical Computer Science and 'computing' as it is performed by *physical* objects. I will codify this point by introducing an *equivalence principle* between physical (*semantic*) information and computation. I will then explain how one can exploit physical properties to build *digital* (hence *scalable*) machines, and why *time non-locality* (memory) is such an advantageous feature to solve hard combinatorial optimization problems.

- In the Prologue, I will argue that *any* physical system performs some computation, with us, the observers, the interpreters of such a computation. I will elevate this argument to an *equivalence principle* between physical (*semantic*) information and computation.

- The way in which 'computing' is presently understood (à la Turing) is discussed in Chapter 2.

- In Chapter 3, I will show that *non-quantum* dynamical systems have several advantages compared to quantum systems for what concerns *computation*. I will also discuss the ideal features the *phase space* of dynamical systems ought to have in order to compute combinatorial optimization problems.

- In Chapter 4, I will discuss the usefulness of *time non-locality* (memory) in computing, and will introduce the physical systems (*memelements* and *memprocessors*) I will employ to define Mem-Computing machines.

The Physics of computing and computing with Physics

1

An experiment is a question which Science poses to Nature, and a measurement is the recording of Nature's answer.
Max Planck (1858−1947)

The theory of computation has traditionally been studied almost entirely in the abstract, as a topic in pure Mathematics. This is to miss the point of it. Computers are physical objects, and computations are physical processes. What computers can or cannot compute is determined by the laws of Physics alone, and not by pure Mathematics.
David Deutsch (1953−)

Takeaways from this Prologue

- 'Computing' means much more than just using modern computers.
- *Any* physical system performs some type of computation, with us, the observers, providing *meaning* (*semantics*) to such a computation.
- This suggests an *equivalence principle* between physical (*semantic*) information and computation.
- A physical system 'does not care' about the mathematical classification or difficulty of computational problems.
- If it is *designed* to solve a specific problem, it will solve it because 'it is in its nature'. It cannot do otherwise.
- MemComputing machines are *designed complex systems*.
- They satisfy some general criteria.

1.1   What does it mean 'to compute'?

Before delving deeper into the various aspects of MemComputing, I think it would be useful for the reader to clarify in this Prologue what I mean by 'computing' in its broader sense. Hopefully, this *viewpoint* will facilitate the understanding of the relation between *physical* concepts and *mathematical* statements I will discuss in the rest of the book.

Scientific paradigm
Any scientific enterprise is based upon some *viewpoint*, which is the guiding principle of future developments in the field. This viewpoint and the practices it entails constitute what is sometimes called a *scientific paradigm* (Kuhn, 1962).

Modern computers

The word 'computing' is nowadays synonymous with using devices we call 'computers' (Fig. 1.1), such as a laptop, a cluster of central processing units (CPUs), or graphic processing units (GPUs), or even a pocket calculator, to solve problems that are tedious, very difficult, or practically impossible to do by hand. We employ these devices for innumerable applications, oftentimes without even realizing it.

For instance, we use computers to 'calculate'—from the Latin 'calculus', or small pebble used in antiquity in place of arithmetic numbers—the amount of electricity that needs to be distributed across a power grid, or to estimate the best routes trucks need to follow to deliver their goods. We employ them to calculate the trajectory of a missile, a planet, or a star.

In summary, we use them for anything that is beyond reach of our immediate human mind capabilities. Essentially, anything that is beyond simple mathematical or logical operations.

Fig. 1.1 Sketch of modern computers.

As I will discuss in more detail in Chapter 2, the very notion of computing has been formalized by introducing the concept of a Turing machine (Turing, 1936), an *ideal* 'device' that attempts to define *mathematically* what it means to compute, and classify problems according to how easily they can be solved by such a machine. However, there is a *big* difference between mathematical statements and physical reality, as we intend it in the Natural Sciences. (Natural Sciences encompass those disciplines, such as Physics, Chemistry, and Biology, that rely on *experiments* as the ultimate validation of a scientific statement.)

Mathematical statements only need to be logically consistent as built upon a set of foundational axioms (cf. Section 1.4), *irrespective* of whether they correspond to any physical reality.[1] Physical reality is instead what we probe with our senses or instruments that extend the reach of our senses (Di Ventra, 2018).

In the context of computing, therefore, such a mathematical idealization seems to overlook (or at least to relegate as an afterthought) a very important aspect of computing: its *physical nature*. In fact, it is not an exaggeration to say that:

The goal of a computing machine is to compute!
This statement seems an obvious tautology, and, in fact, a trivial one. But it is worth remembering that the *goal* of a computing machine is to solve problems that are challenging for us, humans. If a computing machine, quantum or not, does *not* accomplish this goal, then it is more of academic than of practical interest (cf. Section 13.9).

[1] For instance, the concept of infinity is well defined in Mathematics, but we cannot *measure* infinite quantities because our instruments are finite (Di Ventra, 2018).

[2]A similar viewpoint is shared by physicist David Deutsch (Deutsch, 1997)—see quote at the beginning of this Chapter—and was advocated by the late Rolf Landauer (1927−1999) (Landauer, 1991).

> **Computing is first and foremost a physical process.**

It is not something that 'lives' only in the abstract world of Mathematics. Rather, the 'act of computing' itself can be, and ultimately *is* done by physical, material objects.[2] To better clarify this important point let me make the following considerations.

1.2   The Physics of computing

Example 1.1

An astronomical object

Suppose you are given a telescope to observe some object, e.g., a planet in our solar system (Fig. 1.2). For simplicity, you can think of this object as a point in the sky.

The first day you have the telescope in hand, you record the position of the object. You then perform this observation for several days in a row at some time during the day (even for a year if you are patient enough).

By doing this experiment you can then trace the object's trajectory, with respect to you, the observer, during the time that has elapsed from the first day of observation to the last. The experiment can, in principle, be refined by decreasing the time between observations.

What have you really done with this experiment? You have assigned a position (a set of real numbers) to the planetary object at some intervals of time. The position as a function of time defines the orbit of this object with respect to a reference frame (you).

This experiment then really answers the question: what is the position of the planetary object after some time has elapsed from my initial observation, the 'initial condition' of my experiment?

In fact, we are answering this same question at every observation: we determine the new position of the object from the last recorded one.

However, you could have equally asked the question: can you *calculate* the position of the object after some time has elapsed from the initial position?

To answer such a question, you would most likely think of applying Newton's equations of motion,[3] and solving them numerically on our modern computers.

If you actually did this, you would have replaced the act of experimental observation with an *algorithmic* procedure[4] you can carry out on our modern computers. But is there truly a difference between the previous two questions in regard to the *information* we want to extract?

Fig. 1.2 A man and his telescope.

[3]Or Einstein's equations, if you feel brave.

[4]See Section 2.3 for a formal definition of *algorithm*.

If you think about it, whether you determine the trajectory of the planetary object by using our modern computers, or by direct observation, namely by employing as a 'computing device' the planetary object itself, does not really change the *ultimate goal*, which is to find the object's position after some time has passed from an initial point in the sky.

You can then *interpret* the results of the measurement, the assignment of the numbers representing the object's position, as a type of 'calculation'. And you could call the device that does such a calculation a 'computer', in its broader etymological sense.

So, what you have really done with this experiment is to determine or 'calculate' the orbit of the planetary object. This calculation started with the initial observation of the position of the object, the *input* of the calculation, and ended with the last recorded position, or any other intermediate measurement, the *outputs* of the calculation.

But was is it really you who calculated the orbit, or the object itself? Of course, you, the *observer*—the 'active agent' of the calculation—has to *interpret* the results of the measurements as positions of a physical object in time: you are assigning a (mathematical) *meaning* to what we call 'position as a function of time'.

And you need a device (the telescope or your own eyes) to 'write' the input of the computation and 'read' its output.

However, it was actually the planetary object that 'calculated' what we interpret as its time-dependent position. Incidentally, it did so in the presence of the whole Universe surrounding it!

And the planetary object *does not care* whether such a calculation is easy or not: it is simply 'natural' for this physical system to do what it does. It cannot do otherwise!

In fact, such a calculation would be *impossible* to carry out using our modern computers by numerically integrating Newton's (or Einstein's) equations of motion if we had to take into account the effect of the *entire* Universe on the particular planetary object we are interested in.

The numerical task is made even more challenging by the fact that chaotic dynamics is common in our solar system.[5]

Therefore, in a very literal sense, the physical system (the planetary object) calculates its own orbit *infinitely* better than any model integrated numerically on our modern computers could ever do.

> **Information from Nature**
> It makes no difference, in terms of the *information* we want to extract from the natural world, whether we determine the orbit of a planetary object by calculating it on modern computers, or by measuring it via a telescope. In fact, the experimental procedure determines the orbit *infinitely* better than the numerical computation can ever do, because it takes into account *all* possible physical effects and interactions.

[5] As I will discuss in Chapter 3, chaotic dynamics is an undesirable feature of the equations of motion—and computation in general—since it requires an exponentially increasing precision with increasing problem size.

The conclusions I reached about the previous experiment are not just limited to the particular setting, experimental apparatus, or physical object I have discussed in the previous example.

In fact, I can reach the same conclusions for *any* physical system and *any* physical property we, the observers, may be interested in measuring or calculating. Here is another example.

Fig. 1.3 A thermometer measuring the temperature of a gas enclosed between a cold and a hot surface.

Example 1.2

Temperature of a gas

You are given a thermometer to determine how the temperature of air (or some other gas) varies from a hot surface to a cold one placed some distance apart (Fig. 1.3). The interaction between the thermometer and the air sets the temperature that I read with the thermometer. I can then measure/calculate the temperature as a function of position across the two surfaces.

I *interpret* what I read on the thermometer as the parameter I call 'temperature'. The thermometer acts as an input/output device to extract such information.

But it is truly the air that 'calculates' the physical property we are interested in, and the air has no choice other than to do what it has to do when it is sandwiched between hot and cold surfaces.

Finally, let me provide the following two examples that will come in handy in Chapters 11 and 12 when I will discuss how digital MemComputing machines work.

Example 1.3

The pinball machine

Consider a pinball machine (see schematic in Fig. 1.4). Given the initial position and velocity of the ball at the entrance of the pinball machine, the ball will be driven by gravity to the exit point while bouncing off pop bumpers.

In addition, if there are no regions where the ball could get stuck (local minima in its potential energy) the ball will *always* get out at the bottom. It has no other choice.

The Physics of the machine (gravity plus quasi-elastic collisions with the pop bumpers) dictates what the ball needs to do. Along the way the ball determines the trajectory it needs to follow to go from the entrance to the exit.

We can then say that the ball has 'calculated' its own trajectory!

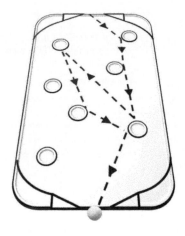

Fig. 1.4 Schematic of a pinball machine. When the ball enters from the top entrance of the machine, gravity will drive it toward the bottom exit point, while bouncing off various pop bumpers (the circles). Along the way the ball 'calculates' the path it needs to follow toward the exit.

The next example is similar, but it will provide yet another analogy of how to intuitively interpret the way digital MemComputing machines find solutions of a given problem (see Chapter 12).

Example 1.4

The waterfall

Consider water that flows down a hill toward an estuary (Fig. 1.5). There may be several paths through which it could *percolate* as driven (directed) by gravity. Here, I used the word 'percolation' on purpose, to mean that the water will naturally flow through all the paths that are available to it, under the effect of gravity.

However, if there are no places on the hill where the water may be absorbed or get stuck, it will always go downhill toward the estuary (the exit point or points, if there are several). It cannot do otherwise: gravity forces the water to go through the paths that lead to the exit.

Fig. 1.5 *Directed percolation* of water down a hill due to gravity. The water 'calculates' all possible paths from its source to the estuary.

Along the way, the water has 'calculated' all possible paths consistent with the constraints of the terrain. In other words, the water flow has determined all available paths going from the 'input point' (the water source) to the 'output point' (the estuary). And it has done so effortlessly, without worrying whether that problem is computationally difficult or not.

Indeed, it has done so in the presence of *all* possible obstacles and imperfections one can think of along its path. And if one of the paths becomes unavailable because we put an obstacle to shut it close, the water will still flow downhill through the other available paths, namely it 'calculates', on the fly, the remaining available paths (it could even find other paths around the blocked one).

Percolation
It is that phenomenon by which a liquid passes through a porous material.

Directed percolation
It is the phenomenon of percolation along a given direction, typically due to the effect of gravity.

I could provide many more examples, but I hope the previous ones are sufficient to support my suggestion that there is indeed a relation between physical systems and 'computation'.

The attentive reader, however, may have noticed that, while discussing this relation, I have somewhat glossed over a *very important* point, which

I now want to expand on.

1.3   Syntactic vs. semantic information

In all the cases I have considered so far, there is always an 'active agent' of the computation, namely an *observer* that provides *meaning* to the results of the measurements (the 'computation' of whatever quantity we are interested in).

This is not a minor point. In fact, without this ultimate step of *interpreting* the results of the measurement, the data we collect from our instruments would have no meaning in terms of the physical quantity we are measuring, or, equivalently, that we are trying to 'compute'.

For instance, the data we collect with the telescope in Example 1.1 are *interpreted* by us as the 'orbit' of the planetary object we are observing: we are assigning real numbers to what we observe, and attribute the meaning of 'position' to those numbers.

However, the light from the object that reaches our telescope (and our own eyes) would behave the same, namely it would do whatever it has to do according to its nature, whether we interpret such a signal as position of the planetary object in the sky or not.

In the same vein, the air molecules in between cold and hot surfaces in Example 1.2, do whatever they have to do, whether we *interpret* what we read on the thermometer as a 'temperature', or something else. It is us, the observers, who *interpret* the read-out on the thermometer as that physical quantity we call temperature of the gas. And so forth, for the other examples as well.

The 'interpretative' side of computation

If you think that this 'interpretative' step applies only to the physical systems I discussed previously, and not to how we actually do computation, as we traditionally intend it, consider again our modern computers (Section 1.1).

These machines are ultimately built out of *physical* devices, such as transistors, resistors, capacitors, and the like (Hennessy and Patterson, 2006). Each one of these physical devices carries an electrical current and/or supports a voltage difference.

When we type on the keyboard of, say, a laptop to input some data (numbers and letters that *we* interpret as numbers and letters), the electrical currents/voltages in the hardware components of the laptop do whatever they have to do. Ultimately, after some elapsed time, these currents/voltages have produced physical changes that we read, say, from the laptop screen, as light signals that *we* interpret as symbols (numbers and letters).[6]

Information

The word *information* (or *knowledge*) may acquire different meanings in different contexts. In its most 'quantifiable' use, it typically indicates a set of *messages* generated by a data source (Cover and Thomas, 2006). In this case, no meaning to such messages is attributed. Hence, this type of information is properly called *syntactic*, to distinguish it from *semantic* information, namely knowledge to which we assign a *meaning*.

Information theory

The study of how messages are quantified, stored, and transmitted is known as *information theory* or *mathematical theory of communication* (Cover and Thomas, 2006). It assigns a value to the messages that are communicated according to how 'surprising' they are to the receiver. The less 'surprising' a message is, the less information it carries, and vice versa (cf. Section 5.3.2). Note that this is *not* a definition of 'information', and indeed it depends on (it is relative to) the state of knowledge (degree of 'surprise') of the receiver. It is then merely a way to quantify such a 'surprise'. That is why 'communication theory' seems a more appropriate term than 'information theory'.

Information channel

A channel that communicates messages from a set of transmitters (senders) to a set of receivers, is known as *information channel*.

---

[6] In this sense, and following Turing's original idea (see Chapter 2), modern computers manipulate *symbols*, not 'information'.

The fact that the 'state' of these currents/voltages is *interpreted* by us, the *observers*, as symbols on a screen, is a step that we do *after* the electrons in the laptop have done their job, and the light of the screen has reached our eyes, according to specific physical laws.

In other words, without us attributing a *meaning* to such symbols on the screen, the result of the computation would be meaningless!

Of course, the electrons transmit some form of information ('knowledge') through the physical 'channels' that link the input (the keyboard where we type) to the output (the screen from which we read). And the light that reaches our eyes also carries information.

This information can indeed be easily quantified (see, e.g., *Shannon's self-information* in Section 5.3.2). However, such information has yet to acquire any meaning. We therefore call it *syntactic*.

In its most abstract terms, syntactic information refers to the transmission of 'strings of symbols' (*messages*), such as 0s and 1s, along an 'information channel', without regard to the *meaning* of such symbols (Cover and Thomas, 2006).[7]

Instead, when we attribute a *meaning* to such information, then we are considering its *semantic* value. We then call it *semantic information*.[8]

This type of information is not so easy to quantify (at least in general) as the syntactic one. Nonetheless, it is fundamental to how we do computation. Indeed, it is so imbued in our way of approaching computation that we rarely think of it when we use our modern computers.

This type of semantic information is what I meant in the preceding Section 1.2 by *interpreting* a measurement as a computation of some physical quantity. We, the observers, provide a *meaning* to the output of the measurement, as we do when we give a meaning to the symbols we read out of the electrical currents/voltages that make our modern computers work.

1.4 The *equivalence principle* of physical information and computation

The previous examples, coupled to the understanding of *semantic information*, invite a generalization.

In fact, we could think of any other type of experiment and physical system, and we would arrive at the same conclusion: the physical system 'calculates' some property that we, the observers, extract with some device and *interpret* accordingly.[9]

I will then consider these examples as expressions of an *equivalence principle* between physical, *semantic* information (as it is interpreted by us, the observers) and computation, namely between the *semantic information* that we extract from experiments, and that which we would

State of a physical system
The *state* of a physical system is the collection of (physical and/or mathematical) quantities required to describe its dynamics (whether deterministic or stochastic). For instance, in Classical Mechanics the (microscopic) state of a system is the collection of all the (physical) positions and momenta of all its particles (Goldstein, 1965). In Quantum Mechanics it is a (mathematical) vector—a 'ray': the equivalence class of all vectors related by an overall complex phase—(or a statistical operator, if its knowledge is not complete) in a Hilbert space (Messiah, 1958). The state of the MemComputing machines (non-quantum dynamical systems) I will discuss in this book is the collection of *all* variables needed to describe their dynamics (e.g., memory variables and voltages in an electrical circuit realization). See Section 3.2.

[7] Arguably, there is *always* some 'meaning' associated with what we call 'symbols', 'messages', 'channels', etc. Therefore, even syntactic information is never truly devoid of semantics.

[8] As a further example, if I wrote the word 'abbanniatu' on a paper, you would see the black ink lines that define the symbols (letters) of the word. The light scattered by the paper ink (the source) that reaches your eyes (the receivers) transmits *syntactic* information about the string of symbols in such a word (the message). However, unless you were versed in Sicilian dialect, you would not understand that the *meaning* of such a word (its *semantics*) is 'well known' (pun intended).

[9] In this sense, 'computation' is a form of *meta-Mathematics*, namely a process that 'transcends' Mathematics, and belongs to the physical world as interpreted by us, the observers (cf. also Section 13.3.3).

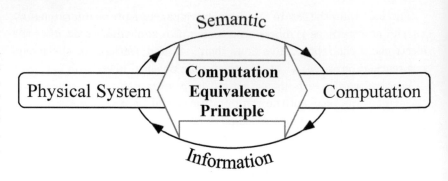

Fig. 1.6 There is an *equivalence* between the information extracted from experiments, as it is interpreted by us, the observers (*semantic information*), and *computation* of various quantities. Physical systems naturally 'compute' such quantities, with the observer providing *meaning* to such a computation.

like to obtain by 'traditional computational means'.

I will then state the following (Fig. 1.6):

> **Computation Equivalence Principle**
>
> *Any* **physical system performs some type of computation, with us, the observers, the ones providing** *meaning* **(***semantics***) to such a computation.**

In other words, it is up to us, the observers, to *write* the input and *read* the output of such a computation, and *interpret* its results, namely the quantity that such a system 'naturally calculates'.[10]

Following this argument, then the act of *measuring* the physical state of a system is equivalent to the act of initializing ('inputting' the initial condition of) the computation.

In the same vein, the act of *measuring* the physical state of the same system, after some time has passed, is equivalent to 'reading' the final result of the computation.

What happens in between the observations of the initial and final states—the actual computation by the physical machine—is dictated by physical laws of motion.[11]

1.5   Computing with Physics

If we take such a broad view of computing, we can then think of new ways in which *physical properties* of actual *physical systems* could be used to solve problems that are difficult to solve with other 'traditional' means.[12]

In fact, this idea is not new and has already been considered for many applications. For instance, the speedometer in our cars employs the Physics of electromagnetism to determine their speed, or a seismograph employs mechanical forces to detect and determine the strength of earthquakes.

Principle vs. axiom

In Physics, a *principle* is a statement whose validity rests on *experimental* data. In other words, it is an *inference*, namely a conclusion, that has been reached on the basis of some experimental evidence, and generalized to a wide class of phenomena for which the principle is enunciated.

In Mathematics, an *axiom* (or 'postulate') is a statement whose truth value is taken to be 'self-evident', and it is foundational to other mathematical statements derived from it.

---

[10]Note that although both the input/output can be written/read by a device (like a remote sensor or a control system) or some other 'machine', the ultimate step of *interpreting* (providing *meaning* to) the data has to be performed by humans. Apart from those portrayed in science fiction movies, we do not have 'machines' that accomplish such a feat (cf. the four waves of Artificial Intelligence in Section 2.5).

[11]Following terminology mostly used in Computer Science, we will call this part of the dynamics, the *transition function*. See Chapter 3.

[12]This reinforces the notion that the *meaning* we attribute to 'computation' resides in the *use* we make of it.

Another example that is closer to the ideas and problems discussed in this book is the use of a quantum feature, known as *entanglement*, to solve a problem like prime factorization (Shor, 1997) that is difficult to solve with traditional algorithmic approaches (see Section 3.2 and Example 7.5).

And the list could go on. We can then summarize the previous discussion by stating that

> **There are many physical properties we employ in our technology to extract/calculate specific quantities of interest.**

This book is dedicated to yet another Physics-inspired approach to computation, MemComputing, that takes advantage of another physical property *–time non-locality*—to solve a wide range of problems.

In particular, when restricted to its *digital* version, MemComputing is specifically designed to solve *combinatorial optimization problems* (these problems will be defined in Chapter 2).

As we will see in the subsequent Chapters, the Physics we will exploit does not necessarily require quantum features and allows us to build *scalable* machines, namely machines that do not need physical resources to increase at an exponential rate with increasing size of the problem.

In fact, if we can *design* a physical system to solve a particular problem (whether the determination of the speed of a car or the solution of a combinatorial problem) that system *has to* solve it, because *it has no other choice*!

It is in the *nature* of that system to do what it has to do. As I have discussed earlier, such a system *does not know* and *does not care* whether a problem is 'easy' or 'difficult' to solve, according to some mathematical classification we, humans, came up with.

In other words, we can say that

> **An appropriately designed physical system has no 'knowledge' of the difficulty of the problem it is designed to solve. It *has* to solve it because it is in 'its nature'.**

In the rest of the book, I will explain how to actually design such systems for the solution of combinatorial optimization problems, that do not require exponentially growing resources with problem size.

Before doing so, however, let me conclude this Prologue with some basic *general criteria* that we would like our MemComputing machines to share. They are taken from (Di Ventra and Pershin, 2013*b*).

Function vs. functional

A *function* is a mapping (relation) between a set (domain) and another set (codomain)—e.g., the set of real numbers and the set of integers—that associates each element of the domain to a single element of the codomain. A *functional* is a mapping between a (vector) space and a field of scalars (e.g., real or complex numbers) (Arfken *et al.*, 1989). In this book, by 'functional' I will mostly intend a mapping between a space of functions and a field of scalars (unless I refer to 'functional polymorphism'; Section 5.3.3). By *functional form* of an equation I will instead mean the (algebraic) form of the equation relating its different variables.

The principle of stationary action

In Physics, we know that the dynamical state of a physical system is the one that renders a quantity, we call *action* (which is a *functional* of the system's coordinates, and often indicated with the symbol S), stationary (Goldstein, 1965; Arnold, 1989). Namely, the variation of such a quantity is zero when evaluated at the actual trajectory between some initial and final times. When we say that a physical system has 'no choice', we mean that it *must* evolve in such a way as to render its action stationary. This is a very powerful *principle*. Although it is typically formulated for *non-dissipative systems*, I will show in Section 11.5, that it can be applied (with an appropriate action) to the operation of digital MemComputing machines, which are *dissipative systems* (see Section 3.8 for the distinction between non-dissipative and dissipative systems).

Some of these criteria will provide a good reference and starting point to think about ways in which MemComputing may be implemented in hardware.

Others turn out to be *emergent properties* (or *epiphenomena*) in MemComputing: they appear 'naturally' from the way in which MemComputing machines are built from their elementary blocks (check Chapters 11 and 12 if you are eager to see which 'emergent properties' I am referring to).

1.6   MemComputing criteria

Some of the criteria I will now outline are similar to those envisioned in the field of Quantum Computation (DiVincenzo, 2000). Indeed, it should not come as a complete surprise that some of them are similar to those required even by our modern computers.

After all, the goal of any 'computing machine' is precisely 'to compute' something. Therefore, some criteria belong to the general class of 'computing machines'.

However, others are specific to the MemComputing paradigm. The criteria are as follows (Di Ventra and Pershin, 2013b).

I. Scalable massively parallel architecture with combined information processing and storage

We envision MemComputing to be realized by appropriate devices with memory, namely devices whose response to a given input depends on their *past dynamics* (time non-locality).

In this book, I will focus on an *electronic* realization of MemComputing, by combining circuit elements with memory and standard electronic components, both *passive* such as resistors, capacitors, and inductors, and *active*, such as transistors, and voltage generators.[13]

However, the notion of MemComputing is much more general and could, in principle, be realized using, e.g., optics or a combination of optics and electronics.

Irrespective, a MemComputing machine will be a collection of elementary units arranged together to solve a certain class of problems. We then expect that all, or at least a large chunk of these elementary units will be *collectively* involved in the computation.

As I will discuss in Chapter 3, this type of *massively parallel* architecture is fundamentally different from the traditional ones, even those we call 'parallel'.

Note also that the combination of information processing and storage has an added practical advantage: it simplifies the *hardware design* be-

---

Complex systems

'Emergent properties' are very common in Nature. They are the hallmarks of *complex systems*, namely systems that showcase phenomena that cannot be easily accounted for only by the properties of their individual constituents. Rather, they 'emerge' from the mutual (and often highly non-linear) interaction among their elementary units (Holland, 1998). MemComputing machines are an example of *designed* and *strongly coupled* complex systems.

---

[13]See Section 4.5 for a precise distinction between passive and active elements.

Information storage

It pertains to the *retention* of (syntactic) information in some property of a physical medium (e.g. electric charge on a capacitor) for a time much longer than the time required to *read* such information. It can also be identified as *long-term memory* (Hennessy and Patterson, 2006).

cause it reduces the amount of components that need to be integrated on a *chip*.

II. Sufficiently long information storage times (of the relevant units)

The elementary units of a MemComputing machine should have sufficiently long information storage times, much longer than the calculation time, namely the time at which the output is read.

However, this long information storage time is not necessary for all units of the machine, only those from which the result of the computation will be read. In fact, the other units may have a very short storage time, so long as, once the computation halts, their loss of information does not affect the units from which the output is read.

Ideally, it is desirable that these fundamental units be realized at such a scale that allows easy integration on a single chip, consume low power, and have relatively short read/write times to facilitate both the writing of the input as well as the reading of the output of the computation.

III. The ability to initialize the states of the machine units

The fundamental units of a MemComputing machine should be initialized before the computation begins. Ideally, we would like to be able to initialize all of them.

If that is impractical, we should be able to initialize at least those that are relevant for the input of the computation.

IV. Mechanisms of collective dynamics

This is the first emerging property I was referring to previously. The MemComputing architecture should (and indeed does) provide a mechanism of *collective dynamics*.

As I will amply discuss in Chapters 11 and 12, during time evolution—from the initial time of the computation, till an output is read— all or a large chunk of the elementary units perform computation collectively—in a *correlated* way—as if the machine were a dynamically 'rigid', 'macroscopic' object.

In fact, I will show in Chapters 11 and 12 that these machines operate at an out-of-equilibrium *correlated state*, without parameter tuning.[14]

V. The ability to read the final result (from the relevant units)

Once the computation is terminated, the read-out of the result should ideally be performed without modifying the states of the units from

Chip
In electronics, by the term *chip* (or 'microchip') we mean a set of electronic circuits (made of transistors, resistors, capacitors, etc.) integrated on a small chunk of some material (typically silicon): an *integrated circuit* (Hennessy and Patterson, 2006).

Computation at the edge of chaos
It has been suggested in the past that computation performed on physical units that operate at the 'edge of chaos', namely *not* in the chaotic regime, but rather at the *critical point* of a *continuous phase transition*, should provide substantial advantages compared to our traditional computers (Langton, 1990). Digital MemComputing machines are, in some sense, a practical realization of this idea: they operate at a (long-range) *ordered state* (see Section 4.3.4 and Chapters 11 and 12).

Dynamical/non-equilibrium phase transition
It is a transition between different phases of matter that occurs *away* from thermodynamic equilibrium (Henkel *et al.*, 2008). See Section 4.3.2 for a quick summary of phase transitions and criticality, and Chapters 11 and 12 for the relation of these notions to MemComputing.

[14]Like the critical state of some sort of *dynamical phase transition*.

Topology and computing

Topology is the branch of Mathematics that deals with the properties of objects that are invariant under *arbitrary continuous transformations* (Fomenko, 1994). Perturbations of a physical system can be described as arbitrary continuous deformations of their dynamics. *Topological invariants* (see Section 11.3 for a precise definition) are not as common as geometric ones (those properties preserved by geometric changes, such as rotations or linear scale changes). However, they are *very robust* against perturbations, much more so than geometric invariants. In fact, topological quantities are fundamentally *non-perturbative*. Therefore, if a machine employs topological features to compute, it is *topologically protected* against perturbations. Only *discontinuous* changes of these topological invariants—e.g., when the system goes through a (quantum) phase transition—would substantially change the dynamics of the machine. *Topological* Quantum Computing is based on such an idea, but has yet to be realized (Freedman *et al.*, 2002). MemComputing is another example of *topological computing*, but with non-quantum systems (see Chapters 11 and 12).

Topological theory

We say that a theory is *topological* if it is *independent* of the choice of *metric* (see Section 6.3.1 for the definition of 'metric' and 'metric space'), namely it is a theory that is insensitive to the distances between points in space-time (or whatever variables the theory is built on). The (supersymmetric) Topological Field Theory I will discuss in Section 11.3 is an example of such a theory.

which the output is read. This is consistent with criterion II, and allows those units to store the result of the computation for a sufficiently long time.

VI. Robustness against small imperfections and noise

This is a very important property that would guarantee stability of the computation against small imperfections of the machine units as well as the unavoidable noise that any physical system is subjected to. In fact, it is not hard to imagine that deviations from the ideal computing architecture can be easily introduced during either the fabrication of the machine or its operation.

In addition, even without invoking specific noise sources, we expect MemComputing machines to operate (at least) at room temperature. Therefore, thermal effects are definitely present.

As I will discuss in Chapters 11 and 12 robustness against small imperfections and noise is yet another important emerging property of (digital) MemComputing machines. It is a consequence of their *topological* nature, namely it is related to the type of dynamics they experience during computation.

Of course, if the architecture is *substantially* changed, e.g., by cutting some connections between elementary units, we do not expect the machine to operate as originally designed.

1.7   Chapter summary

- I have taken a much broader view of what is typically meant by 'computing'. In particular, I have argued that *any* physical system performs some kind of computation, with us, the observers, the 'active agents' that *interpret* the results of such a computation (we provide *semantic* value to such information).

- Although this view may not be what most people have in mind when discussing computation, it is, in fact, not new and it is being employed in various scientific and technological applications.

- The *equivalence principle* I envision between physical experiment and computation will guide us in the quest to find the appropriate physical features of actual physical systems that can be exploited to compute efficiently a wide class of problems, in particular combinatorial optimization ones.

- I have also listed the general criteria MemComputing machines should share for their hardware implementation. As I will show in

Chapters 11 and 12, some of these criteria are *emerging properties* of these machines.

– MemComputing machines can then be classified as *designed complex systems*.

– In Chapters 11 and 12, I will show that they operate at a *long-range correlated state*, and their dynamics are 'topologically protected'.

– Before proceeding further, however, I need first to introduce the notion of Turing machine, define rigorously the problems we want to solve, and discuss how our modern computers realize, in an *approximate* but quite faithful *physical* way, the mathematical concept of a *deterministic* Turing machine.

Computing: The Turing way

2

Fig. 2.1 Modern computers execute logical operations and arithmetic on binary numbers.

A man provided with paper, pencil, and rubber, and subject to strict discipline, is in effect a universal machine.
Alan Turing (1912−1954)

Takeaways from this Chapter

- 'Computing' according to Turing.
- The algorithmic/combinatorial representation of a problem.
- What is a Boolean problem?
- Does a *physical* Turing machine really exist?
- Digital vs. analog machines.

2.1 Traditional computer operations

As I have explained in Chapter 1 our present understanding of computing refers primarily to the use of devices ('modern computers') that allow us to solve mathematical problems that are difficult to do by hand. To do this, we first transform the problem that needs to be solved into a set of mathematical instructions. These are then fed into the computing device, which, given an input of the problem, executes them, and spits out a set of symbols we can read and interpret.

The basic mathematical instructions that a computer executes are *arithmetic* and *logic* operations (Fig. 2.1). Arithmetic operations are addition, subtraction, multiplication and division of integers (belonging to the set $\mathbb{Z}$), typically expressed in the binary numeral system.

Logic operations are all those that relate two or more variables, such as A AND B, or A XOR B, etc., where A and B are *variables* that can take the values of 'true' or 'false' ('1' or '0', respectively, if we use a binary value), and 'AND' and 'XOR' are *logic gates* that, given a set of input truth values, give out particular truth values according to specific rules (*truth tables*).

Why do we do calculations this way? And how do we translate this mathematical construct into actual, *physical* devices?

2.2 Turing machines

The general idea behind the way we do calculations nowadays is known as *Turing paradigm*, from the name of the mathematician Alan Turing (Fig. 2.2) who formalized it in 1936 (Turing, 1936). Turing starts his original paper by saying:

> The 'computable' numbers may be described briefly as the real numbers whose expressions as a decimal are calculable by finite means.

and, later in the paper, he writes:

> According to my definition, a number is computable if its decimal can be written down by a machine.

He also extends such a definition to *functions* of computable numbers, hence stating that 'a function is computable *if* it can be written down by a machine using only finite means'.[1]

In other words, Turing *defines* as *computable* only those mathematical objects that are calculable by *finite* means. He claims the rationale behind this is that 'the human memory is necessarily limited' and, therefore, unable to calculate quantities that require *infinite* means (Turing, 1936).

We see immediately that Turing's original goal was *not* to specify a *general* model of computation. Rather, he was seeking to formalize the way we, humans, perform computations,[2] or, more generally, to put into a mathematical language some (but not all) aspects of what constitutes computation.[3] He was not, in 1936, referring to our modern-day computers, since they had yet to be invented!

We also see that this program leaves out an enormous amount of *non-computable numbers* and, hence, *non-computable functions*. This is because the set of computable numbers is *countable*, while the set of real numbers is *uncountable*, therefore almost all real numbers (and functions on them) are *not* computable according to Turing. This seems indeed quite a strong limitation, and as we will see in this book it has quite a few implications for the non-Turing paradigm of MemComputing.

2.2.1 Universal Turing machine

Let's now make the previous statements a bit more rigorous. It is not my goal to provide a complete discussion of Turing machines in the present

Fig. 2.2 Alan Turing (1912–1954). Sketch by the author.

[1] Note that what really matters here is that there is some *algorithm* (see Section 2.3 for a precise definition of 'algorithm') that produces such a number, even if the latter can only be written as an infinite (but well-defined) decimal expansion.

[2] How a 'human computer' solves decision problems (cf. Section 2.3.4).

[3] More precisely, *automatic* computation or *a-machines*, as Turing defined them, namely those machines whose state at every iteration *completely* determines the next step to take without regard to the *history* of the machine's evolution (Turing, 1936).

Countable and uncountable sets
A set is said to be *countable* (or *denumerable*) if it has the same *cardinality* (number of elements) as some subset of the set of natural numbers, $\mathbb{N}$. It could be finite or infinite. An infinite set is *uncountable* (or *nondenumerable*) if its cardinality is larger than that of natural numbers.

Hypothesis
A *hypothesis* (in Mathematics often called a *conjecture*) is a statement we make about a physical phenomenon or a mathematical proposition of which we do *not* know its truth value *a priori*. Hence, hypotheses are not necessarily facts.

Theorem
A *theorem* is a mathematical statement whose truth value (whether true or false) has been logically proved starting from a set of axioms (cf. Section 1.4) or from previously proved theorems.

Physical theory
A *physical theory* is a set of statements (mathematical or not) based on some hypotheses, aimed at *describing* a limited set of natural phenomena, and *predicting* new ones (see, e.g., (Di Ventra, 2018)). I will make use of several physical theories in this book to describe MemComputing machines and predict their dynamics.

---

[4]In Physics and Engineering parlance the tape head is an *active* element, see Section 4.5.1. In fact, it can be identified with the processors of our modern computers.

Fig. 2.3 Schematic representation of a Turing machine. It consists of an infinite tape and a read/write head (an *active* device). The state of the machine is a collection of finite-string symbols (e.g., the *bit* values 0s and 1s) stored in memory. The head reads the state of the machine, and according to a set of instructions, it writes a symbol on the tape, and either moves its tape head to the left or to the right, or not at all, till the computation halts.

book. The reader is better off consulting the following books (Turing, 2004), (Garey and Johnson, 1990), and (Arora and Boaz, 2009) for a comprehensive analysis. Here, I will discuss only those features that set them apart from MemComputing machines.

Let's start by defining a machine that is now called *universal Turing machine* (UTM). The term 'universal' does not mean that it can do any type of computation (see previous discussion regarding what a Turing machine can and cannot calculate).

It simply means that its mathematical structure is such that it can *simulate* any other Turing machine, and hence it is, in some sense, more 'general' than any other arbitrary Turing machine. This, however, does *not* imply that it is *the* most general computing paradigm one can come up with–although, thanks in part to the Church-Turing hypotheses I will enunciate in Section 2.2.2, this is what is sometimes implied in some textbooks or articles on the subject.

In simple terms a UTM is a *mathematical* model of an object that manipulates *symbols* on an infinite tape according to a set of rules (instructions). The machine then has access to an *infinite memory*, although, at any given time, it never uses it all to work.

The *state* of the machine is the collection of symbols that can be recorded on the tape. According to a finite set of instructions that are provided to the machine by a *control unit* (something or someone that tells the machine what to do), the machine then reads the state on its tape via a tape 'head'. According to the state read, and the set of instructions, it then writes a symbol on the tape, and either moves its tape head to the left or to the right, or not at all. The computation then proceeds till it halts.

The fundamental ingredients of a UTM are then the *tape* itself, a *tape head*[4] to read and write on the tape, a *state register* that stores the states of the UTM (including the initial and final states), and finally a *finite table* of instructions that tells the machine what type of problem to solve and how (see schematic in Fig. 2.3).

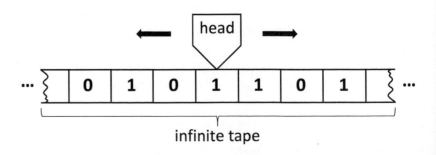

Formally, a UTM is defined as the seven-tuple

$$UTM = (Q, \Gamma, b, \Sigma, \delta, q_0, F),\qquad(2.1)$$

where Q is the set of possible states that the UTM can be in, Γ is the set of symbols that can be written on the tape, $b \in \Gamma$ is the 'blank' symbol (and is not part of the input), Σ is the set of input symbols, $q_0 \in Q$ is the initial state of the UTM, $F \subseteq Q$ is the set of final states, and[5]

$$\delta : Q\backslash F \times \Gamma \to Q \times \Gamma \times \{L, N, R\},\qquad(2.2)$$

is the *transition function* where L, N and R are 'Left shift', 'No shift', and 'Right shift', respectively. The transition function determines precisely the mapping between the initial and final states.

We can then summarize what a Turing machine is:[6]

Turing machine
A mapping, δ, from a *finite* string of symbols to a *finite* string of symbols:

$$\{finite \text{ string of symbols}\} \xrightarrow{\delta} \{finite \text{ string of symbols}\},$$

in *discrete time.*

From this formal definition it can be readily seen that *no* explicit information about any physical system or its properties is involved in a UTM. In other words, the UTM is just an abstract mathematical object, through which it is possible to determine the computational *complexity* of an *algorithm* (see also Section 2.3). This means that given $N = \text{cardinality}[\Sigma]$ (how big the size of the input is), and the type of transition function, δ, we can determine the number of operations (steps) the UTM should perform in order to find the final state F.

How general is this definition in terms of what can be computed within this Turing paradigm? To answer this question some *conjectures* have been proposed. Note that a 'mathematical conjecture' is simply a *hypothesis*. It is *not* a mathematically demonstrated statement, or theorem.

Indeed, none of these conjectures has been proved yet. It is then unfortunate that quite often these hypotheses are taken as facts. That should not be the case.

2.2.2 The various Church–Turing hypotheses

There are several *hypotheses* (not theorems!) that are attributed to Church and Turing, even though the majority of them have never been formulated by these two mathematicians. Here, I will mention only two, which are mostly discussed in the literature.

Binary digit or bit
A logical state that can acquire one of two possible values is called a *binary digit* or simply a *bit.*

Binary numeral system
Any number can be represented by a sequence (possibly infinite) of bits. This method defines the *binary numeral system.*

[5]The symbol '\' in eqn (2.2) means 'excluded', while '×' means 'Cartesian product' between sets. For instance, the Cartesian product of two sets X and Y is the set of all ordered pairs (x, y), with x in X and y in Y.

[6]The string of symbols is typically a collection of bit values 0s and 1s, but that is not necessary.

Physical, discrete, central processing unit (CPU), and wall times
The word 'time' has different meanings according to the context. *Physical time* (or *continuous time*), as defined in Physics, is 'that quantity which is measured by a clock'. In Chapter 3, I will introduce dynamical systems representing MemComputing machines. In that case, 'time' is truly physical time. The *discrete time* of a Turing machine is instead a way to count the *number of steps* the machine needs to complete its calculation (or *number of operations* of an algorithm). *CPU time* is the time taken by the central processing unit to execute a given program or task. *Wall time* is the actual physical time that has elapsed from the start of a computing task to its end, as performed by an actual modern computer. Wall time then depends on many factors, such as the computer architecture, number of processors that compute in parallel, the input/output time, etc.

[7]It is not even a statement on the *resources* required by a given machine to simulate another Turing machine or vice versa, namely how different computing models are related to each other.

[8]To be precise, when θ is not strictly an integer, we should say that the complexity is a *power law*, not a polynomial. However, a power law can always be bounded by a polynomial of appropriate degree. We then typically say 'polynomial scalability' whenever the algorithm scales as $O(N^\theta)$, whatever is θ.

I. The 'original' Church–Turing hypothesis

A function on the natural numbers can be calculated by an 'effective' (namely mechanical or algorithmic) method[a], *if* and *only if* it is computable by a Turing machine (Arora and Boaz, 2009).

[a]See Section 2.3 for a definition of 'algorithm'.

Note that this conjecture is *not* a statement on the *complexity* with which a given machine can solve a given problem, namely how many *resources* a machine needs to solve the problem.[7] To address the issue of resources, yet another hypothesis has been introduced.

II. The 'extended' Church–Turing hypothesis

All 'reasonable' models of computation yield the same class of problems that can be computed in a time that grows *polynomially* with input size (Arora and Boaz, 2009).

Here, by 'reasonable' it is typically meant those models of computation that do not require infinite precision. Therefore, machines whose input or output (or both) cannot be represented with finite precision (we call them *analog computers*, see also Section 2.8) would not be considered 'reasonable' models of computation (I have given a few examples of these machines in Chapter 1).

By 'time' we really mean 'resources', since these are the ones that ultimately decide whether a computation is efficient or not. Finally, by 'polynomial growth' we mean that if we let the size of the input, $N = \text{cardinality}[\Sigma]$—where Σ is defined in eqn (2.1)—grow, then the resources (time) grow(s) as a function that is bounded by a polynomial.

It is common to indicate the complexity of an algorithm running on a given machine with the big-O notation (Arora and Boaz, 2009). For example, to indicate that an algorithm scales polynomially on a Turing machine, we would simply write that it scales as $O(N^\theta)$, with θ some number.[8]

In fact, in the literature it is typically common to define

An *exponential* complexity would instead be represented as $O(2^N)$ (or sometimes as $O(e^N)$ according to the chosen base). A *logarithmic* scaling as $O(\log(N))$, and a *factorial* (which is super-exponential) as $O(N!)$, and so on.

2.3   Algorithmic solution of a problem

How is this Turing paradigm implemented in practice? To answer this question let's look at a specific problem. Consider, for instance, the problem known as the *subset-sum problem* (see margin note).

For the sake of simplicity, let's forget for a moment about the precision p (number of bits) with which we represent the N numbers of the set G. Let's then try to answer the question posed by the subset-sum problem.

A straightforward (*algorithmic*) way to do this is as follows. Let's first calculate the total number of subsets that we need to check. Their size, k, ranges between 0 and N, and the number of ways of picking k unordered outcomes from N values is $\begin{pmatrix} N \\ k \end{pmatrix}$. So, there are in total

> **Subset-sum problem**
> Consider a set G of N integers ($N \in \mathbb{Z}$), each represented with at most p bits (precision of each number). Consider another integer $s \in \mathbb{Z}$.
> Question: is there a subset of G whose sum is s?

$$\sum_{k=0}^{N} \begin{pmatrix} N \\ k \end{pmatrix} = \sum_{k=0}^{N} \frac{N!}{k!(N-k)!} = 2^N \qquad (2.3)$$

possible subsets.

Since for each of the 2^N subsets we need to sum at most N numbers, this brute-force algorithm would require a number of operations ('steps') on the order of $O(N2^N)$, namely it would require an *exponential* running time (steps) to halt the program, or, according to the discussion in Section 2.2.2, we say that its complexity is exponential.[9]

This simple example summarizes the following notion:[10]

> **Algorithm (or 'mechanical procedure')**
> A procedure to find the solution of a given problem requiring
>
> (i) a *finite* number of steps,
>
> (ii) the first *and* last steps of the computation can be identified with *finite* means,
>
> (iii) given an input, a *unique* output is determined.

Note that the last requirement is true for *non-probabilistic* algorithms. In the case of *probabilistic* algorithms, the output is determined by assigning a *probability distribution* for the transition function to determine the next step of the computation.

Also, since we have tried all possible 'combinations' of the sub-set sum problem, we may equivalently say that we have solved that problem in a *combinatorial* way. Many problems we will discuss in this book are traditionally solved in a combinatorial way, so the words 'algorithmic' and 'combinatorial' may be used interchangeably for these cases. However, the notion of 'algorithm' as expressed previously is much more general.

[9] The big-O notation is only concerned with the leading scalability term of an algorithm as a function of the input size. Since an exponential dominates a linear function, $O(N2^N)$ can be simply replaced by $O(2^N)$. In addition, the big-O notation tells us nothing about the 'pre-factor' in front of such a function. Although the pre-factor is not important for purely theoretical considerations, it can make a *substantial* difference in the practical run time of an algorithm.

[10] Note that other authors may define an algorithm somewhat differently, but the general idea holds.

2.3.1   Deterministic vs. non-deterministic Turing machines

If we want to represent the steps a UTM needs to perform to solve a problem, such as the subset-sum I discussed in Section 2.3, then we could imagine drawing a *solution tree*, like the one in Fig. 2.4, where the branches represent the various possibilities in going from one iteration of the computation to the next.

At each iteration step of the computation, the machine needs to explore a growing number of possibilities. If the algorithm is such that these possibilities grow exponentially with input size, then the corresponding tree grows accordingly.

Fig. 2.4 Typical solution tree for a problem whose difficulty grows exponentially with input size. **Left:** a *deterministic* Turing machine needs to explore the whole tree at each computational step to find the solution. **Right:** a *non-deterministic* Turing machine is able to 'guess' correctly the next step to take, thus 'cutting through' the original exponential number of possibilities to find the correct solution.

Deterministic Turing machine

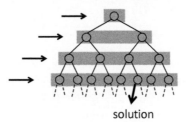

solution

Non-deterministic Turing machine

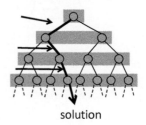

solution

A Turing machine—on which such algorithm is run—that needs to explore this solution tree one possibility at a time is called a *deterministic Turing machine*. It is often said that our modern computers, like the laptop I am using to write this book, are such machines. However, this is strictly *not correct*! Our modern computers are the closest *physical* realizations of such a machine, but they are definitely *not* an ideal (*mathematical*) deterministic Turing machine (see also Section 2.8).

Suppose now that the machine, at each step of the computation, is able to 'guess' correctly the best next move to reach the solution. In other words, suppose you were given a machine that via some internal mechanism unknown to the user—an *oracle* (Arora and Boaz, 2009)—is able to always *guide* the machine toward the correct solution without exploring an exponentially growing number of possibilities to solve the problem. Rather, this machine is somehow able to choose the correct path, at each computational step, toward the solution.

Equivalently, one may think of this machine as branching out into the many paths of the solution tree, each of them following one of the possible transition functions (Garey and Johnson, 1990). This second interpretation is that of a *parallel* machine with an *exponentially* growing number of processing units at each level of the solution tree (Garey and Johnson, 1990).

Decidable and undecidable problems
A problem is called *decidable* if it can be solved by a Turing machine. A problem that cannot be solved by a Turing machine is called *undecidable* (Arora and Boaz, 2009). See Section 5.4 for an example of an undecidable problem.

Either one of the previous equivalent interpretations defines a machine called *non-deterministic Turing machine*,[11] and would be able to solve a difficult problem, like the subset-sum problem, much more efficiently. In fact, it is not too difficult to convince oneself that such a machine would need only a *polynomial* number of steps to find the correct solution of the subset-sum problem (Garey and Johnson, 1990). This is because at every step of the computation, it just needs to determine one location of the corresponding branch, not all of them.

2.3.2 The NP vs. P question

Of course, such a machine seems impossible to realize in practice, but its definition allows the exploration of many interesting mathematical questions. The most famous of these (which is still unsolved) is whether we can find an algorithm running on a *deterministic* Turing machine that can solve a difficult problem, such as the subset-sum problem, in polynomial time (resources).

Such a question is sometimes referred to as the 'NP vs. P problem' better formulated in the margin note—where 'NP' stands for 'Non-deterministic Polynomial', and 'P' for 'Polynomial'. It is like asking whether there is indeed a way to realize in practice a non-deterministic Turing machine, since, as we have seen in Section 2.3.1, such a machine would indeed solve difficult problems, like the subset-sum, in polynomial time.[12]

2.3.3 NP, NP-complete, and NP-hard problems

In fact, the subset-sum problem is a special type of problem, known as *NP-complete* (Arora and Boaz, 2009). This means that if we found an algorithm that solves it in polynomial time on a deterministic Turing machine, any other problem in the NP class can be solved in polynomial time by the same machine.

The reason is because one can convert ('reduce') *any* NP problem into an NP-complete problem, and this mapping (which is an algorithm executable by a deterministic Turing machine) requires only polynomial resources. I will provide an explicit example of such a mapping in Section 9.2.2.

An important remark

The notion of NP-completeness only pertains to the *existence* of a transformation—requiring only polynomial resources—from one NP-complete problem to any other in the NP class. It does *not*

[11]Do not confuse the word 'non-deterministic' with 'probabilistic' Turing machines. A *probabilistic Turing machine* is one that at each step of the computation determines the next step to follow according to some probability distribution (Garey and Johnson, 1990).

Is NP = P?
Consider the set of all problems whose answer can be *found* by a Turing machine in polynomial time. Call this set (class) P. Consider now the set of all problems whose solution can be *checked* in polynomial time, but it cannot necessarily be solved in such a time. Call this set (class) NP. Question: is the set NP the same as P?

[12] Note that, given a problem class, we need to consider the *worst-case scenario*: some instances of NP problems could be 'easy' to solve even with deterministic Turing machines.

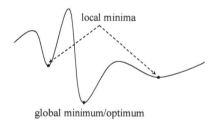

local minima

global minimum/optimum

Fig. 2.5 A *non-convex* function has several hills and valleys with many local minima (non-optimal or *sub-optimal solutions*) and one or more global minima/*optimal solution(s)*.

[13]More precisely, a problem X is NP-hard if any problem in the class NP can be reduced in polynomial-time to an instance of X.

'Solving' an optimization problem?
Properly speaking, we do not 'solve' optimization problems, because we cannot check that, whatever assignment of the variables we obtain within a certain solution time, that assignment is the optimum. I will nonetheless use this word to mean that the 'solution' of such a problem is the best assignment found within the prescribed time interval (see also Section 7.9).

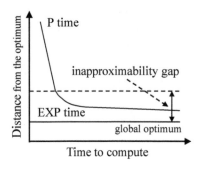

Fig. 2.6 *Inapproximability gap* of APX-hard problems. The solution to these optimization problems can be approximated in polynomial (P) time up to a 'distance' from the optimum. After that, exponentially (EXP) increasing time is required to further refine the solution.

---

establish the *degree* of such a polynomial. From a strictly *theoretical* point of view, the existence of such a transformation is a very strong result. However, from a *practical* point of view, if the degree of the polynomial is large, it may not always be convenient to transform one NP problem into another. In other words, the *representation* of a problem may considerably affect its time to solution (Garey and Johnson, 1990). See also Section 13.8.4 for a discussion of this issue in MemComputing.

A problem is said to be *NP-hard* if it is 'at least as hard to solve' as any problem in NP (Garey and Johnson, 1990).[13] It is then NP-complete if it is NP-hard and also in the class NP.

2.3.4   Optimization, decision, and search problems

Some problems, most notably *optimization problems* (e.g., those that require finding the optimal solution in a vast *non-convex* landscape dotted with hills, valleys, and saddles of varying heights) are NP-hard (Fig. 2.5). I will tackle various problems in this class in Chapters 9 and 10.

For these types of problems, even if we did find some solution, it would be hard to check in polynomial time that such a solution is indeed the *optimum* ('global minimum'), because we would have to compare it with the (typically) large number of sub-optimal solutions ('local minima').

In fact, some of these problems are so difficult that even if we found an approximation to the optimum, further refining such an approximation takes a time growing exponentially with input size (Fig. 2.6) (Arora and Boaz, 2009). These problems, also called *APX-hard*, are known to have an *inapproximability gap*.

This means that, if NP $\neq$ P, no algorithm can overcome, in polynomial time, a fraction of the optimal solution. In other words, one can approximate the problem in polynomial time with respect to the size, N, of the input only up to a certain 'distance' from the optimum. To go beyond this limit an exponentially growing time in N is always needed, unless NP = P.

Finally, the question we have posed for the subset-sum problem requires a simple binary answer: yes or no. In other words, we only asked whether a solution *exists*. This is what we would call a *decision problem*.

Many times, however, we would like to *find* a solution to the problem, if it exists. This is a different question than the previous (decision) one. We then say that we are trying to find the solution to a *search problem*. Many search problems can be formulated as decision problems and the complexity of the two types of problems is, oftentimes, the same for the same problem (Arora and Boaz, 2009).

In Chapters 9 and 10 we will encounter other examples of NP-complete and NP-hard problems, and we will show how MemComputing machines tackle them. At this stage of the book, however, a general understanding of what these classes of problems are is more than enough. The reader interested in a more in depth discussion is urged to consult the books I have already referenced (Turing, 2004); (Garey and Johnson, 1990); and (Arora and Boaz, 2009).

2.4   Boolean problems

Once we have understood the difference between the various Turing machines, and the types of problems they can solve, let's see how these problems are represented in practice.

We have already mentioned that we, humans, need to interact *meaningfully* with any computing machine, implying that we should be able to feed into it a *finite* set of symbols, and read the result of the computation *unequivocally* with *finite* means.

Again, 'finite' here means that the way we write the input and read the results cannot, in any way, require exponential resources with increasing input size. This requirement means that we want to solve a problem whose input and output are *digital*, or at least it can be written this way (maybe as an approximation).

Manipulation of digital symbols is best performed by *Boolean algebra*, that is the set of logical propositions that relate the truth value of two or more *variables* represented in a binary form, namely, whether their value is 'true' (logical 1) or 'false' (logical 0) (Rautenberg, 2009). For instance, an AND gate is a logical proposition that takes in two truth values of two input *literals* (variables or their negation), and gives out the corresponding truth value of the only output literal. The OR gate has a similar structure but different truth values at the output.

The symbols of the AND and OR gates and their *truth table* in the most common binary representation, which relates input and output variables, are given in Fig. 2.7. Note also that a *multi-terminal* OR gate, namely one that has n input variables and one output, will always give 1 as output, provided *at least one* input variable is 1. When all input variables are 0, then its output will be 0.

The proposition 'A AND B' that relates the two variables A and B is what we call a *clause*. Let's then summarize these Boolean concepts for future use:

(i) **Variable**: logical symbol that represents the truth value of 'true' (logical 1), or 'false' (logical 0), of a quantity.

(ii) **Literal**: logical assignment of a variable: either the value of

Boolean algebra rules
Let's use the symbols $\vee$ = OR, $\wedge$ = AND, and $\neg$ = negation. The following rules (from which any other Boolean relation can be derived) are valid between variables A, B, and C (Rautenberg, 2009):

Idempotence:

$$A \wedge A = A,$$

$$A \vee A = A.$$

Commutativity:

$$A \wedge B = B \wedge A,$$

$$A \vee B = B \vee A.$$

Associativity:

$$A \wedge (B \wedge C) = (A \wedge B) \wedge C = A \wedge B \wedge C,$$

$$A \vee (B \vee C) = (A \vee B) \vee C = A \vee B \vee C.$$

Distributivity:

$$A \wedge (B \vee C) = (A \wedge B) \vee (A \wedge C),$$

$$A \vee (B \wedge C) = (A \vee B) \wedge (A \vee C).$$

Duality:

$$\neg A \vee \neg B = \neg(A \wedge B),$$

$$\neg(A \vee B) = (\neg A) \wedge (\neg B).$$

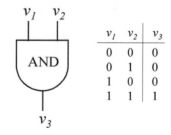

v_1	v_2	v_3
0	0	0
0	1	0
1	0	0
1	1	1

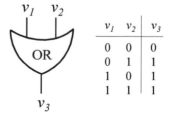

v_1	v_2	v_3
0	0	0
0	1	1
1	0	1
1	1	1

Fig. 2.7 Left: symbols of an AND gate and an OR gate. **Right:** their truth table in a binary representation.

the variable itself, or its negation.

(iii) **Clause**: logical proposition (formula) that relates variables.

A clause is also sometimes called a *constraint*. For later use, we also note that *any* propositional formula can be converted into an equivalent form known as *conjunctive normal form* (CNF).

This form defines a very important problem known as *satisfiability* (SAT) *problem* or *constraint satisfaction problem*, where the variables have to satisfy a set of constraints.

Satisfiability (SAT) problem

It is the problem of finding an assignment of the Boolean variables, such that their CNF formula evaluates to true. Equivalently, we can say that each clause should have at least one literal that is true under the assignment (for a disjunctive clause). Such a formula and all its clauses are then said to be 'satisfied'. If there is no assignment satisfying all clauses, the CNF formula is said to be 'unsatisfiable'.

An example of a CNF formula relating the three variables v_1, v_2, v_3 is:

$$\varphi(\mathbf{v}) = (\neg v_1 \vee v_2 \vee v_3) \wedge (\neg v_1 \vee \neg v_2 \vee v_3) \wedge (v_1 \vee \neg v_2 \vee \neg v_3) \wedge (\neg v_1 \vee v_2 \vee \neg v_3), \tag{2.4}$$

where $\mathbf{v} = (v_1, v_2, v_3)$, the symbols $\vee$ and $\wedge$ represent the logical OR and AND, respectively, while the symbol $\neg$ represents negation. In this formula, there are four clauses/constraints, each a disjunction of three literals. The four clauses are then related by logical ANDs.

The particular CNF formula (2.4) representing a SAT problem in which all clauses contain three distinct literals, of which none is a negation of the others, is called an *instance* of a 3-SAT problem. The 3-SAT is also an NP-complete problem, like the subset-sum problem (Arora and Boaz, 2009). As such, any other problem in the NP class can be reduced to it in polynomial time. The 2-SAT (when all clauses have 2 literals) is instead in the class P of problems that can be solved in polynomial time.

If the CNF formula cannot be satisfied by any set of literals, we could still try to solve a very important *optimization* problem: we could look for the variables assignment that *maximizes* the number of *satisfied* clauses. This problem is called a

Maximum satisfiability (MAX-SAT) problem

It is the problem of finding an assignment of the Boolean variables

---

Conjunctive normal form

A logic formula is in conjunctive normal form, *CNF*, if it is a conjunction ('ANDs') of disjunctions ('ORs') of literals. In this case, the disjunctions are the clauses or constraints.

Instance of a problem

An instance of a problem is just one of its possible realizations. For example, eqn (2.4) is a particular instance of the general 3-SAT problem.

Weighted MAX-SAT

Consider a MAX-SAT CNF formula in which we assign weights $w_j \in \mathbb{R}^+$ to each clause j. The NP-hard problem of finding the assignment of its variables that maximizes the sum of the weights of the satisfied clauses is called *weighted MAX-SAT*. The MAX-SAT is a special case of the weighted one, in which all clauses have weight 1.

Partial MAX-SAT

It is a MAX-SAT problem in which some clauses in the CNF formula are *always* satisfied, no matter the variable assignment (namely they have *infinite* weight).

Weighted partial MAX-SAT

It is a partial MAX-SAT for which an assignment is sought that maximizes the sum of the clauses with finite weight.

that *maximizes* the number of *satisfied* clauses—equivalently, *minimizes* the number of *unsatisfied* clauses—in a given logic formula.

Interestingly, unlike its SAT counterpart, even if all clauses have two literals, the corresponding maximum satisfiability problem (called MAX-E2SAT) is NP-complete (Garey and Johnson, 1990). This shows the fundamental difference between satisfiable and unsatifiable problems.

2.4.1   Boolean circuits

A CNF formula has a simple *Boolean circuit* representation (Arora and Boaz, 2009). An example of such a circuit is shown in Fig. 2.8 for the 3-SAT represented by the formula (2.4).

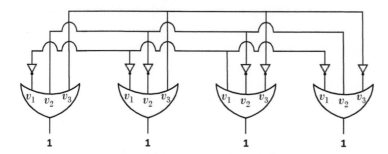

Fig. 2.8 Example of a Boolean circuit for a 3-SAT representing the CNF formula (2.4). The output of each 3-terminal OR gate is set to 1 (true), and the problem is to look for an assignment of the variables v_1, v_2, and v_3 that satisfies each clause. The triangle symbols indicate negations of the variables.

Let's then assume that our problem can be written in the form of a Boolean circuit (whether in CNF or not), namely it can be written as a set of *logic gates* whose input is a set of truth values (in a binary representation, a collections of 0s and 1s), and whose output is also a collection of truth values.

Let's then define the following general problems that we will mostly deal with in this book:

Boolean problem
Let $\mathbf{y} \in \mathbb{Z}^n$ and $f(\mathbf{y}) : \mathbb{Z}^n \to \mathbb{Z}^{n_o}$, a system of Boolean statements (with n, n_o integers) that accept as input the Boolean variables $\mathbf{y}$. Solving the problem defined by f and $\mathbf{b} \in \mathbb{Z}^{n_o}$ means that the machine must return a possible input $\mathbf{y}$ of f, if it exists, that satisfies $f(\mathbf{y}) = \mathbf{b}$.

However, this class of problems, although vast in its own, does not exhaust those that digital MemComputing machines can tackle. I will then introduce a much larger class of problems in Section 2.5.

The waves of Artificial Intelligence

Artificial Intelligence (AI) (also called 'Machine Intelligence') is the field of Science that studies any machine or device able to receive input from its environment and take appropriate actions to maximize its chances of achieving its goals successfully (Poole *et al.*, 1998). Although the following timeline is not super-precise, the field (whose name was coined at a conference in 1956) can be said to have experienced these (somewhat overlapping) 'waves':

First wave (1960s–2000s)— Rule-based learning, whereby the machine follows a set of fixed rules (*handcrafted knowledge*). Our modern computers operate that way.

Second wave (2000s–2020s)— Deep neural networks (*statistical models*) that are good at learning features from a set of data with labels ('supervised learning') without a rule-based programming, but have a hard time to generalize without labels ('unsupervised learning'); see Chapter 10. This wave was possible because of the availability of 'big data', as well as hardware and methods advancements.

Third wave (2020s–?)— Machines that generalize very efficiently from a very small set of data, so that they can adapt to the context in which they operate and learn without much supervision (*contextual adaptation*). It requires efficient 'unsupervised learning'.

Fourth wave (?)—Human-level intelligence? I am not even sure there is a unique agreed-upon *definition* of 'intelligence' (or even 'thinking') in this case.

2.5   Combinatorial optimization problems

In addition to problems that involve logic propositions, there are others that necessitate other types of relations or constraints between variables. For instance, in Chapter 9 I will discuss a few examples of an *algebraic* problem known as *Integer Linear Programming* (Arora and Boaz, 2009). This is a very important optimization problem (belonging to the NP-hard class) in which algebraic relations (inequalities) need to be satisfied between integer (not necessarily Boolean) variables (see Section 9.2.5 for a precise definition of this problem and its solution using DMMs).

In fact, this problem can be further extended by allowing some variables to acquire *non-integer* values. It is then called *Mixed-Integer Linear Programming* and finds a lot of applications in industry (Arora and Boaz, 2009).

In addition, the majority of problems in the field of *Machine Learning* (a sub-field of *Artificial Intelligence*) require tackling some type of optimization. I will discuss some examples of these problems in Chapter 10 and show how to employ MemComputing effectively in such a field, and in *Quantum Mechanics* (Messiah, 1958).

In this book we will deal with all these types of problems, whether they are just combinatorial in nature, such as the sub-set sum problem (Section 2.3), or also of an optimization type. It is then useful to classify them even more generally as follows.

Consider a problem $\mathcal{P}$, and a function $\varphi(\mathbf{v})$ of a set of variables $\mathbf{v}$ with cardinality n whose target space is the real axis. In this book, these variables typically take only integer values, but we can extend them to take any real value. The set $\mathbf{v}$ then belongs to $\mathbb{R}^n$.

The function $\varphi(\mathbf{v})$ is then $\varphi : \mathbb{R}^n \to \mathbb{R}$. We call φ an *objective function*.

Suppose also that the variables $\mathbf{v}$ are subject to additional constraints, namely $\mathbf{v} \in \Omega$, where Ω is a closed subset of $\mathbb{R}^n$.

We will then define the following general category of problems:

Combinatorial optimization problems

Those problems $\mathcal{P}$ that require the maximization (or minimization) of an *objective function* $\varphi(\mathbf{v}) : \mathbb{R}^n \to \mathbb{R}$ subject to the *constraints* $\mathbf{v} \in \Omega \subseteq \mathbb{R}^n$.

These are precisely the problems that *digital* MemComputing machines have been invented for (Traversa and Di Ventra, 2017; Di Ventra and Traversa, 2018a).

2.6 Why are some problems so difficult to solve (with algorithms)?

Having discussed the types of problems we will be mostly concerned in this book and how to represent them, I can now provide an intuitive understanding of why some of them are difficult to solve with traditional algorithms. This will also anticipate the reason why digital MemComputing machines are instead very efficient in solving these same problems. For clarity, let's refer to the 3-SAT problem represented in Fig. 2.8 and eqn (2.4).

We see from that figure (or the right-hand side of eqn (2.4)) that the variable v_1 appears in the first, second and fourth clauses negated, and in the third clause as is (I am numbering the clauses from left to right). On the other hand, the variable v_2 appears as is in the first and fourth clauses, negated in the other two. And so on for the third variable. The goal now is to find an assignment of all variables that renders that formula satisfied: $\varphi(\mathbf{v}) = 1$ ('true').

Let's now make an analogy that we will indeed use when we will discuss DMMs in Chapter 6. Imagine the circuit in Fig. 2.8 as an actual *physical* circuit (e.g., an electric one) so that the lines in that figure represent transmission or *information channels* relating the truth values of the variables at each terminal of all gates. Being now a physical circuit, you can think of it distributed in *real space*, namely the different terminals of the different gates are *physically separated* (and the variables can be conveniently thought of as voltages at the corresponding terminals).

Suppose now that we initiate the calculation by assigning a 1 to the variable v_1 and a 0 to the variable v_2. Since the clauses are 3-terminal OR gates, according to their truth table, we must have $v_3 = 1$ for the first clause to be satisfied. However, if we do that, then we realize that the fourth clause is unsatisfied.

I could instead assign the logical 0 to v_1, so that I have the freedom to assign either 0 or 1 to both v_2 and v_3. I would then have to check that all clauses are satisfied with either of those choices, in an *iterative* way, namely one after the other.

This process can be done quite easily in a problem that has only three variables and four clauses. But suppose you have to do it for several variables and several clauses (in the thousands or even millions!), then you immediately realize that it becomes a challenging combinatorial problem, that requires an enormous amount of checks along the way. The difficulty is precisely that we need to check each clause *iteratively*, by varying the assignment of each variable *one by one*.

An algorithm that accomplishes this task would then check *all* possibilities, and thus *guarantee* whether there is an assignment such that

Machine Learning
The sub-field of Artificial Intelligence that studies algorithms that learn features from a set of data to infer additional features, without having been directly trained on them, is called *Machine Learning* (Goodfellow, Bengio and Courville, 2016). I will dedicate the entire Chapter 10 to discuss how to use MemComputing in this field.

Branch and bound algorithm
The way in which we solve a combinatorial optimization problem as I have described in Section 2.6 can be classified as the simplest realization of the *branch and bound algorithm* (Arora and Boaz, 2009). The idea is to simply go through different 'branches' of the *solution tree* (Fig. 2.4). Each branch is then checked against some estimated *bounds* on the optimal solution. The branch is discarded if it does not give rise to a better solution than the best one presently found. If no bound is provided, the search is exhaustive, namely one checks all possibilities.

Complete algorithm
An algorithm able to verify whether a combinatorial optimization problem is satisfiable, and if not, to report the *global optimum*.

Incomplete algorithm
An algorithm that does *not* guarantee that it will eventually find a satisfying assignment or declare that a given problem is unsatisfiable.

Heuristic algorithm
An algorithm that employs a series of approximations to achieve a solution of a problem in a reasonable time, without regard for optimality and completeness. Arguably, this definition does not differ much from the one of 'incomplete algorithm'. In fact, the two terms are often used interchangeably.

Short vs. long length scales
The discussion in Section 2.6 shows that traditional algorithms can easily assign the truth value of the literals appearing in 'nearby' ('locally coupled') clauses. In Physics terms, we would say they can easily explore/detect the *short* length scales (*local* information) or *high energies* (*ultraviolet limit*) of a problem. Instead, they have a hard time assigning literal values belonging to 'far-away' clauses, namely they struggle to explore/detect the *long* length scales (*non-local/global* information) or *low energies* (*infrared limit*) of the problem. As I will discuss in this book, MemComputing machines efficiently explore *both* length (energy) scales. In fact, the exploration goes from short scales (high energies) to long scales (low energies). See also the discussion in Section 13.7.5.

$\varphi(\mathbf{v}) = 1$ or not. This type of algorithm could also be used, in principle, to find the *global optimum* of an optimization problem (e.g., when $\varphi(\mathbf{v}) = 0$, one could ask what the *maximum* number of satisfied clauses is) and verify whether the problem is not satisfiable ($\varphi(\mathbf{v}) = 0$). Such algorithms (sometimes called *solvers*, when one refers to their software implementation) are then called *complete* or *exact* (Arora and Boaz, 2009).

Of course, such a 'brute force' search for the solution of the problem is extremely time consuming. Therefore, one often resorts to *heuristic* approaches, that attempt to first satisfy a much smaller subset of all clauses (typically chosen at random), and then iteratively try to use that information to satisfy all of them, employing various strategies (Arora and Boaz, 2009).

For instance, one could perform a *local search* in the space of configurations of the variables by iteratively moving from one configuration to another differing by one or a few literals, and comparing the 'quality' of such configurations, e.g., by means of some energy or *cost function*, namely a function that measures the 'cost' of that move (see Chapter 10 for an example of cost function in the context of Machine Learning). These algorithms/solvers do not typically guarantee convergence to the correct solution (hence the name *incomplete* or *heuristic*).

Much research has gone, and still goes into finding efficient algorithms, both exact and heuristic, that improve the time to solution. Note, however, that such algorithms when applied to hard problem instances in the NP-complete/NP-hard classes still require exponential time to converge, when increasing the number of variables (Arora and Boaz, 2009).

2.7   The need for *non-perturbative* methods

We then see that the difficulty of these combinatorial optimization problems arises from the fact that variables appear in distinct, 'far-away' clauses (in the physical sense as described previously). In order to solve such problems efficiently one would then need *correlated* assignments of the variables at different clauses that could be 'physically' far away from each other.

For instance, one would need to know *simultaneously* that the assignment $v_1 = 1$, $v_2 = 0$, and $v_3 = 1$ for the CNF formula (2.4) does not satisfy the first and the fourth clauses, without checking one clause first and then the other. It is like having an 'oracle' (see Section 2.3.1) that through some mechanism, unknown to the user, would recognize at each step of the computation how to make this type of *global* assignments of variables (as if the machine were able to 'see' the whole problem *at once*).

In other words, in order to compute efficiently hard problems one needs some sort of *non-local, collective* information during the solution search.[14] As I have discussed in Section 1.6 (criterion IV of MemComputing), this collective behavior may be the one that characterizes 'computation at the edge of chaos' or at the *critical state* of a continuous phase transition (Langton, 1990). We will see in Chapters 11 and 12 that DMMs do employ such non-local, collective information in the form of physical trajectories, called *instantons* (Di Ventra *et al.*, 2017; Di Ventra and Ovchinnikov, 2019), which mimic the behavior of an oracle by 'choosing' the correct path to follow to find the solution.

[14]In physical language, non-local effects are characterized by *long wavelengths*. In turn, this implies *low energies*, and vice versa (cf. Section 11.6). As I will discuss in Chapters 11 and 12, *instantons* are the long-wavelength, low-energy dynamics of digital MemComputing machines.

An important remark

Both 'complete' and 'local-search' algorithms operate essentially by *perturbing* the value that the literals of the given problem have at each iteration, namely by changing one or a few literals out of the many in the problem specification. According to the previous discussion, to solve hard problem instances efficiently, one needs a *non-perturbative* approach, that provides a *collective* (correlated) assignment of many literals. As I will explain at length in Chapters 11 and 12, DMMs employ *instantons* to compute. Instantons belong to the class of *non-perturbative* physical phenomena (like *quantum tunneling*, to which they are related, see Section 11.6), namely phenomena for which *perturbation theory* is inadequate (Coleman, 1977). In fact, DMMs have features of *strongly coupled systems*.

Strongly coupled systems and the failure of perturbation theory

When the elementary units of a physical system are *strongly coupled*, one cannot describe their dynamics by considering first a 'simpler' version of the system (e.g., a collection of non-interacting units), and then *perturbing* such a system by adding *small* terms ('small' compared to the characteristic energy scale of the system) describing their interactions. We then say that *perturbation theory* fails for such systems. An example is *Quantum Chromodynamics*, where the coupling between *quarks* and *gluons* becomes too large at low energies for perturbation theory to hold (Peskin and Schroeder, 1995). The digital MemComputing machines I will discuss in this book showcase features of *strongly coupled systems*: their low-energy dynamics are *non-perturbative* (see Section 11.10).

Once we have rigorously defined the problems we want to solve, and explained why some of them are difficult to solve, let's now look at the way an actual *physical* digital machine could work.

2.8 A *physical* 'Turing machine'?

By definition (2.1), Turing machines map a *finite* string of symbols into a *finite* string of symbols, and do this in steps ('discrete time'). In modern parlance, this is what we call *digital* machines.

This follows Turing's original idea, in the sense that we, humans, can only *interpret* a finite number of symbols at a time. In other words, although we did come up with the *mathematical concept* of infinity, we can neither visualize it nor represent it with any *physically* existing object, and we cannot measure infinities since our instruments are finite (Di Ventra, 2018).

For the sake of simplicity, and because this is what is employed in modern computers, from now on, the symbols that we will use to represent the input and output of a computation are just 0 and 1, namely we use a binary representation. The logical 'false' is represented by 0 and the logical 'true' by 1. The finite precision of input and output provides an incredible advantage to these machines: they are *scalable*. The word 'scalable' means that, no matter the size of the input, and without the need of expending exponentially growing resources, the machine can still function and provide the answer in the form of a finite output, namely we do not need any approximation, or extra interpretative tools to understand the values representing the output.

This scalability comes from the fact that the two logical states represented by the numbers 0 and 1 are 'sufficiently distinct' that even small errors in their assignment will not 'scramble' them. For instance, even if we had, say, a 10% error in representing the 0 or 1 states, we can still say that they are two different symbols representing the logical 'false' (0) and the logical 'true' (1).

Note that in this discussion of scalability of a digital machine I have said nothing about its *transition function*, namely the function that maps the input symbols into the output symbols, see eqn (2.2). In fact, as I will discuss later in the book, the transition function can be chosen at will so long as the input and output can be interpreted with finite means, and the *form* of the transition function is not substantially modified by noise or perturbations so as to 'scramble' the output.[15]

We can then define:

[15] As I will explain in Chapters 11 and 12, this is the case for *digital* MemComputing machines whose input and output are digital, and the transition function is of *topological* character, namely employs topological features to map input to output. Therefore, the transition function of digital MemComputing machines is robust against noise and perturbations (see Section 12.7).

Digital machine
A machine whose input and output can be written and read, respectively, with an error that is *independent* of the size of the machine or the problem they attempt to solve. Such a machine is then *scalable* in terms of resources.

If the input or output (or both) were, say, real numbers, then an arbitrary real number requires an *infinite* number of bits (0s and 1s) to be represented. Therefore, (almost all) real numbers cannot be read or written with finite means (they are not 'computable' according to Turing, see Section 2.2).

This poses a significant challenge for the scalability of such a machine. This is easy to understand.

If the input and output states of the machine are represented by real numbers that do not differ much from each other, then in order to maintain the same level of precision in calculations, as the size of the problem grows, such a machine would require an increasingly large number of resources (typically growing exponentially as the size of the problem scales linearly) to be able to distinguish the different states.

This is because even tiny perturbations in the assignment of those states can render them indistinguishable from each other. Such a machine would then be called *analog*, to distinguish it from a digital one, and would not be easily scalable.

We then define:

> **Analog machine**
> A machine whose input or output (or both) *cannot* be written and read, respectively, with an error that is independent of the size of the machine or the problem they attempt to solve. Such a machine is *not* easily scalable in terms of resources.

I have already provided some examples of analog machines in the Prologue (Chapter 1), although I did not classify them as such.

For instance, the planetary object that calculates its orbit, together with the telescope is such a machine. The position of the object is defined by real numbers.

Similarly, the speedometer of the car (in its most 'natural' version) moves its pointer in a continuous way, hence assigning real numbers to the car velocity. And so on, for all the other cases in which the input or output, or both, of the computation are real numbers.

Note again that I did not say anything about the transition function (2.2), namely the mapping that relates input to output. I want to stress once more that what is mostly important for scalability is the ability to write/read input/output with finite precision, *not* the type of transition function (provided the latter is not susceptible to noise so as to destroy the digital structure of the input/output).[16]

This is a very important point, that allows us to build actual scalable machines: our own modern computers.

Our modern computers are *not* deterministic Turing machines
So, how do we actually *build* in hardware a digital (deterministic Turing) machine? Since we want to map integers into integers, it should be easy, right? Not quite.

It may come as a surprise to some, but the way modern-day computers function at a practical level, makes them the closest realization of a deterministic Turing machine, but *not exactly* the same thing. By modern computers, I again mean those that are actually built in hardware, not those existing only in the abstract world of Mathematics.

In fact, in the real, *physical* world there is *no* such a thing as a true Turing machine! Let me explain.

The bits that represent the input and the output of a given operation are physically represented in hardware as high or low current (or voltage)

[16] For instance, in (Ercsey-Ravasz and Toroczkai, 2011), the transition function relies on *chaotic* behavior, which is highly sensitive to noise. So, even if the input and output were digital, a chaotic transition function would be *detrimental* to scalability.

> **A common misconception**
> Strictly speaking, our modern computers are *not* deterministic Turing machines! This is because they are *physical* objects and, hence, cannot represent an exact *mathematical* concept. However, they are the closest *physical realization* of the mathematical idea of deterministic Turing machines, if we do not worry about the fact that modern computers do *not* have infinite memory like a Turing machine, and operate in *continuous* time, not in steps (see Section 2.2.1).

[17]Note that, to reduce power consumption, in actual computer architectures, complementary transistors are typically used to represent the logical bits and functions (cf. Section 3.3, and in particular the definition of CMOS).

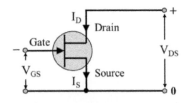

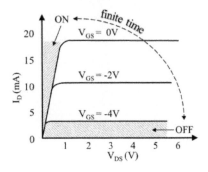

Fig. 2.9 Top panel: circuit diagram of a *transistor*. The drain current, I_D, is controlled by the gate-source voltage, V_{GS}, at a fixed drain-source voltage difference, V_{DS}. **Bottom panel:** Schematic of the I_D-V_{DS} characteristics of a transistor for different values of V_{GS}. When I_D is within a fixed threshold, ϵ_{OFF}, the transistor is in the OFF state (logical 0). Within another threshold, ϵ_{ON}, it is in the ON state (logical 1). It takes a *finite time* to switch from the ON to the OFF state, and vice versa.

[18]For example, the range $0 \div 0.5$V is sometimes used to represent the logical 0, while the range $2.4 \div 2.9$V, the logical 1. Any voltage value in between is not a valid logical state (Horowitz, 2015).

[19]And, of course, *we*, the users, attribute some *meaning* to the physical representation of the bits (cf. Section 1.3).

values of devices (e.g., *transistors*; Fig. 2.9), or some other physical property, such as patterns of magnetization in a magnetizable material, or charges in capacitors, etc. (Hennessy and Patterson, 2006).

Say, a high transistor current represents a logical bit 1, while a low current a logical bit 0.[17] The transition function itself is implemented by physical objects such as transistors, and other circuit elements.

This physical representation comes with the unavoidable baggage of *any* physical system: *noise*, circuit *imperfections*, and materials *defects* of the electronic components of the full circuit. This unavoidable noise and imperfections/defects do not allow the machine to have, at any given time, a well-defined current (or voltage) for all the elements representing the bits.

In other words, *there is no way* that, at any given time, an actual *physical* computer could represent its bits *exactly* as ideal mathematical 1s or 0s.

In addition, it takes some physical (continuous) *time* for a transistor to switch from a high to a low current, and vice versa, namely, unlike for a Turing machine, the switching from a physical representation of a logical 1 to a logical 0, and vice versa, is *not* instantaneous (Fig. 2.9).

Therefore, our modern computers operate in *continuous time*, not in steps, as a Turing machine!

In this respect, since the physical devices in our modern-day computers can only provide real numbers, and not the exact mathematical values of 1 and 0, they seem to have the characteristics of an analog machine.

Nevertheless, we would never conclude that, due to these features, our computers are not digital, hence not scalable. So, where is the catch?

The reason is that, *within a certain threshold*, which, importantly, is *independent of the size* of the machine or the input of the problem to solve, we can always attribute to those physical properties (such as the current), that represent the states of the input and output, the mathematical value of 1 or 0 (Fig. 2.9). That threshold is *fixed* by a single unit of the machine (say a transistor) and is *independent* of how many units (transistors) we put together.[18]

This is the only thing we require of our modern computers: to be able to write and read input and output, respectively, within a certain threshold *independent* of the size of the problem.[19]

We then still (rightfully) call our modern computers *digital*, not analog, because they do not need an exponential increase in the number of resources when we scale the size of the problem we want them to solve.

Of course, for practical reasons, related to how small the transistors can be fabricated reliably, the heat generated during calculations, costs of production, etc., our modern computers *are* limited by the number of transistors we can fit in a CPU, or the memory they hold, etc.

However, these practical, albeit important, limitations do not invalidate the fact that our modern computers are digital.

2.9   Toward a *physical* digital machine

Once we have understood this important point regarding what constitutes a scalable machine, we are then free to play with the transition function at will. This is where Physics ideas will come in handy.

In fact, instead of trying to force physical systems, such as transistors, to approximate the features of a deterministic Turing machine, we may employ other physical properties and systems that allow us to compute, while keeping the output and input digital.

This brings us *away* from the Turing paradigm of computation but opens up a vast number of possibilities. In fact, as I have anticipated in Chapter 1, physical systems can perform complex computations effortlessly, and they do that without 'caring' whether such computations can be easily performed by Turing (or other mathematical) machines.

As an illustration, let's go back to the planetary example of Chapter 1.[20] Suppose you want to simulate the orbit of Earth, as measured from the Sun, using Newtonian mechanics.

If we used an algorithm running on our modern computers, we would need to *discretize* both space and time, and then solve Newton's equations of motion within those approximations. The result we obtain is again an *approximation* of the actual physical trajectory.

In addition, numerical errors accumulate in time, and, if we do not perform the simulations with particular care, what should be a simple elliptic orbit may soon mold into an unrecognizable set of lines as time progresses. And yet, Earth performs this motion effortlessly, not just alone with the Sun, but also in the presence of all the other planets and objects in the entire Universe!

The actual physical system (Earth) does not care whether a Turing machine or its approximation (our modern computers) can calculate its exact orbit: it does so irrespective of how we, humans, *interpret* its motion.[21]

[21]See Section 1.3 for a discussion on this *semantic information* step of the computation.

Of course, as I have already mentioned, the 'computing machine' Earth (plus telescope) would represent a truly analog computer, as nowhere during the execution of its trajectory, we can write a finite input or read a finite output: any point of the Earth's trajectory in space is represented by a collection of real numbers.

The goal is then to find a *physical* system that performs in *continuous time* (and continuous space as well) some form of computation, while preserving the *digital* structure of the input and output. This way, we may take advantage of the tremendous power physical systems offer in performing complex calculations, while being able to interpret their results with *finite means*. This is the starting point of digital MemComputing machines.

2.10 Chapter summary

- In this Chapter we have discussed the way computing is presently done and interpreted. A problem is tackled *algorithmically*, and, typically, no physical information is used to help in the computation.
- This way of computing goes back to the ideas introduced by Turing to formalize how we, humans, solve problems.
- The Turing machine is an ideal *digital* device that maps a *finite* set of symbols into a *finite* set of symbols in *discrete time* (steps).
- Keeping the *finite* structure of input and output is the only requirement of a *scalable* machine, provided, of course, noise is not strong enough so as to 'scramble' the output.
- The machine itself can perform *continuous-time* and *continuous-space* computation, so long as it reads in and writes out a set of symbols that can be *interpreted* by us, humans, with *finite* means. As I will explain in this book, this is what a *digital* MemComputing machine does.

Dynamical systems picture of computing

The difficulty lies not so much in developing new ideas as in escaping from old ones.
John Maynard Keynes (1883–1946)

A new scientific truth does not triumph by convincing its opponents and making them see the light, but rather because its opponents eventually die, and a new generation grows up that is familiar with it.
Max Planck (1858–1947)

Takeaways from this Chapter

- Major differences between computing with quantum and non-quantum systems.
- 'Classical' does *not* necessarily mean 'Turing'.
- Difference between 'intrinsic' and 'standard' parallelism.
- The long-time behavior of dynamical systems.
- The notion of 'integrability' of dynamical systems.
- Which phase space is ideal for computing?

In the last Section 2.9 of Chapter 2, we have seen that if we found a physical system that, while solving a problem, still preserves the *digital* (finite) structure of the input and output, we may take advantage of its *continuous time* (and space) features, and, at the same time, be able to interpret the results of the computation with finite means.

Unlike a Turing machine, 'time' here is not simply 'counting steps' of an algorithm. Rather, it is a well-defined *physical variable*[1] that labels the evolution (dynamics) of the system (see also Section 2.2.1). This is the reason we call these physical systems, *dynamical systems* (Perko, 2001).

[1] Let's recall that *physical time* is that quantity which is *measured* by a clock.

3.1   Quantum dynamical systems

A dynamical system can manifest quantum or non-quantum properties, or both. Since the machines I will discuss in this book do not really need quantum features, such as *entanglement*, to operate, I will only consider *non-quantum* dynamical systems.[2]

This is not a minor point. Suppose, in fact, you needed a quantum feature such as entanglement to compute. This property, which implies that the state of a given particle in a collection of particles cannot be described independently of the others, *irrespective* of their spatial separation, is prominent in the computing paradigm known as *Quantum Computing* (Nielsen and Chuang, 2010). It is this feature (together with *quantum interference*) that may allow, for instance, the factorization of numbers into primes in polynomial time on a (*ideal*) quantum computer versus the exponential time of traditional algorithms running on our traditional computers (Shor, 1997).

However, in order to describe the dynamics of the state of a system with such a feature, one needs to solve the *partial* differential equation known as Schrödinger's equation (or its 'mixed-state version' if the system is not in a pure state; the interested reader is urged to look into the book (Messiah, 1958)). The state itself is a vector in a topological (complex) vector space known as *Hilbert space*, whose dimension grows *exponentially* with system size (Nielsen and Chuang, 2010).

To make this point even more explicit, suppose you are given the fundamental unit of a quantum computer, known as a 'quantum bit' or *qubit*. This is a quantum-mechanical system with two states, e.g., the two states of the spin of an electron. To the qubit is associated a two-dimensional Hilbert space spanned by two vectors, we can indicate as $|0\rangle$ and $|1\rangle$.[3]

Now, unlike a *classical* bit, that can be in only one of two states (Section 2.2.1), a qubit can be, at any given time (before an actual measurement is performed on the system!), in a state $|\Psi\rangle$ which is a *linear superposition* of $|0\rangle$ and $|1\rangle$: $|\Psi\rangle = \alpha|0\rangle + \beta|1\rangle$, with α and β some complex numbers such that $|\alpha|^2 + |\beta|^2 = 1$.

If we now put together n such qubits, we can form a much larger Hilbert space of the collection of these n qubits which is the *tensor product* of the Hilbert spaces of the individual qubits (Nielsen and Chuang, 2010). However, the dimension of this new n-qubit Hilbert space is not n, but 2^n, namely *exponential* in the number of qubits (Fig. 3.1).[4]

It is this exponential growth that makes it difficult to *simulate* quantum computers efficiently on our traditional computers. For instance, for a meager 50-qubit system, whose computational power can be over-

---

[2]Quantum effects may still affect the physical properties of some devices used in actual MemComputing machines, e.g., spin-based devices (Pershin and Di Ventra, 2008). However, their *macroscopic* dynamics do not typically require the full *microscopic* description based on, e.g., *state vectors* in Hilbert spaces. In other words, even these 'quantum-mechanical' devices can be described by ordinary differential equations of the type (3.1).

Quantum entanglement
It is a quantum feature of systems with at least two degrees of freedom (cf. Section 3.3) that allows their individual elements to 'correlate' with each other at very long distances, as if the whole system were 'rigid': a perturbation in (measurement of) one of its parts, would immediately affect other parts arbitrarily far away. In Section 4.3.1 we will see that entanglement can be viewed as a type of *ideal long-range order*, one in which the correlations do not decay spatially.

[3]Following Dirac, a general state vector, Ψ, in Hilbert space is indicated as $|\Psi\rangle$, called 'ket' (Messiah, 1958).

Hilbert space
A topological vector space endowed with the operation of *inner product* among its elements (that associates each pair of vectors to a scalar) is called a *Hilbert space* (Arfken *et al.*, 1989).

[4]The space of *states* of our traditional computers contains also 2^n possible states, with n the number of, e.g., transistors representing bits. However, unlike a quantum computer, a traditional computer can be, at any given time, in one and only one of such states.

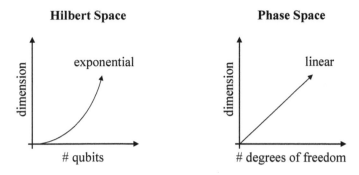

Fig. 3.1 The dimension of the Hilbert space of a quantum computer increases *exponentially* with the number of qubits. Instead, the dimension of the *phase space* of a non-quantum system grows *linearly* with the number of degrees of freedom (dynamical variables describing the system). See Sections 3.2 and 3.3.

shadowed by our traditional computers, we would need to keep track of $2^{50} - 1 \simeq 10^{15}$ complex numbers![5]

Therefore, in order to take advantage of a quantum computer, one is forced to build it in *hardware*. However, the hardware realization of quantum computers faces its own challenges; see, e.g., (Dyakonov, 2019).

If entanglement needs to be maintained for a reasonable time for an actual computation to occur, then quantum computers typically need to operate at extremely low temperatures (a few milli-Kelvins, 10^{-3}K).[6] This is because entanglement is very sensitive to environment fluctuations, a phenomenon known as *decoherence*, which makes the quantum system behave more closely to a classical one as time goes by.

Operating at such extreme temperatures (much smaller than the cosmic microwave background radiation of about 2.7 K) makes the hardware realization of these machines very difficult to scale.

In fact, to counteract for the unavoidable decoherence, for each qubit, one typically needs a lot (possibly thousands) of other qubits that provide *error correction* for the given qubit, thus guaranteeing the necessary *fidelity* (quality; cf. Section 10.7) in the operation of quantum gates.[7]

3.2 Non-quantum dynamical systems

The limitations and difficulties that plague the practical realization of quantum computers I have outlined in the previous Section 3.1, *do not* apply to *non-quantum* dynamical systems, for at least three reasons.

However, before discussing these reasons, let me define more precisely what I mean by non-quantum dynamical systems, as those I will consider in this book. These are *dynamical systems* described by *ordinary differential equations* (ODEs) of the type:[8]

Dynamical system

$$\dot{\mathbf{x}}(t) = F(\mathbf{x}(t)); \quad \mathbf{x}(t = t_0) = \mathbf{x}_0, \tag{3.1}$$

where $\mathbf{x}(t)$ is a D-dimensional set of *dynamical (state) variables* that de-

[5]The -1 is due to the state normalization constraint, but it is irrelevant at that size. Note also that the memory requirement for such a 50-qubit simulation can easily exceed tens of terabytes (10^{12} bytes) of random access memory (RAM), if not petabytes (10^{15} bytes)!

[6]Entanglement of photons may provide a path to much higher (even room) temperature(s), but it is severely limited by the size of the instrumentation (i.e., the lasers) used.

Tensor product
Given two vector spaces A and B, their *tensor product* is a vector space $A \otimes B$, endowed with a bi-linear map (a function that is linear in both arguments) from the Cartesian product $A \times B$ (see Section 2.2.1) to $A \otimes B$ (Arfken *et al.*, 1989).

[7]Note also that a quantum computer is a *probabilistic* machine, namely the calculation of a given problem needs to be repeated an *ensemble* of times on an equally prepared system, so that an *average* of the results can be collected. This is an *intrinsic* feature of Quantum Mechanics whereby the measurement of the state of a quantum system only provides one outcome out of a probability distribution of outcomes (Messiah, 1958). The MemComputing machines I will introduce later in the book are instead *deterministic* machines, and they are fundamentally *non-quantum*.

[8]Equations (3.1) define also an *initial value problem*, and the symbol $\dot{\mathbf{x}}(t)$ stands for $d\mathbf{x}(t)/dt$.

[9]Sometimes, the state $\mathbf{x}(t)$ is also called a *trajectory* in the phase space.

[10]The flow vector field, $F(\mathbf{x})$, 'lives' on the *tangent space*, $TX_{\mathbf{x}}$, of X, at each point $\mathbf{x}$. It is a section of the *tangent bundle*, TX. See Section 11.3.

Topological space

There are different ways to define a topological space. Intuitively, it is a space of *points* and *neighborhoods* of such points that are related to each other via specific axioms. The interested reader should consult, e.g., (Fomenko, 1994) for a rigorous definition.

Manifold

It is a topological space of some dimension D that at each point 'resembles' Euclidean space. More precisely, *any* point in the space has an open neighborhood which is *homeomorphic* to ('homeo' 'same', 'morphic' 'shape'; namely it is topologically equivalent to) an open subset of Euclidean D-dimensional space. We will consider D-manifolds that are also *smooth* (with a *differential structure* that allows for differential calculus on the manifold).

Phase space

It is the topological manifold $X \subset \mathbb{R}^D$ of all states of a given *non-quantum* dynamical system and its dimension, D, is equal to the number of *degrees of freedom* of the system (Goldstein, 1965; Arnold, 1989). Unlike a Hilbert space, the phase space is typically *non-linear* and endowed with just a *metric*, which is a way to measure distances between points in the space (Arfken *et al.*, 1989). The metric induces a topology, which is also a *Hausdorff space* (a space where for each pair of distinct points in X, there exist two disjoint open sets to which these points belong). See also Section 6.3.1 and Fig. 7.21.

fines the *state* of the system at time t.[9] It belongs to some D-dimensional *topological space*, X, called *phase space*, and F is a vector (called the *flow vector field*) representing the laws of temporal evolution of $\mathbf{x}$.[10]

For instance, in classical Newtonian mechanics, if $\mathbf{x}(t)$ is the velocity of a particle of unit mass, F is the sum of all forces acting on that particle. In Hamiltonian dynamics, $\mathbf{x}(t)$ is the collection of generalized coordinates and momenta of the particles, and F is related to the derivative of the Hamiltonian with respect to coordinates and momenta (Goldstein, 1965). In Chapter 7, F will represent digital MemComputing machines that can be realized in *hardware*. For now, it suffices to say that it is some continuous functional of the state vector, $\mathbf{x}(t)$.

Of course, we may also envision MemComputing machines whose fundamental units are described by *partial* differential equations, such as those used in the description of classical *fields* (Landau and Lifshitz, 1975).

However, unless there is a clear physical advantage, for simulation purposes (cf. Chapter 8), it is obviously better to consider only dynamical systems described by ODEs, as I will do in this book.

Note also that eqns (3.1) are D *coupled* ODEs, where $\mathbf{x} \equiv (x_1, \ldots, x_D)$ and $F \equiv (F_1, \ldots, F_D)$. We can then equivalently write eqns (3.1) as:

$$\begin{pmatrix} \dot{x}_1(t) \\ \dot{x}_2(t) \\ \vdots \\ \dot{x}_D(t) \end{pmatrix} = \begin{pmatrix} F_1(\mathbf{x}(t)) \\ F_2(\mathbf{x}(t)) \\ \vdots \\ F_D(\mathbf{x}(t)) \end{pmatrix} ; \quad \begin{pmatrix} x_1(t = t_0) \\ x_2(t = t_0) \\ \vdots \\ x_D(t = t_0) \end{pmatrix} = \begin{pmatrix} x_{1,0} \\ x_{2,0} \\ \vdots \\ x_{D,0} \end{pmatrix} \quad (3.2)$$

The coupling between the different ODEs comes precisely from the fact that the different components of the flow vector field F depend on the whole vector of variables $\mathbf{x}$. This is a very important point. In fact, in Section 3.6 I will show that such a feature leads to a type of parallelism (*intrinsic parallelism*) that is not shared by our modern 'parallel computers'.

For later use, let's also remark that eqns (3.1) can be equivalently written as

$$T(t)\mathbf{x}(t_0) = \mathbf{x}(t_0) + \int_{t_0}^{t} F(\mathbf{x}(t'))dt' \quad (3.3)$$

with $T : \mathbb{R} \times X \to X$, also called a *flow field* or just *flow*.

Finally, let me mention that the ODEs (3.1) are called *autonomous*, because they do *not* depend *explicitly* on time.

If the dynamics were *explicitly* dependent on time, namely the equations would be of the type $\dot{\mathbf{x}}(t) = F(\mathbf{x}(t), t)$, we would call them *non-autonomous* (Perko, 2001). As already mentioned, we will only deal with *autonomous* dynamical systems in this book.

3.3 Advantages of non-quantum systems (for computing)

With these definitions in mind, we can now enumerate the main advantages of non-quantum dynamical systems over quantum systems for what concerns *computation*.

Flow of a dynamical system
The *flow* $T : \mathbb{R} \times X \to X$ is a C^r *semigroup* with $r \geq 1$, and *Lipschitz continuous*. It is a semigroup because it does not necessarily have the inverse property of a group. C^r means that all its derivatives up to the r-th exist and are continuous. Lipschitz continuity means that the field is limited on how rapidly it may vary in time (cf. Section 8.5.1). See (Arfken *et al.*, 1989) for a more precise definition of all these terms.

An important remark
Note that I am only discussing advantages in the field of *computation*. Of course, quantum systems showcase phenomena that do *not* have a classical counterpart, and which we employ heavily in our modern technology, such as in lasers, transistors, and atomic clocks, to name just a few. Somehow, these successes give the impression that anything 'quantum' must be useful. However, the question of 'usefulness' in *any* technology should be first addressed in terms of its *end goal*. As I have already mentioned in Section 1.1, in the case of computing, the end goal is to solve problems that are quite challenging for us, humans. If a machine is 'quantum', but does *not* accomplish that goal, then it may be academically interesting, but it has little practical utility (cf. Section 13.9).

I. First advantage of non-quantum systems

First of all, as I mentioned previously, the equations of motion of the non-quantum systems I will discuss in this book are ODEs, and ODEs *can* be efficiently simulated using our traditional computers. For instance, according to whether one uses *explicit* methods of integration or *implicit* ones (see the definition in Section 8.4.1), the numerical overhead of simulating them (in terms of integration steps) can vary from linear (e.g., forward Euler) to some other polynomial (e.g., Runge-Kutta 4th order) (Press *et al.*, 2007).

The *memory* (RAM) requirement for such simulations instead always scales *linearly* with the number of *degrees of freedom*. This linear RAM requirement is because at any given step of the computation only the *present state* of all variables in the integration must be stored. I will

---

Degrees of freedom
The state variables, x_i, in eqns (3.1), are also called *degrees of freedom* (Goldstein, 1965). This nomenclature comes from the fact that they already take into account any constraint on the system. Therefore, they are the maximum number of 'free' variables that characterize the dynamics of the system. Their number is equal to the dimension, D, of the phase space, X.

show several examples of this numerical advantage in Chapters 8, 9, and 10, when simulating the equations of motion of digital MemComputing machines.

CMOS

Complementary metal-oxide-semiconductor (CMOS) technology is the modern fabrication process to integrate field-effect transistors (cf. Fig. 2.9) on chips, namely to fabricate *integrated circuits*. In particular, it employs complementary pairs of *p*-type and *n*-type metal–oxide–semiconductor field-effect transistors to implement logic functions. A *p*-type transistor operates such that when the gate has positive voltage no source-drain current flows, while when the gate voltage is zero, source-drain current flows. An *n*-type transistor is complementary, namely when the gate has positive voltage current flows, when it has zero voltage, no current flows. When put together, a logical one is represented by one transistor ON and the other OFF, and vice versa for a logical zero. This way, they use power only when they switch, and not when they hold a state. Any type of traditional logic gate can be implemented using complementary transistors (Hennessy and Patterson, 2006).

II. Second advantage of non-quantum systems

Another important difference between quantum and non-quantum systems can be immediately seen from the previous definition (3.1). Unlike the Hilbert space of a quantum system, the *dimension* of the phase space grows *linearly* with the number of *state variables* (degrees of freedom); see Fig. 3.1. This is by no means a minor point. In fact, coupled with the polynomial convergence of integration methods for ODEs, this means that the numerical cost of *simulating* a non-quantum system described by eqns (3.1) should grow *polynomially* with the number of state variables. As we will see later in the book (Chapters 8, 9, and 10), unlike quantum computers, all this makes it possible to *simulate* digital MemComputing machines *efficiently* with our modern computers.

III. Third advantage of non-quantum systems

Unlike quantum systems, non-quantum systems are *not* sensitive to decoherence effects (Section 3.1). Therefore, if one can design a non-quantum machine (like a MemComputing machine) that can sustain reasonable amounts of noise and perturbations, then such a machine can operate at *room temperature* ($\sim$ 298 K). This makes its hardware realization much easier than in the case of a quantum computer. In fact, if standard electronics are employed (as I will consider in this book), then one can take full advantage of the well-established *complementary metal-oxide-semiconductor* (CMOS) technology that underlies the success of our modern computers (see also discussion in Section 4.5) (Hennessy and Patterson, 2006).

3.4   Deterministic vs. predictable dynamics

It is important to stress that the equations of motion (3.1) (like their quantum counterpart, the Schrödinger's equation) require the assignment of *initial conditions* so that at the initial time of the dynamics, the system is in some state $\mathbf{x}_0$. Without noise, given this initial condition, eqns (3.1) would then determine *uniquely* the state $\mathbf{x}(t)$ at a later time t. The system is *deterministic*.

Note, however, that *deterministic* dynamics does not necessarily mean that it is easily *predictable*. For instance, *deterministic chaos* is a term indicating that the system dynamics evolve in a deterministic way, but by no means the dynamical trajectories of the system can be easily predicted at later times (Gilmore, 1998) (see Section 3.9). The digital MemComputing machines I will consider in this book are *non*-chaotic *deterministic* dynamical systems (Di Ventra and Traversa, 2017a).[11]

3.5 On the word 'classical'

Here, I want to pause for a moment and make a clarification that I think is very important. It has been proved mathematically that an *ideal*[12] quantum computer, if built in hardware, can solve, e.g., *prime factorization* in polynomial time vs. the exponential time of 'classical machines' (Shor, 1994). This is sometimes referred to with the unfortunate moniker of 'quantum supremacy' over classical computers.

What is meant by 'classical' in this context is either our modern digital computers or their idealized version, the deterministic Turing machine (see Chapter 2). Non-quantum physical systems are indeed (correctly) called 'classical' in the Physics literature. However, it would be a misrepresentation to characterize the non-quantum, classical physical systems that realize the concept of digital MemComputing machines as deterministic Turing machines. As we will see in Chapter 5, MemComputing machines, whether digital or not, are *not* Turing machines!

It would then be really naïve to put them in the same category. This is the reason why I have used the term 'non-quantum' rather than 'classical' all along, and I will continue to do so when necessary for the remainder of the book.

3.6 Intrinsic vs. standard parallelism

After these preliminaries, suppose now you could actually build a machine as described by eqns (3.1). In other words, suppose you have a physical system that can be mathematically described by eqns (3.1), with some flow vector field, F.

We call the degrees of freedom of this system *coupled* or *dependent* on each other. These are the physical situations of interest.

(In fact, if the degrees of freedom were *independent*, each one of them would evolve separately from the others, without ever 'knowing' about each other's existence.)

This system, through its dynamics, performs some computation (e.g., a planetary object tracing its trajectory around the Sun, as I mentioned in Chapter 1). In the spirit of what we have discussed for Turing machines (see Chapter 2), we can then interpret eqns (3.1) as follows.

If the system starts at a time t_0 in a (input) state $\mathbf{x}(t_0)$, then after an interval of time Δt, it will be in a new (output) state $\mathbf{x}(t_0 + \Delta t)$, as

[11]Nevertheless, as I will discuss in Chapters 11 and 12, their dynamics are still not so easily predictable *a priori* due to the presence of *instantons*. However, as we will see, this lack of predictability of the *microscopic dynamics* does not preclude the machines from solving the problems they are designed to tackle. This is because several (even quite distinct) trajectories can lead to the problem solution(s).

[12]Namely, *decoherence-free* (cf. Section 3.1).

Classical does *not* mean Turing!
In this book I will consider only *non-quantum* systems to compute. Although these are often called 'classical' in the Physics literature, they are by no means synonymous with Turing machines, or with our modern computers implemented with, e.g., the von Neumann architecture (Section 5.2).

Independent degrees of freedom
If each component of the flow vector field, F, in eqns (3.1) depends on only one state variable (*degree of freedom*), x_i, namely eqns (3.1) are:

$$\dot{x}_1 = F_1(x_1),$$
$$\vdots$$
$$\dot{x}_D = F_D(x_D),$$

then the degrees of freedom are *independent*. This is not a particularly interesting case, and I will not consider it in this book.

[13] In physical terms, the transition function acts as the *flow* of the dynamics (3.3).

[14] Here, I am assuming that there is no noise in the system. In reality, noise is always present in *any* physical system. I will discuss the role of noise in Section 12.7.

[15] Note that, despite this collective behavior, the *correlations* among different elements in the system can still decay exponentially, both spatially and temporally. I will show in Chapters 11 and 12 that digital MemComputing machines are particular dynamical systems that develop *dynamical long-range order*, namely the correlations (in both space and time) of their degrees of freedom follow a *power law*, not an exponential one, despite *local* interactions between their units.

determined by eqns (3.1).

In other words, there must be a mapping, or better a *transition function*, δ, that relates the initial and final states, the input and the output, respectively, of the computation performed by this physical system (cf. eqn (2.2)).[13]

We can then write eqns (3.1) in a form similar to eqn (2.2) that we used to define a Turing machine:[14]

Intrinsic parallelism

$$\delta(\mathbf{x}(t_0)) = \mathbf{x}(t_0 + \Delta t). \tag{3.4}$$

Now, this seemingly trivial equation represents a very important feature for a machine whose computation is described by it.

In fact, consider only one of the D elements of the state vector $\mathbf{x}(t)$. Let's now perturb it slightly. Then, in view of the flow vector field F in eqn (3.1), which is common to all variables in $\mathbf{x}(t)$, we see from eqn (3.4) that *every* element of the vector $\mathbf{x}(t)$ is, at any given time, affected by the dynamics of *all* the other elements.

We would then say that the dynamics described by eqn (3.4) are *collective*: all elements of the physical system participate in the dynamics *at once*.[15]

Since we are discussing dynamical systems in the context of computing, we call this property *intrinsic parallelism*. It is *fundamentally different* from the *standard parallelism* we intend when we refer to our modern computers (see Fig. 3.2).

Standard Parallelism

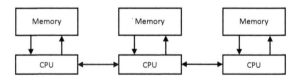

Fig. 3.2 Top panel: representation of a *standard parallel machine*. The different central processing units (CPUs) have their own memory and do not interact with each other until the end of each *clock cycle*. **Bottom panel:** representation of an *intrinsically parallel machine*. All the elements (seven in the figure) of the state, $\mathbf{x}(t)$, evolve *collectively* under the action of the common flow vector field.

Intrinsic Parallelism

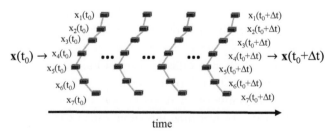

To see this difference, let's recall that our 'parallel computers' are made of several central processing units (CPUs), each with their own memory or accessing a shared memory, according to the architecture. The most important point is that if a program is executed on all the CPUs, each one of them performs a particular task during a *clock cycle*.

However, the tasks executed by each CPU are *independent* of the tasks executed by the other ones. In other words, it is *only* at the end of the clock cycle that all the CPUs share their results, not during the cycle interval (the CPUs are *synchronized*). Within a cycle each CPU has *no information* on what the other CPUs are doing.

We can make this analysis more rigorous by describing what I just wrote in terms of dynamical systems theory. This is similar to the discussion in (Di Ventra and Traversa, 2018a).

Suppose we have n_s CPUs with n_s memory units. The states of the n_s CPUs, and the symbols written on the total memory formed by the n_s memory units, can be compactly written as vector functions $\mathbf{s}(t) = [s_1(t), ..., s_{n_s}(t)]$ and $\mathbf{m}(t) = [m_1(t), ..., m_{n_s}(t)]$, respectively.

During each clock cycle Δt the CPUs perform computations *independently*. Therefore, treating each CPU as a dynamical system, we can define n_s *independent* flow vector fields describing the dynamics of each CPU:

$$(s_j(t + \Delta t), m_{j_w}(t + \Delta t)) = \phi_{\Delta t}^j(s_j(t), \mathbf{m}(t)), \quad 1 \leq j \leq n_s, \quad (3.5)$$

where m_{j_w} is the memory unit written by the j-th CPU, and $\phi_{\Delta t}^j$ is a function of its arguments. In principle, $\phi_{\Delta t}^j$ could be different for the different CPUs.

Now, the j-th CPU reads the memory $\mathbf{m}(t)$ at only the time t, and not during the clock interval $\mathcal{I}_{\Delta t} =]t, t + \Delta t]$. Since it does not perform any change on this memory, apart from the unit m_{j_w}, the evolution of the entire state during the clock cycle $\mathcal{I}_{\Delta t}$ is fully determined by the set of *independent equations*:

Standard parallelism

$$(s_j(t' \in \mathcal{I}_{\Delta t}), m_{j_w}(t' \in \mathcal{I}_{\Delta t})) = \phi_{t'-t}^j(s_j(t), \mathbf{m}(t)), \quad 1 \leq j \leq n_s. \quad (3.6)$$

As anticipated, we call this *standard parallelism* to distinguish it from the one described by eqn (3.4). The distinction between standard and intrinsic parallelism is visually represented in Fig. 3.2, and boils down to the ability of employing some form of *collective* dynamics to compute.

It should then not come as a surprise that an intrinsically parallel machine may provide considerable advantages compared to a standard parallel one, provided we maintain the digital structure of its input and output.[16] In order to do this, however, we need to understand a few

Clock cycle
In modern computers, CPUs perform elementary operations (such as accessing their memory bank, fetching instructions, or writing data) within a single electronic pulse, called a *clock cycle* or just *cycle* (Hennessy and Patterson, 2006).

Synchronous computation
Computation in which the different processing units of the machine are *synchronized* after a well-defined period of time (a 'cycle' of the computation) is called *synchronous*. This means that every unit has to wait for all the others to finish their computation during a cycle, before moving on to the next cycle. Our modern parallel computers operate this way.

Asynchronous computation
Computation in which all processing units of the machine compute and exchange information *simultaneously*, without the need to wait for a predetermined period of time, is called *asynchronous*. Neural networks and MemComputing machines perform such a computation. In fact, *artificial neural networks* are a sub-class of the more general concept of universal MemComputing machine (cf. Section 5.1).

[16]If we are interested in solving the combinatorial optimization problems I defined in Section 2.5.

more things about dynamical systems. In particular, we want to isolate the main features such systems ought to have to compute efficiently.

3.7   Critical points

Let's first note that when the right-hand side of eqns (3.1) is zero, the system is temporarily at a 'stand-still'. We call the particular points in the phase space, X, where this happens, *critical points*.

Mathematically, we say that a point, $\mathbf{x}_{cr}$, in the phase space is critical if:

Critical point

$$\mathbf{x}_{cr} \to F(\mathbf{x}_{cr}) = 0. \tag{3.7}$$

Uniqueness of the trajectory
Given an initial condition, $\mathbf{x}(t = 0)$, the ODEs (3.1) admit a *unique solution* at time t if the flow vector field F is *Lipschitz continuous* (see Section 8.5.1 and (Arfken et al., 1989)). This immediately rules out the intersection of two trajectories in phase space, or the self-intersection of a single trajectory, except at a critical point.

Critical points and nontrivial topology
Euclidean D-dimensional space, $\mathbb{R}^D$, has a *trivial* topology, meaning that it can be continuously *contracted* to a single point. Since the phase space manifold is *locally homeomorphic* to Euclidean space, it is *topologically nontrivial* only if these local Euclidean spaces cannot be 'glued' together without obstructions (the phase space is *locally* contractible, but *not globally*). This happens when critical points are present. Therefore, the presence of critical points indicates a nontrivial topology of the space (Fomenko, 1994). All the dynamical systems I will discuss in this book span a phase space with nontrivial topology.

As I will discuss in Chapters 11, and 12, critical points are critical (pun intended) for the proper operation of MemComputing machines. In addition, they always appear in the typically *non-convex landscapes* defined by the hard combinatorial optimization problems I will consider in Chapters 8, 9 and 10.

Each critical point is characterized by an *index* which is of a *topological* nature, in the sense that it is related to the 'shape' (the *topology*) of the phase space, rather than its local (geometric) features (Fomenko, 1994). This is true also for the *number* of critical points of smooth functions on the phase space: this number is constrained by the topology of phase space (Fomenko, 1994).

As we will see in Chapters 11 and 12, the fact that DMMs are robust against noise and perturbations can be ultimately traced back to their use of critical points, and trajectories connecting them. These trajectories, called *instantons*, are also of topological character.

To understand what a critical point index is let's first make the following analysis. In a neighborhood of any critical point we can perform *linear stability analysis* (Perko, 2001), namely we can expand eqns (3.1) to linear order:

$$\dot{\mathbf{x}} \approx \mathbf{J}(\mathbf{x}_{cr})(\mathbf{x} - \mathbf{x}_{cr}), \tag{3.8}$$

where $\mathbf{J}$ is the Jacobian matrix

$$[\mathbf{J}(\mathbf{x}_{cr})]_{ij} = \left.\frac{\partial F_i(\mathbf{x})}{\partial x_j}\right|_{\mathbf{x}_{cr}}, \tag{3.9}$$

evaluated at the critical point $\mathbf{x}_{cr}$. We can now determine the (typically complex) eigenvalues, λ_i, of this matrix for the given critical point.

The linearized eqns (3.8) provide the local trajectories

$$\mathbf{x}(t) \approx \mathbf{x}_{cr} + \sum_i \mathbf{u}_i e^{\lambda_i t}, \tag{3.10}$$

where the sum is over the eigenvalues λ_i and the associated eigenvectors $\mathbf{u}_i$ of the Jacobian matrix.

$$\dot{x} = F(x_{cr}) = 0$$

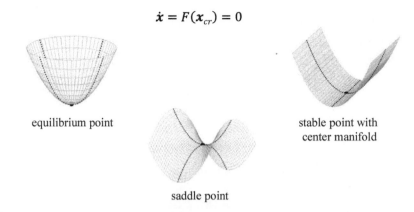

equilibrium point

stable point with
center manifold

saddle point

Fig. 3.3 Critical points, solution of $F(\mathbf{x}_{cr}) = 0$. An equilibrium point has all eigenvalues, λ_i, of the Jacobian with negative real part: Re $\lambda_i < 0$ (stable directions). A saddle point has both stable (Re $\lambda_i < 0$) and unstable (Re $\lambda_j > 0$) directions. In addition to stable and unstable directions, critical points may also support center manifolds or 'flat' directions (Re $\lambda_i = 0$).

The eigenvectors corresponding to Re $\lambda_i < 0$ and Re $\lambda_i > 0$ define the vector spaces tangent to the *stable* and *unstable manifolds*, respectively, at each critical point. In other words, for a negative (real part) eigenvalue, from eqn (3.10), it is clear that the dynamics tend to bring the system back to the critical point; that direction is *attractive*.

Instead, a positive (real part) eigenvalue indicates that the corresponding direction in the phase space is *repulsive*. If a critical point has only stable directions then it is an *equilibrium* or *fixed* point (whether *local* or *global*) (see Fig. 3.3). If it has both stable and unstable directions it is called a *saddle point* (Fig. 3.3).

All eigenvectors with Re $\lambda_i = 0$ span a *flat* manifold tangent to a *center manifold* (Fig. 3.3). It is clear from eqn (3.10) that center manifolds are somewhat irrelevant to the dynamics, since in those manifolds the system does not 'move'. A dynamical system can then have an equilibrium point with most directions being stable and a few flat ones. The stability of that point would still be governed by the stable directions.

We call *index* of a critical point the number of eigenvalues with *positive* real part (number of *unstable* directions). The reader interested in learning more about these concepts should consult the book (Fomenko, 1994) that also provides a visual account of these important notions.

Stable and unstable manifolds

The *stable manifold*, $M_s(\mathbf{x}_{cr})$, of a critical point $\mathbf{x}_{cr}$ is the collection of all points in the phase space that flow under eqns (3.1) toward $\mathbf{x}_{cr}$ for $t \to +\infty$. The *unstable manifold*, $M_u(\mathbf{x}_{cr})$, of a critical point $\mathbf{x}_{cr}$ is the collection of all points in the phase space that flow under eqns (3.1) toward $\mathbf{x}_{cr}$ for $t \to -\infty$.

Index of a critical point

The number of *unstable* directions of a critical point $\mathbf{x}_{cr}$ is called its *index*—indicated with ind$(\mathbf{x}_{cr})$ (Fomenko, 1994). The index is then the *dimension* of the unstable manifold, $M_u(\mathbf{x}_{cr})$, at the critical point: ind$(\mathbf{x}_{cr}) =$ dim$M_u(\mathbf{x}_{cr})$. It measures how stable a critical point is. Maxima have the largest index, minima the lowest (zero), and saddle points some integer in between.

3.8 Dissipative vs. conservative dynamical systems

Let me make another important remark. Dynamical systems can be typically grouped into two categories, according to whether they conserve or not volumes in phase space.

To make this point clearer, consider at time $t = 0$ a $(D-1)$-dimensional

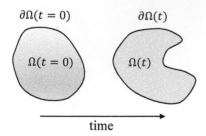

$\partial\Omega(t = 0)$ $\partial\Omega(t)$

$\Omega(t = 0)$ $\Omega(t)$

time

Fig. 3.4 Variation in time of a volume of phase space Ω enclosed by the surface $\partial\Omega$.

surface $\partial\Omega(t = 0)$ that encloses a *volume* of phase space $\Omega(t = 0)$. Let's then take all points on this surface and let them evolve according to eqns (3.1). After some time, t, the initial surface $\partial\Omega(t = 0)$ then evolves into $\partial\Omega(t)$ and it encloses the volume $\Omega(t)$ (Fig. 3.4). We then classify dynamical systems as:

Conservative systems
Dynamical systems that conserve phase space volume:

$$\Omega(t) = \Omega(t = 0).$$

Non-conservative systems
Dynamical systems that do not conserve phase space volume:

$$\Omega(t) \neq \Omega(t = 0).$$

Typical conservative systems are those whose dynamics can be described by a Hamiltonian function (Goldstein, 1965). Examples of non-conservative systems instead are those that involve, for instance, some type of friction among particles.

Of particular interest for the remainder of this book are those systems in which the phase space volume *contracts* during dynamics. We call these particular non-conservative systems:

Dissipative systems
Dynamical systems such that the phase space volume contracts:

$$\Omega(t) < \Omega(t = 0).$$

With this classification, we can further determine whether a dynamical system is dissipative or conservative according to the *divergence* of the flow vector field, F, in eqns (3.1).

To see this, we can compute the time variation of the volume of phase space, $\Omega(t)$, which is equal to the outgoing flux of the field F through the surface $\partial\Omega(t)$ of such volume:

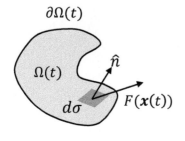

$\partial\Omega(t)$

$\Omega(t)$ $\hat{n}$

$d\sigma$ $F(\boldsymbol{x}(t))$

Fig. 3.5 Schematic of the outgoing flux of the field F at a time t through the surface $\partial\Omega(t)$ enclosing a volume of phase space $\Omega(t)$, where $d\sigma$ is an infinitesimal element of $\partial\Omega(t)$, and $\hat{n}$ the outgoing unit vector normal to the same surface.

$$\frac{d\Omega}{dt} = \int_{\partial\Omega(t)} F(\mathbf{x}(t)) \cdot \hat{n}\, d\sigma = \int_{\Omega(t)} \nabla \cdot F(\mathbf{x}(t)) d\mathbf{x}, \qquad (3.11)$$

where $\hat{n}$ is the outgoing unit vector normal to the surface $\partial\Omega(t)$, $d\sigma$ an infinitesimal element of the same surface (Fig. 3.5), and

$$\nabla \cdot F(\mathbf{x}(t)) = \sum_{i=1}^{D} \partial F_i(\mathbf{x}(t))/\partial x_i, \qquad (3.12)$$

with F_i the components of the vector field F, eqn (3.2).

In the second equality of eqn (3.11) I have used the divergence theorem to relate the flux of F through the closed surface $\partial\Omega(t)$ to its divergence in the volume enclosed by that surface (Arfken, Weber and Harris, 1989).

From eqn (3.11) we see that if the divergence is zero, $\nabla \cdot F(\mathbf{x}(t)) = 0$, then the dynamics preserve volumes. We say that these are *conservative* systems.

If the divergence is negative, $\nabla \cdot F(\mathbf{x}(t)) < 0$, in some region of phase space, then volumes contract in that region and we have a *dissipative* system.[17]

The distinction between these two classes of systems is very important because conservative systems do *not have attractors* (Section 3.9 below), namely bounded subsets of the phase space to which clusters of initial conditions will asymptotically tend.[18]

Since we want to map the set of solutions of combinatorial optimization problems to some set of points in the phase space, and ideally we want the dynamical system to reach that set *irrespective* of the initial conditions, then dissipative systems are the ones we are interested in.

Let's then discuss the different types of long-time behavior we should expect of dissipative dynamical systems.

3.9   Attractors

The output of a physical machine has to be read after some time has passed from the initial time at which the computation has started. We are then interested in the 'long-time' properties of dynamical systems.

In principle, from a strictly *mathematical* point of view, 'long-time' should properly mean 'infinite time'. However, this is not *physically* necessary.[19]

In fact, as I have discussed in Section 2.8, it is enough that we wait some time beyond which we can determine that the system trajectory is within a *finite* distance from an attractor, with that distance *independent* of the size of the problem we are trying to solve (see also Chapter 6).

With this clarification in mind, let's note that equations of motion like (3.1) can show several features in the long-time limit. The simplest one is what we call *equilibrium point*.

These states are such that, as their name suggests, they do not vary by application of the flow vector field F, and as I have explained in the Section 3.7, they are critical points with all stable directions (or possibly with also center manifolds).

More precisely, we call the state $\mathbf{x}_{eq}$ an equilibrium point *if*, and *only if*, for every time t (see Fig. 3.6)

[17] In general, there could also be dynamical systems for which volumes *expand* in some parts of the phase space (positive divergence of the flow vector field). Examples are systems with *unstable* periodic orbits or chaos that *repel* nearby trajectories. These regions of phase space are called *repellers*. We will not be concerned with these types of systems in this book because DMMs with fixed points do not support periodic orbits or chaos (Di Ventra and Traversa, 2017a; Di Ventra and Traversa, 2017b).

[18] Equivalently, we can say that conservative systems have *no transient dynamics*.

Indistinguishable trajectories
In actual physical systems, when two trajectories differ by a small amount, $\mathbf{x}'(t) - \mathbf{x}(t) = \epsilon$, with the volume in phase space spanned by ϵ smaller than the measurement apparatus accuracy, then they are *physically indistinguishable*. Even if the accuracy of the measuring instrument were zero, and we could operate at zero absolute temperature (0 K)—which we cannot—then the *uncertainty principle* of Quantum Mechanics (for the appropriate observables) would set a minimal scale over which two trajectories are indistinguishable (Messiah, 1958).

[19] Incidentally, this is also true in Quantum Mechanics. Due to the uncertainty relation between energy, E, and time, t, $\Delta E \Delta t \sim \hbar$, with $\hbar$ the reduced Planck constant, we would need an *infinite* time to measure an energy state with zero error ($\Delta E = 0$). Instead, we are satisfied when we reach the measurement apparatus accuracy.

Fig. 3.6 Sketch of the three possible phase-space trajectories of a dynamical system in the long-time limit (*attractors*): equilibrium points, (quasi-)periodic orbits, and chaos. (In the latter case, the line can self-intersect only at critical points for a continuous-time system of dimensionality $D \geq 3$.) In all cases, there may be several initial conditions $\mathbf{x}(t = t_0)$ that lead to the same attractor. The set of these initial conditions is the *basin of attraction* of the attractor.

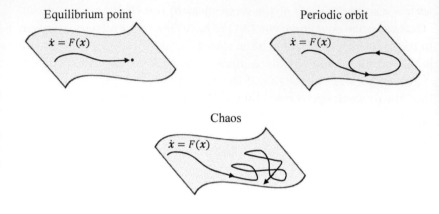

Equilibrium point

$\dot{x} = F(x)$

Periodic orbit

$\dot{x} = F(x)$

Chaos

$\dot{x} = F(x)$

Equilibrium point

$$F(\mathbf{x}_{eq}) = 0, \quad \forall t. \tag{3.13}$$

Equivalently, in terms of the semigroup $T(t)$, eqn (3.3), the equilibrium point satisfies the condition

Equilibrium point

$$T(t)\mathbf{x}_{eq} = \mathbf{x}_{eq}, \quad \forall t. \tag{3.14}$$

[20]The term 'fixed point' should be properly used for *discrete* dynamical systems, i.e., *maps* (Hilborn, 2001). I will abuse this terminology and use it interchangeably with the term 'equilibrium point'. The reason is that when a continuous-time dynamical system is *simulated* on a computer, it is necessarily transformed into a map.

[21]In fact, the word 'equilibrium' does *not* necessarily mean *thermodynamic equilibrium* (Goldenfeld, 1992). The system may be *far* from thermodynamic equilibrium and still satisfy (3.13). To stress this difference further, we could also say that the equilibrium point is a *steady state* of the system, when the latter is not at thermodynamic equilibrium (Di Ventra, 2008). Since all the MemComputing machines I will discuss in this book are such systems, there cannot be any confusion, and therefore I will not distinguish between 'equilibrium point' and 'steady state'.

Note that, in the literature, equilibrium points are sometimes called *fixed points*,[20] or (if they do *not* represent the thermodynamic equilibrium) *steady-state solutions*.[21]

Equations (3.1) may have one equilibrium point, many, or none at all. If equilibrium points are present, they are the ideal candidates for us to map them into the *solutions* of a particular problem, since once the system has reached them, it will stay put forever after.

The equilibrium points we mainly consider in this book are properly called *asymptotically stable* equilibrium points, meaning that at those points, the real parts of the eigenvalues of the Jacobian are negative (see eqn (3.10)).

Note that some of the eigenvalues of the Jacobian of some of the equilibrium points we will consider, may have zero real part, hence giving rise to *center manifolds*. As we have already discussed, their presence does not affect the stability of the equilibrium point (Section 3.7).

In addition to equilibrium points, eqns (3.1) may support also *periodic orbits* (in two dimensions these are often called *limit cycles*) (Perko, 2001). These are solutions that repeat themselves at specific intervals of time. More precisely, we call a state $\mathbf{x}_P$ periodic, if, and only if, there exists a constant $T_P > 0$ (the orbit's *period*) such that (after a possible transient time) the state satisfies the following (see Fig. 3.6):

Periodic orbit

$$\mathbf{x}_P(t) = \mathbf{x}_P(t + nT_P), \quad n \in \mathbb{N}. \tag{3.15}$$

It is clear that such orbits 'reside' on hyper-surfaces of multi-dimensional *invariant tori* (Fig. 3.7).

However, a dynamical system may also have *quasi-periodic orbits* as attractors (Perko, 2001).[22]

These are orbits that contain a finite number of *incommensurate* frequencies (namely frequencies that are not rational multiples of each other). Therefore, they never really come back *exactly* on themselves, like periodic orbits. Rather, they *densely cover* the hyper-surfaces of multi-dimensional invariant tori (Fig. 3.7).

A few remarks on chaos

Finally, eqns (3.1) may show a highly irregular pattern at long times (Fig. 3.6), which, despite being *deterministic*, is difficult (practically impossible) to predict. We call a system displaying this dynamical pattern *chaotic*. For *continuous-time* dynamics it can only occur if $D \geq 3$.

There is no agreement in the literature about what is an appropriate definition of deterministic chaos. However, there is a fairly general understanding that one can classify deterministic chaotic dynamics as the one with the following two (*topological*) properties (Devaney, 1992; Aulbach and Kieninger, 2001).

Consider a physical system whose dynamics are described by the flow field T in eqn (3.3) and an *invariant* subset $\mathcal{A}$ of the phase space ($T\mathcal{A} = \mathcal{A}$). We then say that the system is chaotic if (Devaney, 1992):[23]

Deterministic chaotic dynamics (à la Devaney)

(i) T is *topologically transitive* in $\mathcal{A}$ (see margin note);

(ii) the set of *periodic orbits* of T is *dense* in $\mathcal{A}$.

Given properties (i) and (ii), one can prove (Banks *et al.*, 1992) that if a dynamical system satisfies them, then it follows that $T|_{\mathcal{A}}$ depends *sensitively* on initial conditions. This means that if a system described by eqns (3.1) is chaotic, at sufficiently long times, its dynamics are highly *irregular* and *unpredictable*, and two initial conditions arbitrarily close to each other typically lead to completely different trajectories.

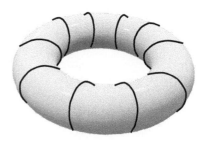

Fig. 3.7 A dynamical system may have both *periodic* and *quasi-periodic* orbits as attractors. *Periodic* orbits 'live' on hyper-surfaces of multi-dimensional invariant tori. *Quasi-periodic* orbits, instead, *densely cover* the hyper-surfaces of invariant tori.

[22]Quasi-periodic orbits can immediately be ruled out if the dynamical system is *globally* dissipative. This is because any volume of phase space would shrink as time goes by, till it reaches a region of zero volume (see eqn (3.11) with $\nabla \cdot F(\mathbf{x}(t)) < 0$). Since quasi-periodic orbits cover the hyper-surface of a multi-dimensional torus, the volume occupied by the initial points inside the torus must shrink to zero, and the torus will eventually disappear (Hilborn, 2001).

[23]This definition of chaos does *not* hold for *stochastic* dynamical systems, as in eqn (12.6). See (Ovchinnikov and Di Ventra, 2019) for details.

Invariant sets
A subset, $\mathcal{A}$, of the phase space is said to be *invariant* under the dynamics described by the flow field T, eqn (3.3), if $T\mathcal{A} = \mathcal{A}$. An equilibrium point (see eqn (3.14)) or a periodic orbit are examples of invariant sets.

Dense set
Consider a topological space X. A subset $\mathcal{B}$ of X is *dense* in X if every point in X either belongs to $\mathcal{B}$, or is a limit point of $\mathcal{B}$. For instance, the rational numbers are a dense subset of the real numbers.

Topological transitivity

Topological transitivity means that, given any two points in the phase space, we can find an orbit that comes arbitrarily close to both. More precisely, we say that the flow is topologically transitive in $\mathcal{A}$ if for any pair of non-vacuous open sets U, V $\subset \mathcal{A}$, there exists a positive time t' such that $T|_{\mathcal{A}}(t', U) \cap V \neq \emptyset$, where $T|_{\mathcal{A}}(t', U) = \{t' \in \mathbb{R}^+; \mathbf{x}(0) \in U | T|_{\mathcal{A}}(t')\mathbf{x}(0) : U \to \mathcal{A}\}$ (Gottschalk and Hedlund, 1955). A dynamical system with a closed invariant set with nonempty interior, and different from the whole space cannot be transitive. This is the case of DMMs, even without equilibrium points (Di Ventra and Traversa, 2017*a*).

Spontaneous symmetry breaking

It is the phenomenon by which the *ground* (or equilibrium) *state* of the system has *less* symmetry than the system's equations of motion. For *continuous* (global) symmetries, their spontaneous breakdown is accompanied by the emergence of *massless* (zero-energy) Goldstone modes (Peskin and Schroeder, 1995). See also discussion in Chapter 11.

Explicit symmetry breaking

It is the phenomenon by which a symmetry is broken explicitly by an *external* perturbation that does not preserve the symmetry.

[24]To make this point clearer, in the literature, *chaotic attractors* are often called *strange attractors*, although 'strange' truly means that the attractor is not piece-wise differentiable (Gilmore, 1998). An attractor can then be strange but not necessarily chaotic and vice versa. However, for most cases involving differential equations (not maps), the geometrical property of 'strangeness' appears in tandem with chaotic dynamics (Ott, 1993).

However, sensitivity to initial conditions is *not* enough for the dynamics to be chaotic. In fact, one can easily find cases in which there is sensitivity to initial conditions, and yet the system is not chaotic. An example is an *ergodic* system (see Section 12.3 for the definition of 'ergodicity'), where the trajectories of nearby points diverge algebraically, while in a chaotic system they diverge exponentially (see, e.g., (Ott, 1993) and Section 3.10).

Incidentally, the properties (i) and (ii) imply that only *topological* (i.e., global) features are enough to characterize a flow as chaotic, without the need of introducing a *metric* (namely a distance between points, which is a *local* property). In fact, in Chapter 11 I will discuss that chaos can be ruled out in digital MemComputing machines on topological grounds only.

For the purpose of this book, the previous discussion about chaos is more than enough. The reader interested in learning more about this fascinating phenomenon and its various features can consult the books (Hilborn, 2001; Ott, 1993) or the references in the introduction of (Ovchinnikov and Di Ventra, 2019), where chaos is also presented as a *symmetry-breaking* phenomenon, whether *spontaneous* or *explicit*.

In view of all these preliminaries we then see that, given an initial condition of eqns (3.1), the system may end up exploring some parts of the phase space, X, that have very different long-time behavior. We call

Attractor

The set of all points in the phase space to which a *dissipative* dynamical system converges in the long-time limit.

For instance, the attractor of an equilibrium point is that equilibrium point itself. The attractor of a periodic orbit is the set of points in the phase space traced by eqns (3.1) that satisfy condition (3.15). For a quasi-periodic orbit, it is the *dense* portion of the hyper-surface of the multi-dimensional torus the solution covers (Fig. 3.7). Finally, for chaotic dynamics the attractor is a subspace of the phase space with non-trivial (*fractal*) geometry.[24]

Therefore, according to the initial conditions we choose for eqns (3.1), we may end up into any of these attractors. We can then define the following important notion:

Basin of attraction

The set of *initial* points in the phase space that evolve according to eqns (3.1), and whose trajectories end up in a given attractor.

In general, then, a phase space clusters into sub-spaces, which are the different basins of attraction.

With regard to computation, it is clear that neither (quasi-)periodic orbits, nor chaotic behavior are good features if we want our physical machine to compute the solution of a particular problem. In both cases, the solution or output would still change in time, in the chaotic case, even unpredictably. Hence, we would like to avoid them, if possible.

Instead, I will show in Chapter 7 that we can map the output of the computation by a digital MemComputing machine into the fixed points of the corresponding dynamical system. The phase space of such machines is then the union of the basins of attraction of the equilibrium points which represent the solutions of the problem they are designed to solve.

Note that, in principle, one may still conceive of using periodic orbits as representations of the solutions/output of a given problem. In that case, one needs to make sure that no fixed points and chaos coexist. I will not discuss this idea in this book.

As for chaotic dynamics, it is difficult to imagine it could represent the output of a machine that is also scalable: any small perturbation on the trajectory of the system would render the future dynamics unpredictable.[25] This is equally true in *hardware* as well in the *software* simulation of such a machine. According to the terminology used in the extended Church-Turing hypothesis, such machines are *not* 'reasonable' (see Section 2.2.1).

[25] In fact, in (Ercsey-Ravasz and Toroczkai, 2011), a chaotic dynamical system has been designed to tackle SAT problems. It was then shown that chaotic dynamics require an energy expenditure that increases *exponentially* with problem size.

3.10 Integrable vs. non-integrable dynamical systems

Finally, let me linger on a notion that is very important for dynamical systems: the issue of *integrability*. Here, I need to make a major distinction between conservative and non-conservative (dissipative) systems (these definitions are in Section 3.8).

Compact manifold (space)
It is a manifold (space) which is both *closed* (contains all its limit points) and *bounded* (all its points are within a fixed distance from each other) (Fomenko, 1994). For instance, any closed and bounded interval of the real axis, $[a, b]$, with $a, b \in \mathbb{R}$, is compact, while $]a, b]$ is not, since it is not closed on the left boundary. And $[a, \infty)$ is not compact because it is not bounded.

Conservative systems

For *conservative* systems, in particular those described by a Hamiltonian (Goldstein, 1965; Arnold, 1989), integrability is well defined but requires quite a few background notions, which are not necessary for this book, in view of the fact that digital MemComputing machines are *not* conservative systems. Nonetheless, for completeness I will outline what this notion means in this case as well.

A dynamical system with D degrees of freedom is said to be integrable if there exist D *first integrals of motion* (constants of motion),[26] $I_j(\mathbf{x})$ with $j = 1, \ldots D$, that are in *involution* with each other, *independent*,

[26] Since a (Quantum) Field Theory has an *infinite* number of degrees of freedom, it is integrable if and only if it has an *infinite* number of constants of motion.

and the *manifold* defined by the points $\mathbf{x}$ in the phase space such that $I_j(\mathbf{x}) = c_j$, with c_j constants, is *compact*.

Although integrable systems are the exception rather than the rule in conservative systems, their importance lies in the fact that their dynamics are highly *constrained* in the compact manifold defined by the first integrals, and one can start from them to study other types of dynamical systems as *perturbations* of integrable ones (Goldstein, 1965).

Non-conservative systems

Having discussed (very briefly) the notion of integrability in conservative systems, let's now turn to the dissipative ones, namely the ones we are really interested in this book. For these, the definition of integrability does not rely directly on the existence of integrals of motion. Rather, it is strongly tied to the *absence of chaos* (Gilmore, 1998). To understand this connection let's proceed as follows.

Recall from Section 3.7 that critical points in the phase space may have *unstable* directions. For each unstable direction there is an *unstable manifold*, so that points on those manifolds tend to move *away* from the critical point. In fact, if I take two infinitesimally close points on those manifolds, their trajectories will tend to diverge as time goes on. This seems similar to a feature of chaos, but it is not enough to produce chaos (see again discussion in Section 3.9).

What is needed for true chaotic behavior is that those trajectories 'fold' on each other and 'stretch' in the phase space (Gilmore, 1998), see Fig. 3.6.[27] These are the mechanisms that truly create a *strange* (chaotic) attractor. If that is the case, then there are no well-defined *global*[28] unstable manifolds in the phase space, rather some 'objects' that are sometimes called *branched manifolds* (Gilmore, 1998).

If this situation occurs, then we say that the flow vector field of the dynamical system is *non-integrable*:

Non-integrable dynamical system

A dynamical system whose flow is such that its *global unstable manifolds* of *all* dimensionalities are *not* well-defined sub-manifolds of the phase space.

The reverse then defines *integrability* in a general dissipative dynamical system:[29]

Integrable dynamical system

A dynamical system whose flow is such that its *global unstable man-*

First integral

First integral or *constant of motion* is any quantity that is *invariant* (constant) during time evolution.

Poisson bracket

Given two functionals, $f(\mathbf{q}, \mathbf{p})$ and $g(\mathbf{q}, \mathbf{p})$, of the canonical coordinates, $\mathbf{q}$, and momenta, $\mathbf{p}$, their *Poisson bracket* is (sum over all degrees of freedom D):

$$\{f, g\} = \sum_{i}^{D} \left(\frac{\partial f}{\partial q_i} \frac{\partial g}{\partial p_i} - \frac{\partial f}{\partial p_i} \frac{\partial g}{\partial q_i} \right).$$

Involution

Constants of motion, $I_j(\mathbf{q}, \mathbf{p})$ $(j = 1, \cdots, D)$, are in *involution* with each other when their Poisson brackets are zero: $\{I_j, I_i\} = 0, \forall i, j = 1, \cdots, D$.

Independence

First integrals, $I_j(\mathbf{q}, \mathbf{p})$, are *independent* if any linear combination of them is zero, $\sum_j \alpha_j I_j = 0$, only if *all* coefficients α_j are zero. See (Arnold, 1989) for a thorough discussion of all these concepts.

[27] Equivalently, volumes of phase space *continuously* increase in some directions and decrease in others.

[28] The word 'global' again reinforces the notion that chaotic dynamics can be defined with *topological* features only (Section 3.9).

[29] We also say that, for integrable systems, the unstable manifolds of the flow vector field *foliate* the phase space, similarly to the foliation of the phase space with invariant tori that the first integrals of a Hamiltonian system generate (Arnold, 1989).

ifolds of *all* dimensionalities *are* well-defined sub-manifolds of the phase space.

From the previous discussion we then see that an integrable system has a flow that is not chaotic and vice versa. In other words, we equate integrability to absence of chaos.[30] Digital MemComputing machines are *integrable* systems, namely they do not support chaotic dynamics (see Chapters 7 and 12, in particular Section 12.9).[31]

3.11   Phase-space engineering: The ideal phase space for computing

Let's conclude this Chapter by summarizing the *ideal* features a dynamical system should have in order to represent a 'reasonable' machine,[32] especially for the solution of the combinatorial optimization problems I have defined in Section 2.5.

In fact, from a physical point of view, what I will discuss in the rest of this book is what I call *phase-space engineering*. In other words, we want to 'engineer' the phase space of the MemComputing machines, so that its (topological) features will aid in the efficient solution of combinatorial optimization problems.

These features are as follows:

(i) The dynamical system has to be *dissipative*, so that it has *attractors*.

(ii) Of all the attractors, we choose *equilibrium points* as representations of the *solutions* of a given problem. Say, we map one equilibrium point to one solution.

(iii) This means that we do *not* want (quasi-)periodic orbits or chaos. Absence of chaotic dynamics means that the dynamical system has to be *integrable*.

(iv) However, this is *not* enough since the system dynamics may get 'stuck' in some local minimum in the phase space. Therefore, we need a phase space that, apart from the equilibrium points representing the solutions of the problem, can *only* have as additional critical points, *saddle points*, and no additional minima. This way, the system may end up into those saddle points, and still get out of them.

(v) Finally, we want to *guide* ('direct') the system toward the equilibrium points, namely the system has to be *constrained* to always reach such points.

[30] Note, however, that 'integrability' does *not* mean that the equations of motion (3.1) can be solved *analytically*. Integrable dynamical systems for which an analytical solution of their equations of motion is known are rare (Goldstein, 1965).

[31] Their phase space is *foliated* with the *instanton manifolds* (Section 11.4).

[32] By 'reasonable' I mean a machine that does not need exponentially growing resources with increasing problem size (cf. Section 2.2.2).

Integrability and constrained dynamics
As in the case of conservative systems (see previous Section 3.10), the property of integrability in dissipative ones already sets strong *constraints* on the state trajectory in the phase space. In other words, trajectories of integrable systems cannot *wander* aimlessly in the phase space, as it happens for non-integrable (chaotic) ones (Ott, 1993).

Fig. 3.8 *Ideal* phase space for computing. Given an initial condition, the state of the system evolves from time $t = 0$ through a landscape that has only saddle points and the equilibrium points, x_{eq}, representing the solutions of the given problem. The system is *guided* (driven) toward the equilibria like the ball toward the exit in a pinball machine (Example 1.3), or the water down a hill toward the estuary (Example 1.4). Different initial conditions may lead to different equilibria.

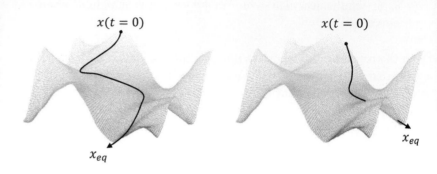

If we engineered a phase space as described previously (see schematic in Fig. 3.8), and a physical system that is forced to explore it until it finds one of the equilibrium points/solutions (according to the initial conditions we set), we would then have a similar situation as the one depicted by the pinball machine in the Example 1.3, or the waterfall I have discussed in the Example 1.4.

In the present case the 'ball' or the 'water' are described by the state $\mathbf{x}(t)$ in phase space at any time, t, and such a state *has to* reach the equilibrium points because it *has no other choice*: it is in its nature to do so.

We will see in Chapters 11 and 12 that, indeed, *digital* MemComputing machines satisfy these properties, and their dynamics are similar to the percolation of water down a hill or a ball traveling down a pinball machine, but they still map a *finite* string of symbols to a *finite* string of symbols.

3.12   Chapter summary

– Putting physical information into a given problem, even if it is in the combinatorial optimization class, means that we need to deal with *dynamical systems*.

– Dynamical systems also support a type of parallelism I named *intrinsic*, to distinguish it from the 'standard' parallelism of our modern 'parallel computers'. Intrinsic parallelism means that all elements of the physical system participate in the dynamics *collectively*.

– In order to read and write the input and output with finite means (*digital machines*), requires us to represent the solutions of the problem with some well-defined long-time features of the corresponding dynamical system. We are then interested in *dissipative* dynamical systems, which support *attractors*.

- In this book, I will only consider as attractors equilibrium points to represent the solutions of a given problem. However, it is quite possible to use periodic orbits as representation of the solutions, while chaotic behavior seems difficult to reconcile with the digital requirements of the output.

- We will also *engineer* dynamical systems such that their phase space does *not* have any minima other than those representing the solutions of a given problem.

- Finally, we will need a mechanism to *guide/direct* the system toward the equilibrium points/solutions of a problem, similar to what water does when it percolates downhill, or a ball when it exits a pinball machine, under the effect of gravity.

- After all these preliminary general remarks, let's try to understand why *time non-locality* (memory) is so advantageous in computing.

Memory, memelements, active elements, and memprocessors

4

Sometimes you will never know the value of a moment until it becomes a memory.
Dr. Seuss (1904−1991)

Takeaways from this Chapter

- The meaning of *time non-locality* (memory).
- The notion of long-range order.
- Memory induces spatial non-locality.
- Difference between standard circuit elements and memelements.
- MemComputing is *not* 'memristive computing'!
- The *necessity* of active elements to compute efficiently.
- Definition of memprocessors.
- MemComputing is *not* just 'in-memory computing'!

In Chapter 3 we have seen that a physical system described by eqns (3.1) is *intrinsically parallel*, in the sense that any of its elements 'knows', through physical interactions of some sort, what the other elements are doing. However, this feature alone is not enough to build a machine that can solve efficiently (as defined in Section 2.2.2) specific problems, in particular the combinatorial optimization problems I introduced in Section 2.5.

In fact, without any additional physical property, a machine described by eqns (3.1) is just a generic 'analog' machine. And we know that analog machines are not easily scalable (see Section 2.8).

In this Chapter, I will introduce an extra physical property, namely *time non-locality* (memory), and provide a general understanding of why such an ubiquitous feature—that is available to both quantum and non-quantum systems—may be advantageous for computing. However, before doing that, let me clarify a few things about the word 'memory'.

4.1 What is memory?

The word 'memory' may carry different meanings according to the context in which it is used.

Fig. 4.1 Human memory can be classified as our ability to encode and retain information that we can retrieve for later use.

Memory in Biology

For instance, when we refer to humans we typically refer to the following (Fig. 4.1):

> **Physiological memory**
> The ability to *encode* and *retain* information that we can *retrieve* for later use (Sherwood, 2015).

We call it 'physiological' because it is related to the way in which that part of the human brain *functions*.

Incidentally, this definition is not limited to humans. It could easily be extended to describe memory in those animals that showcase similar features.[1] Most importantly, this definition presupposes three ways in which *information*[2] is processed as far as memory is concerned:

> **Memory and information processing**
> - *Encoding* information.
> - *Storage* of such information.
> - *Retrieval* of such information at a later time.

Encoding and storage of information can be done on a variety of physical media (for instance, by distributing it across *neurons* in *nervous systems* (Sherwood, 2015); see also Section 12.10). Retrieval of such information requires some other process, and its exact mechanism, as far as nervous systems are concerned, is still not fully understood.

Memory in modern computers

Irrespective, the three main processes I have just outlined also apply to our modern computers. In those, we *encode* information with strings of symbols (0s and 1s), we *store* it on some media via some physical quantity (magnetic features in hard disk drives, charges in transistors of flash memories, etc.), and we *retrieve* it when necessary via a central processing unit (CPU) (Hennessy and Patterson, 2006).

[1] Note that this memory behavior does not necessarily require a nervous system. For instance, it was shown that unicellular organisms, such as *amoebas*, can encode and retain information to adapt their motion to external stimuli (Saigusa *et al.*, 2008), and such a behavior can even be modeled with electronic circuits (Pershin *et al.*, 2009).

[2] I am referring here to *syntactic* information (see Section 1.3).

> **Nervous system**
> The collection of all *neurons* (nerve cells) and *glial cells* in an organism that transmit information to the various parts of its body, according to sensory inputs, is called the *nervous system* of such organism (Kohonen, 1989).

However, this understanding of memory—whether in its physiological aspects or in its realization in modern computers—is not complete, or at least it is not general enough for our purposes.

In fact, this definition somewhat conceals the *dynamical* nature of the processes involved in the manipulation of information. A more physical understanding of 'memory' is thus necessary.

4.2   Memory is equivalent to time non-locality

In Physics, we use the term 'memory' to mean *time non-locality*, namely that the *state* of a physical system (as I have defined it in Section 3.2) at any given time, t, depends on the *history* of its dynamics at previous times. This is the significance of the term 'memory' I will consider in this book.

I can then provide the following general definition:

Physical definition of memory (time non-locality)
A physical system has memory when its state at time t depends on the collection of *previous* states through which the system has evolved (it is *time non-local*).

In other words, the *response* of the system to some perturbation at a time t depends on the state of the system at previous times: the response is *non-local* in time.

Response functions
When a physical system is subject to some external perturbation, $v(t)$, (e.g., an electric field) it responds so that one (or more) of its properties, $u(t)$, (e.g., its current density) changes. The *response function*, g, quantifies how much the perturbation affects the physical property of interest (Kubo, 1957). Typically, this response function is *non-local* in *both* space *and* time. For instance, if we focus only on time non-locality (memory), then the relation between perturbation and response is:

$$u(t) = \int_{t_0}^{t} g(t,t')v(t')dt',$$

where t_0 is the time at which the perturbation is switched on, and the response function, $g(t,t')$, depends on two different times.

For instance, if we apply an electric field $E(t)$ (in one direction) to a conductor, a current density $J(t)$ develops of a magnitude that depends on the conductivity, σ, of the conductor; see, e.g., (Di Ventra, 2008). The conductivity, which is a *response function*, is not simply dependent on time t, but depends on two times, $\sigma(t,t')$ (and indeed on two spatial positions as well, namely it is also non-local in space, see Section 4.4.2).

This dependence on two times is the essence of time non-locality. If we then focus on just the time dependence of the response, the current density at time t is:

$$J(t) = \int_{t_0}^{t} \sigma(t,t')E(t')dt', \quad \textit{time non-local response}, \qquad (4.1)$$

where t_0 is the time when the electric field is switched on. From eqn (4.1) we see that $J(t)$ depends on the *history* of the response (states) at all times: from the initial time when we switched on the field, E, till the time we measured the current density, $J(t)$.

If the conductivity were *local in time* then eqn (4.1) would read:

$$J(t) = \sigma(t)E(t), \quad \textit{time-local response}, \qquad (4.2)$$

namely $J(t)$ responds *instantaneously* to the external field $E(t)$ at every time t.[3]

Note that the previous example, and indeed the very definition of memory I have given at the beginning of this Section, applies to *both* quantum *and* non-quantum systems (Kubo, 1957). In fact, time non-locality is a physical property that is shared by *all* physical systems. In other words, *no* physical system can respond instantaneously to a given perturbation. Therefore, its present state *always* depends, to a certain extent, on the past states through which it has evolved (Di Ventra and Pershin, 2013*a*).

Of course, in many cases this time non-locality is so short that it is difficult to detect with available instruments, but this does not invalidate the point I am making here, which is that memory, intended as time non-locality, is an ubiquitous phenomenon.

4.3 Why time non-locality?

Our next quest is to understand why the physical property of time non-locality, which is easily found also in non-quantum systems, may help in computing difficult problems—like the ones I discussed in Chapter 2—efficiently. For this scope, let's first consider again the Quantum Computing paradigm (Nielsen and Chuang, 2010).

4.3.1 Quantum entanglement revisited

As I have already explained in Section 3.1, quantum computers may employ a feature, known as *entanglement*—which does not have a classical counterpart—to solve prime factorization in polynomial time (Shor, 1997). Again, entanglement is a property that does not allow an independent description of a single particle in a collection of many particles, no matter how far apart we place them.

To understand this point better, consider as an example two qubits (see Section 3.1). Each one of them could be represented by, e.g., the spin states of an electron (Fig 4.2).

To each qubit is associated a Hilbert space with basis states $|0\rangle$ and $|1\rangle$ (or 'spin up' and 'spin down', respectively).[4] The Hilbert space of the two qubits is the *tensor product* (cf. definition in Section 3.1) of the Hilbert spaces of each individual qubit, and has four possible basis states. A general state, $|\Psi_1\rangle$, in the Hilbert space of qubit-1 is a *linear combination* of the type $|\Psi_1\rangle = \alpha_0|0\rangle + \alpha_1|1\rangle$, where α_0 and α_1 are some complex numbers such that $|\alpha_0|^2 + |\alpha_1|^2 = 1$.

The same is true for qubit-2: a general state, $|\Psi_2\rangle$, in the Hilbert space of this qubit is a linear combination, $|\Psi_2\rangle = \beta_0|0\rangle + \beta_1|1\rangle$, where β_0 and β_1 are complex numbers such that $|\beta_0|^2 + |\beta_1|^2 = 1$.

[3]Which is a physical impossibility.

Observable
In Physics, an *observable* is an *experimental procedure* to determine a physical quantity, e.g., position, velocity, voltage, etc., of a system. In Quantum Mechanics observables are represented by *self-adjoint operators*; see, e.g., (Messiah, 1958).

Quantum (projective) measurement
Consider a two-state quantum system and an observable $\hat{O}$ with two eigenstates $|0\rangle$ and $|1\rangle$, so that $\hat{O}|0\rangle = \epsilon_0|0\rangle$, and $\hat{O}|1\rangle = \epsilon_1|1\rangle$. If the state of the quantum system is $|\Psi\rangle = \alpha_0|0\rangle + \alpha_1|1\rangle$ *before* a measurement is performed (with α_0 and α_1 some complex numbers such that $|\alpha_0|^2 + |\alpha_1|^2 = 1$), then the probability that a measurement of $\hat{O}$ gives ϵ_0 is $|\alpha_0|^2$. The probability that a measurement of $\hat{O}$ gives ϵ_1 is instead $|\alpha_1|^2$. However, if the measurement finds, say, the state $|0\rangle$, then the original state of the system *collapses* into $|0\rangle$, and an identical, *immediate* measurement on the same system produces the same state $|0\rangle$ (unless the system is 'destroyed' by the measurement, such as in the case of the absorption of a photon by a photodetector). For an in-depth discussion of these concepts see (Messiah, 1958).

[4]Note that in Quantum Computing and Quantum Information, it is common to employ the convention that the vectors $|1\rangle = (0, 1)^T$ and $|0\rangle = (1, 0)^T$, where T is the transpose.

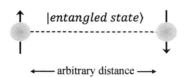

Fig. 4.2 Entangled state of two particles. A measurement on one of them affects the state of the other, *irrespective* of their distance.

Product states

A *joint state* of the 2-qubit system could then be:

$$
\begin{aligned}
|\Psi_1\rangle \otimes |\Psi_2\rangle &= (\alpha_0|0\rangle + \alpha_1|1\rangle) \otimes (\beta_0|0\rangle + \beta_1|1\rangle) \\
&= \alpha_0\beta_0|0\rangle \otimes |0\rangle + \alpha_0\beta_1|0\rangle \otimes |1\rangle \\
&\quad + \alpha_1\beta_0|1\rangle \otimes |0\rangle + \alpha_1\beta_1|1\rangle \otimes |1\rangle \\
&\equiv \alpha_0\beta_0|00\rangle + \alpha_0\beta_1|01\rangle + \alpha_1\beta_0|10\rangle + \alpha_1\beta_1|11\rangle,
\end{aligned}
\tag{4.3}
$$

where in the last line I have indicated the tensor products of the individual states with the shorthand notation, $|00\rangle$, $|01\rangle$, $|10\rangle$, $|11\rangle$, (e.g., $|00\rangle \equiv |0\rangle \otimes |0\rangle$, where the symbol $\otimes$ means tensor product).

These states are known as *product states*. If the 2-qubit system is prepared in one such state, it means that its two qubits are *independent* of each other. The quantum features of the whole system appear *only* in the linear superposition of states of the *individual* qubits, and nothing else.

Entangled states

Instead, let's now form the following 2-qubit state:

$$
|\Psi\rangle = \frac{|0\rangle \otimes |0\rangle + |1\rangle \otimes |1\rangle}{\sqrt{2}} \equiv \frac{|00\rangle + |11\rangle}{\sqrt{2}},
\tag{4.4}
$$

which is also a valid state of the 2-qubit system.[5]

The *normalization* factor, $1/\sqrt{2}$, is important since it guarantees that $\langle\Psi|\Psi\rangle = 1$.[6]

We then see that the state (4.4) *cannot* be written as the tensor product of two separate states, $|\Psi_1\rangle$ in the Hilbert space of qubit-1, and $|\Psi_2\rangle$ in the Hilbert space of qubit-2:

$$
|\Psi\rangle = \frac{|00\rangle + |11\rangle}{\sqrt{2}} = |\text{entangled state}\rangle \neq |\Psi_1\rangle \otimes |\Psi_2\rangle,
\tag{4.5}
$$

no matter how we choose the complex numbers α_0, α_1, β_0, β_1 in eqn (4.3), with the constraint, of course, that quantum states have to be normalized to have a statistical interpretation (check it!). We call state (4.4) *entangled*.

Now, *any* physical system (e.g., the 2-qubit one in the previous example) that is *prepared experimentally* in such a state, or it is *naturally* in such a state (e.g., most states of multi-electron atomic or molecular systems) is such that its individual constituents are strongly dependent on each other, irrespective of how far apart they are in space.[7]

This means that, *irrespective* of their spatial separation, the dynamics of the individual elements of the system are *correlated*, in the sense that they are *intrinsically dependent* on (or *associated* to) each other.

[5] This is also called a *Bell state* (Nielsen and Chuang, 2010).

[6] The symbol $\langle \cdot | \cdot \rangle$ indicates the *inner product* of states in Hilbert space. The normalization is necessary because the states of quantum systems have a statistical interpretation (Messiah, 1958). This means that $\langle 0|0\rangle = \langle 1|1\rangle = 1$. The single states of each qubit are also orthogonal to each other, namely $\langle 0|1\rangle = \langle 1|0\rangle = 0$. Therefore, $\langle 00|00\rangle = \langle 11|11\rangle = 1$ and $\langle 00|11\rangle = \langle 11|00\rangle = 0$. From eqn (4.4) we then get: $\langle\Psi|\Psi\rangle = \langle 00|00\rangle/2 + \langle 11|11\rangle/2 = 1$.

[7] In fact, there is *no* length scale in the state (4.4).

Although it may not seem intuitive, this phenomenon gives rise to a sort of *spatial non-locality* such that a *measurement* of one element (part) of the system, would be immediately 'felt' in other parts arbitrarily far away (Fig. 4.2).[8] Note that this *does not* mean that information can be transferred between distant parts of the system faster than the speed of light!

In fact, the information on the entangled state is *already* present in the physical system itself, whether the system has it naturally, or it has been prepared in such an entangled state by the experimental procedure. It simply means that the whole system acts as a unique, *rigid* entity.

Although this is not how it is typically presented in the literature, we could say that since the correlations do not decay spatially, entanglement realizes some sort of *ideal long-range order* (Di Ventra and Traversa, 2018*a*).

4.3.2 Long-range order/correlations

The word 'order' I just used reminds us of systems whose fundamental units are arranged in an ordered fashion (imagine, for instance, the atoms in a solid). 'Long-range' means that such an order extends over the entire spatial (or temporal) dimension of the system, and even to infinity, if we extend the system to its *thermodynamic limit*.

Given a physical system, some of its properties (e.g., the spins of individual elements of magnetic materials; see the following example) may showcase this type of order. In fact, what will be mostly useful for our purposes is that the *correlations* between the different degrees of freedom of the system are *long ranged*.

Although entanglement is a feature that is proper to quantum mechanical systems, long-range order/correlations is also shared by non-quantum ones. For instance, it appears in many natural phenomena; the onset of *continuous phase transitions* being a typical (and important) example (Goldenfeld, 1992).

Consider, for instance, a ferromagnetic material in which the spin domains are all aligned in some direction (Fig. 4.3). By varying some parameter, say increasing the external temperature, the material goes from one *phase* of matter (ferromagnet) to another (paramagnet) in which the spin domains have random orientations (Fig. 4.3). In proximity to the (*critical*) point in between the two phases, the correlations of the spin fluctuations decay as a *power law*, meaning that different spins can correlate at arbitrary distances from each other.

To quantify what I just wrote we can calculate the (2-point) *correlation function*, $G_c(\mathbf{r})$, at a distance $\mathbf{r}$ between spins $\mathbf{s}(\mathbf{r})$ (vectors with a given magnitude and orientation) at different locations, averaged over all possible spin configurations (we use the symbol $\langle \cdots \rangle$ for the *ensemble*

[8]Note that one can design 'short-ranged' entangled states that can be smoothly deformed into a product state, without a phase transition (Wen, 2013). These are *not* the entangled states I consider in this book.

Thermodynamic limit
Consider a system with N particles occupying a volume V. In Statistical Mechanics we call *thermodynamic limit* the one in which *both* the number of particles *and* the volume go to infinity, while their ratio (the *density*, ρ) is constant:

$$\lim_{\substack{N \to \infty \\ V \to \infty}} \frac{N}{V} = \rho = \text{constant}.$$

Ensemble average
Consider a physical system and 'imagine' an *ensemble* of identical copies of it, all prepared with similar initial conditions. Let all these copies evolve in time, and consider a physical property, $O(t)$. The *ensemble average* of such a quantity, at fixed time t, is its mean over the ensemble, and is indicated with $\langle O(t) \rangle$. In the thermodynamic limit, the mean is independent of the ensemble chosen.

Phase transitions
Any transition between *phases* (or states) of matter (e.g., from solid to liquid, or liquid to gas, etc.) is called a *phase transition*. It is typically induced by changes of a physical parameter such as temperature or pressure. *Continuous phase transitions* are those transitions such that at the *critical* value of that parameter, namely at the point of transition between phases, and in the thermodynamic limit, an infinite *correlation length* and a power-law decay of correlations develop. See, e.g., (Goldenfeld, 1992) for a thorough discussion of all these concepts (see also Chapters 11 and 12).

Ferromagnet

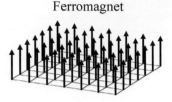

Paramagnet

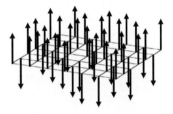

Fig. 4.3 Cartoon of spins on a two-dimensional square lattice, where all spins can take only two directions ('up' or 'down'); this could be a model of an 'easy-axis ferromagnet'. We call *ferromagnetic phase* the *ordered* (or *rigid*) one in which all spins are aligned in one direction (say 'up'). (The other *ground state*, with all spins down, has the same energy as the one with all spins up.) We call *paramagnetic phase* the *disordered* one in which the spins are randomly oriented either 'up' or 'down'. We can go from one phase to the other by, e.g., varying the external temperature.

[9]This is also called $\mathbb{Z}_2$ symmetry (Goldenfeld, 1992).

Ground state(s)

The lowest energy state of a physical system is called the *ground state*. It may be *degenerate*, meaning that there may be different (ground) states with the same energy. For instance, in the ferromagnetic case of Fig. 4.3, the two (ground) states with all spins up or all spins down have the same energy (cf. also Fig. 10.11).

---

[10]According to its previous *history*.

average).

Away from the critical point, we expect the correlations to be *local/short-ranged*, meaning that they typically decay *exponentially*:

$$G_c(\mathbf{r}) \equiv \langle \Delta \mathbf{s}(\mathbf{r}) \Delta \mathbf{s}(0) \rangle \sim \exp(-|\mathbf{r}|/\zeta), \quad |\mathbf{r}| \to \infty, \qquad (4.6)$$

where the *fluctuations* $\Delta \mathbf{s} = \mathbf{s} - \langle \mathbf{s} \rangle$, and ζ is the *correlation length*, namely the characteristic length over which the spin fluctuations correlate. At *criticality* (near the critical point), instead, we find that the correlation function behaves *algebraically* as (Goldenfeld, 1992)

$$G_c(\mathbf{r}) \equiv \langle \Delta \mathbf{s}(\mathbf{r}) \Delta \mathbf{s}(0) \rangle \sim \frac{1}{|\mathbf{r}|^\alpha}, \quad |\mathbf{r}| \to \infty, \qquad (4.7)$$

with α some constant—called the *critical exponent* of the correlation function—that depends on the dimensionality, symmetries of the system, and range of interactions. The correlation length, ζ, instead *diverges* at criticality. (I anticipate here that such a critical behavior entails, for continuous symmetries, *massless/zero-energy modes*; cf. Chapters 11 and 12.)

Since the correlations decay as a power law, by simply scaling the distance $\mathbf{r}$ by some arbitrary factor λ, the correlation function would only change as $G_c(\lambda \mathbf{r}) = \lambda^{-\alpha} G_c(\mathbf{r})$, namely the system is *self-similar* at different scales: *fluctuations occur at all length scales*. We would then call these correlations *scale free* (or *scale invariant*).

4.3.3   More order, less symmetry

Note that in the magnetic example of the preceding Section, the system has different *symmetries* in the two phases, paramagnetic and ferromagnetic. Consider Fig. 4.3. The spins in the paramagnetic phase are randomly oriented either up or down, so even if we reversed them all, the system would 'look the same' as before.

More precisely, the paramagnetic phase is *invariant* with respect to the *spin inversion symmetry* ($\mathbf{s} \to -\mathbf{s}$).[9] The same symmetry is shared by the *Hamiltonian* (or energy function) describing the interactions between spins; see eqn (9.1). However, it is clear that in this phase the spin system is *disordered*.

In the ferromagnetic phase, instead, the spins are *either* all pointing up *or* all pointing down. Their *ground state* does not possess the spin inversion symmetry.

Also, since this ground state does not share the same symmetry as the Hamiltonian (9.1), and the system 'spontaneously' ends up *either* with all spins up *or* with all spins down,[10] we say that a *spontaneous symmetry breaking* has occurred in going from the paramagnetic to the ferromagnetic phase (cf. discussion in Section 3.9). And we would say that the spins in the ferromagnetic phase are in an *ordered/rigid* state.

Order from (spontaneous) symmetry breaking

This argument does not apply just to the previous magnetic example, but to *all* cases in which order in Nature arises from the *(spontaneous) breakdown of a symmetry*.[11]

For instance, if we cool a liquid until it crystallizes, the system transitions from the *disordered* liquid phase (which is invariant with respect to *rotation* and *translation* symmetries) to the *ordered* crystal phase that does not possess rotational and translational symmetries. Therefore, we can say that 'more order' is synonymous with 'less symmetry':

Order and symmetry
If a system transitions from one phase to another, the *ordered/rigid* phase is the one with the *least number of symmetries*:

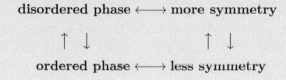

4.3.4   Computation at criticality

We then see that if we could engineer a non-quantum dynamical system so that its correlations are long-ranged, somehow driving it toward the *critical state* of some sort of (continuous) phase transition, its different parts could correlate *strongly* at long distance—while being not completely 'rigid' (as in the fully ordered phase)—making it a good candidate to solve hard problems efficiently.

This is because, as we have seen in Section 2.6, hard problems typically require the *simultaneous* (correlated) assignment of logical values of different variables *anywhere* in the problem specification. For instance, we have seen in Section 2.6 that a conjunctive normal form (CNF) formula, like the one of eqn (2.4), can be viewed as a Boolean *physical* circuit (see Fig. 2.8), with the variables being physical quantities (e.g., voltages) of an actual physical system.

A machine (a collection of elementary units) with long-range correlations can then assign logical values to the variables in a *correlated* way, even if the variables appear in multiple clauses.[12] This is schematically shown in Fig. 4.4.

Incidentally, as I have anticipated in Section 3.3, if our machines had such a property, and were built out of non-quantum systems, they would be able to work even at room temperature (and possibly higher temperatures), not just cryogenic temperatures typical of quantum computers.

[11]There are exceptions to this symmetry-breaking picture of order. A prominent example is the *topological order* (*topological phases*) of some (strongly correlated) electron systems at low temperatures, e.g., quantum Hall systems (Wen, 2013) (see also Section 12.9). Furthermore, there are phase transitions, e.g., the liquid-gas transition such that, at the critical point, no change of symmetry occurs (Goldenfeld, 1992).

Rigidity
In the ordered phase it is difficult to deform the state of the system (we need to pay a large energy penalty to do so). Hence, the latter displays *rigidity* in that phase. For example, the spins in the ferromagnetic phase are *stiff*, or a solid, unlike its liquid (disordered) phase, is mechanically strong against deformations.

Universality classes
There are several *critical exponents* in continuous phase transitions, corresponding to different observables (e.g., specific heat, susceptibility, etc.). These exponents are *universal*, in the sense that physical systems differing by microscopic details, but sharing the same symmetries, dimensionality, and range of interactions, should have the *same* critical exponents. If that is the case, we say that these systems belong to the same *universality class* (Goldenfeld, 1992).

[12]The machine can then discover the *long-wavelength* (*low-energy*) features of the problem it is attempting to solve. These features are the most difficult to find (cf. Section 2.6).

Fig. 4.4 A machine that supports *long-range order/correlations* is able to assign logical values to each variable of a combinatorial optimization problem, with CNF formula $\phi(\mathbf{v})$, in a *correlated* way, even though those variables appear in different clauses. This is because, the different variables can be mapped into quantities (e.g., voltages) of a physical system, with each quantity *spatially* separated from the others—like the spins of the 2D magnet of Fig. 4.3—with the machine correlating such quantities.

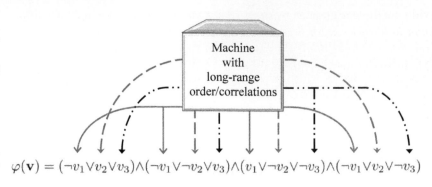

$$\varphi(\mathbf{v}) = (\neg v_1 \vee v_2 \vee v_3) \wedge (\neg v_1 \vee \neg v_2 \vee v_3) \wedge (v_1 \vee \neg v_2 \vee \neg v_3) \wedge (\neg v_1 \vee v_2 \vee \neg v_3)$$

[13] In fact, one can have scale-free distributions *without* criticality (Touboul and Destexhe, 2017). Therefore, long-range correlations would be *enough* to provide an advantage to the machine, even if the latter is not at the critical point of a continuous phase transition.

> **Cellular automata**
> A *cellular automaton* is a model of computation consisting of a (finite-dimensional) grid of cells, each in one of a finite number of states. The computation starts by assigning a state to each cell. In the next step, the new state of a cell is determined by some rules, and depends on the current state of the cell and that of its neighboring cells (Toffoli and Margolus, 1987).

[14] For instance, in the case of the spin model I discussed in Section 4.3.2, one needs to *fine tune* the temperature exactly at the critical value to be at the point of transition between the ferromagnetic and paramagnetic phase. This fine-tuning is *not* desirable for the practical realization of a computing machine. As we will see in Chapters 11 and 12, the collective state of digital MemComputing machines is *attractive*, meaning that they 'naturally fall into' such a state without fine tuning.

[15] In fact, we will see in Chapters 11 and 12 that long-range order has to be a *non-equilibrium* property of the system, not an equilibrium one.

Therefore, computation 'at a continuous phase transition' or, more precisely, *computation at criticality* seems very appealing.[13]

In fact, in Section 1.6 (criterion IV of MemComputing) I have argued that one of the desired criteria of MemComputing machines should be some mechanism of *collective dynamics*. Here, we see more clearly why such a mechanism is so important for the types of problems we want to solve.

As mentioned in the same Section 1.6, a similar idea of computing at criticality was put forward in 1990 by Langton (Langton, 1990), who argued, using *cellular automata* as a model of computing machines, that the optimal conditions for *processing* and *storing* of information are achieved in proximity of a phase transition, in particular continuous ones that show critical behavior. The same author has called this behavior *computation at the edge of chaos*.

A similar conclusion was subsequently reached regarding the operation of the brain, where one observes scale-free features in the firing of neurons at distant locations of the cortex; see, e.g., (Muñoz, 2018). Collective dynamics then seems to be a property that MemComputing machines share with the brain (cf. discussion in Section 12.10).

However, although these ideas are very compelling, even if correct, they do not suggest *how* to actually build a machine that could generate such a scale-free behavior *on its own*, namely *without tuning any external parameters*,[14] and take full advantage of it during computation.[15] This is what I now set out to do.

4.3.5   Time non-locality promotes spatial non-locality

After the previous discussion about long-range order, I would like to show how time non-locality fits to all this. In this Section, I will provide a simple argument to show that *time non-locality* promotes *spatial non-locality*, even for systems that interact *locally* (with short-range interactions). I will demonstrate this result more explicitly and with actual MemComputing machines in Chapter 11.

To see that memory may promote spatial non-locality, let's consider a dynamical system made of only two particles and described by eqns (3.1). Say, one particle moves along the y direction, while another along the z direction of a Cartesian coordinate system (Fig. 4.5).

The particles interact spatially only at the initial time t_0 when they are close to each other, but as soon as they move away, their direct interaction is zero. However, let's assume that somehow they retain memory of such a process via some physical property, e.g., the medium in which they travel.

Non-locality and global information

It is known that *non-local* quantities can 'probe' the *global* (*topological*) features of the space in which the system is embedded. An example of this is the *Wilson loop* which is a *non-local* operator that probes global features of systems described by a *gauge theory*; cf. Section 13.7.3 and (Peskin and Schroeder, 1995). I will show in Chapters 11 and 12 that *time non-locality* is yet another tool to extract global/topological information on a given system.

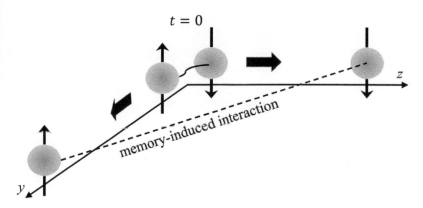

Fig. 4.5 Two particles interact at time $t = 0$ and their interaction is short-ranged, so that, when at $t > 0$, they each move in different directions, they do not interact *directly*. However, the particles have *memory* of their past dynamics. Due to this *time non-locality* they still interact *indirectly* during their dynamics.

This property represents some *internal* (microscopic) *degree of freedom* for the particles and follows its own dynamical equation. We indicate this degree of freedom, or *memory variable*, with $\tilde{x}$.

Equations (3.1) describing the system evolution in the three-dimensional phase space spanned by y, z, and $\tilde{x}$ read

$$\dot{y} = F_y(y, \tilde{x}), \tag{4.8}$$

$$\dot{z} = F_z(z, \tilde{x}), \tag{4.9}$$

$$\dot{\tilde{x}} = F_{\tilde{x}}(y, z, \tilde{x}), \tag{4.10}$$

where F_y, F_z, and $F_{\tilde{x}}$ are some flow fields, whose exact form is not important for this discussion. The flow fields F_y and F_z describe the dynamics of the two particles that depend on the memory variable $\tilde{x}$.

If we integrate each one of these fields we see that they *both* depend on the history of the dynamics of $\tilde{x}$. The latter depends on the positions of each particle at any given time. In fact, from eqns (3.3) which define the flow $T(t)$, and (4.10) we get

$$\tilde{x}(t) \equiv T_{\tilde{x}}(t)\tilde{x}(t_0) = \tilde{x}(t_0) + \int_{t_0}^{t} F_{\tilde{x}}(y(t'), z(t'), \tilde{x}(t'))dt'. \tag{4.11}$$

Internal degrees of freedom

I call *internal* (or microscopic) degrees of freedom, those that are not easily accessible experimentally, e.g., the positions of atomic vacancies in a solid under current flow (Pershin and Di Ventra, 2011*a*).

As time progresses we see from eqn (4.11) that, even if the two particles did not interact directly any longer after $t > t_0$ (namely, there was no direct spatial interaction between them of the type $F_y(y, z)$ or $F_z(y, z)$), they would still interact, *indirectly*, through the time evolution of the internal degree of freedom $\tilde{x}$:[16]

[16]Replace eqn (4.11) into (4.8) and (4.9).

$$\dot{y} = F_y(y(t), \tilde{x}(t_0) + \int_{t_0}^{t} F_{\tilde{x}}(y(t'), z(t'), \tilde{x}(t'))dt')$$
$$= F_y(y(t), history \text{ of } y \text{ and } z), \qquad (4.12)$$
$$\dot{z} = F_z(z(t), \tilde{x}(t_0) + \int_{t_0}^{t} F_{\tilde{x}}(y(t'), z(t'), \tilde{x}(t'))dt')$$
$$= F_z(z(t), history \text{ of } y \text{ and } z). \qquad (4.13)$$

From *time non-locality* (memory) we have then induced *spatial non-locality*, because now the time variation of the position of particle-1 at a given time depends also on the time variation of the position of particle-2 at all previous times, and vice versa.[17]

We can then summarize the effect of memory (time non-locality) as:[18]

> **Shortest-path problem**
> Consider a graph with n vertices (or nodes) and m edges. Each edge may be weighted differently with strength w_{ij}, with $i, j = 1, \ldots, n$. The shortest-path problem is to find the path between two vertices in the graph such that the sum of the weights of the edges of such a path is minimized.

[17]As a further example, consider the spin system of Fig. 4.3. Suppose you change the orientation of one spin in the lattice, but the new orientation depends on the orientation of that spin at previous times, which in turn depends on how all the spins in the system are arranged at any given time. Therefore, even if each spin interacts locally, with just its nearest neighbor spins, due to the presence of memory its orientation will depend on the *collective* state of all spins in the lattice.

[18]It is worth noting that this link between time non-locality and space non-locality appears also (in its abstract form) in the context of information theory (Section 1.3), where transmission of syntactic information in time can be similarly described as the transmission of information in space (Hamming, 1997).

[19]It is similar to leaving pebbles along a path. This is also a type of memory, because it allows you to retrace your own path, or other people to follow it and reach you, at a later time, so long as the pebbles are not displaced (which would represent a 'loss of memory' of the path).

[20]The distribution of the number of connections that each node possesses.

> **Effect of time non-locality**
> *Spatial non-locality* among physical systems may *emerge* from time non-locality (*memory*) alone, even if the physical constituents themselves interact *locally* (with short-range interactions).

A similar conclusion was reached by looking at how ants solve the *shortest-path problem* to find food (Blum, 2005). In that case, ants release pheromones that can be detected by other ants. The pheromones persist for a certain time, so that the shortest path from the nest to the food is the one on which the most pheromones are deposited.

This allows other ants to traverse such a path, and, as a consequence, to keep reinforcing it for other ants to follow.[19] Even though the ants interact only *locally*, the pheromone 'memory trail' produces a spatially non-local type of interaction among them that is exploited by the ants to solve the shortest-path problem.

In support of this, it was further shown that a *scale-free distribution* of the degree of connectivity in complex networks[20] emerges with time non-locality and only local interactions (Caravelli, Hamma and Di Ventra, 2015). In other words, such a scale-free behavior does not need explicit spatially non-local interactions: time non-locality is enough.

All this suggests that memory is indeed a good ingredient to exploit in computation.

4.4 Toward a circuit realization of MemComputing

In view of the previous discussion, it is then natural to look for appropriate physical systems whose response functions are strongly time non-local.

The next question is then: what type of physical systems with memory best realize the concept of MemComputing? In fact, one can think of many such systems, ranging from optical to superconducting.

In this book, I will only focus on non-quantum *electrical circuits*, since they are easy to fabricate and scale up to large sizes (*very large scale integration*). However, the reader should keep in mind that the notion of MemComputing machines can be realized, in principle, with many other types of systems and devices.

4.4.1 Electrical circuits as dynamical systems

Typical electrical circuits are made of standard electronic components with two terminals (such as resistors, capacitors, inductors), and three terminals (such as transistors; Fig. 2.9), assembled to satisfy conservations laws, known as *Kirchhoff's laws* (Horowitz, 2015). These circuit elements are the building blocks of our modern computers.

However, we will use them in different *architectures*, namely assembled in ways that are not typical in our current computers, and, in addition, we will use their *memory properties*, that emerge under specific conditions. To stress that we will be using circuits elements in the regime in which memory is *experimentally* evident, we will collectively call them *memelements*, to distinguish them from the 'standard' circuit elements, those for which memory effects are negligible under the experimental conditions our machines operate.

It is not my goal in this book to delve into all of the details of circuit theory, other than discussing those few features that are strictly necessary. The reader not familiar with circuit theory, and interested in a more in-depth account, is encouraged to read (Horowitz, 2015).

Here, it is enough to know (and I will show explicitly in Section 4.6 and, later on, in Chapter 7 as well) that electrical circuits, with and without memory, can be represented with equations of the type (3.1), and can be built having the properties I will discuss in Section 6.3.

4.4.2 Memelements

I have already pointed out in Section 4.3.5 that *any* physical system subject to external perturbations, such as electromagnetic fields, temperature changes, etc., *cannot* respond instantaneously to such perturbations.

Very large scale integration (VLSI)
VLSI is the industrial process of fabricating *integrated circuits* (cf. Section 1.6) by combining several transistors (nowadays on the order of several billions) and other circuit elements into a single chip (Hennessy and Patterson, 2006).

Kirchhoff's laws
These laws express the *conservation* of charge and energy.

I. Current law: charge conservation
The algebraic sum of currents in a network of conductors meeting at *any* node is zero. If there are n conductors meeting at a node, and we indicate with I_k the current from one of these conductors (with positive or negative sign according to whether the current is directed toward or away from the node, respectively) then the current law states:

$$\sum_{k=1}^{n} I_k = 0. \qquad (4.14)$$

II. Voltage law: energy conservation
The algebraic sum of the voltage differences around *any* closed loop of a circuit is zero. If there are n voltage differences, V_k each, then

$$\sum_{k=1}^{n} V_k = 0. \qquad (4.15)$$

Therefore, the *response* of *any* physical system (e.g., their resistance or capacitance or inductance) *must* depend on its past dynamics, hence it must be memory-dependent/time non-local (Kubo, 1957).

In many cases, we would need to make the perturbations vary so fast to see this memory effect that it may be very difficult to measure in practice. For instance, many resistors that we use on a daily basis have a resistance that is essentially independent of the external current (or voltage) we apply, whether it is a *direct-current* (DC) or a typical *alternating-current* (AC) source.

However, when the frequency of the AC source becomes very high (say, in the range of 100 MHz to 1 GHz for typical resistors), capacitance effects—typically called 'parasitic capacitances'—become observable. These effects give rise to an intrinsic delay in the response of the resistor, effectively creating a *resistor with memory* or *resistive memory* (sometimes called *memristive element* (Chua and Kang, 1976)).

Although in some literature there is the tendency to think that resistive memories are 'fundamental' new circuit elements (Chua and Kang, 1976), I want to stress once more that this is *not* the case: resistors will *always* show memory features if they are appropriately perturbed. Therefore, resistive memories are simply resistors that easily showcase time non-local effects at experimentally accessible frequencies and magnitudes of currents and/or voltages (Di Ventra and Pershin, 2013*a*).

In order to clarify this point further, and to justify the general functional form of memelements that I will provide later in this Section, let me repeat an analysis from (Di Ventra and Pershin, 2013*a*) based on response functions (see definition in Section 4.1). In fact the following derivation is a simple application of *Kubo's response theory* (Kubo, 1957).[21]

Memristive elements are *not* fundamental circuit elements
Resistive memories, sometimes called *memristive elements* (Chua and Kang, 1976), are simply resistors whose response clearly shows time non-local (memory) effects at experimentally accessible frequencies and magnitudes of the applied voltages/currents (Di Ventra and Pershin, 2013*a*). They are *not* novel 'fundamental circuit elements' (Kim *et al.*, 2020)!

[21]The reader not very familiar with the theory of response functions may skip this derivation without much impact on the understanding of the subsequent material.

Memory from microscopic dynamics
Suppose we want to determine the electrical *current density*, $\vec{J}(\mathbf{r}, t)$ of a given conductor when it is subject to an external *electric field* $\vec{E}(\mathbf{r}, t)$. Note that, unlike in eqn (4.1), I am now explicitly including the spatial dependence of both the external field and the resulting current density.

At this stage, in order to apply Kubo's response theory we need a *microscopic* description of the physical system. Whether we work within a quantum or non-quantum framework we would then need the many-body Hamiltonian of the electrons interacting with each other and the ions comprising the conductor.

However, in order to obtain a general understanding of response functions, detailed knowledge of this Hamiltonian is not necessary.

We can then just assume that it is a function $H(\{\vec{R}\})$ of all the ionic positions, $\{\vec{R}\}$, that for simplicity we consider non-quantum.

If the ions can be treated classically, we can then describe their dynamics using Newton's equations of motion. If each ion has a mass M then its dynamics are described by equations of the type:

$$\vec{F}\left(\vec{R}, \frac{\mathrm{d}\vec{R}}{\mathrm{d}t}\right) = M\frac{\mathrm{d}^2\vec{R}}{\mathrm{d}t^2}, \tag{4.16}$$

where $\vec{F}$ is the total force acting on that particular ion.

In eqn (4.16) I have explicitly included in the force $\vec{F}$ also a dependence on the velocity of the ions. This is because, at the microscopic level, the interactions between electrons and ions and between ions themselves, create a 'friction' for an ion to move in the conductor under the action of an external field (Di Ventra, 2008).

Now we apply Kubo's response theory (Kubo, 1957). We then switch on the electric field perturbation at time t_0, and we find, assuming a *linear response* between current density and electric field, the *current-current* response function for each component of the current density ($\mu, \nu = x, y, z$) (see, e.g., (Di Ventra, 2008) for the full derivation)

$$J_\mu(\mathbf{r}, t) = \sum_\nu \int \mathrm{d}\mathbf{r}' \int_{t_0}^t \mathrm{d}t' \sigma_{\mu\nu}\left(\mathbf{r}, \mathbf{r}'; t, t'; \{\vec{R}\}, \left\{\frac{\mathrm{d}\vec{R}}{\mathrm{d}t}\right\}\right) E_\nu(\mathbf{r}', t'), \tag{4.17}$$

where $\{\vec{R}\}$ and $\{\mathrm{d}\vec{R}/\mathrm{d}t\}$ represent the collection of all ionic positions and velocities, respectively.

This is the central result of response theory. As I have anticipated in Section 4.1, it explicitly shows what I have anticipated, namely that the electrical *conductivity*, $\sigma_{\mu\nu}$, is non-local in *both* space *and* time:

$$\sigma_{\mu\nu}(\mathbf{r}, \mathbf{r}'; t, t'; \{\vec{R}\}, \{\mathrm{d}\vec{R}/\mathrm{d}t\}), \quad \textit{space and time non-locality.} \tag{4.18}$$

It is a 2-rank tensor representing the current response in the direction μ under an electric field component in the direction ν.

The non-locality in *space* is represented by the pair of coordinates $\mathbf{r}, \mathbf{r}'$, while the non-locality in *time*, from the pair of times, t, t'.

Since the response in eqn (4.17) is an integral in time from the initial time of the perturbation until the time at which the current is measured (and an integral in space over the entire spatial extension of the system), it depends on the full *history* of the system from the time the perturbation was switched on.

Hamiltonian
The *Hamiltonian* of a dynamical system (whether quantum or non-quantum) is a function of basic dynamical variables, such as positions and momenta of all particles in the system, from which the whole time evolution of its state can be derived. The difference between quantum and non-quantum systems is that in the former case the Hamiltonian is an operator acting on state vectors in a Hilbert space, while in the latter it is a dynamical function in phase space.

Linear vs. non-linear response
A response is said to be 'linear' when the variation of a physical quantity due to a perturbation is *linearly* related to such a perturbation. A response is 'non-linear' when it has a *non-linear* dependence on the perturbation itself.

Let's now write this response in a form that we will use later. First, let's assume that the variation of the electric field is such that it does not induce an appreciable magnetic field. Then the electric field is related to the *electrical potential*, $V(\mathbf{r}, t)$, as $\vec{E}(\mathbf{r}, t) = -\vec{\nabla}V(\mathbf{r}, t)$.

The *total current* that flows across a surface S, is related to the current density as $I = \int_S \vec{J} \cdot d\vec{S}$, with $d\vec{S}$ the infinitesimal surface vector of the surface S through which the current is measured. From eqn (4.17) we thus find

$$I(t) = G\left(\{\vec{R}\}, \left\{\frac{d\vec{R}}{dt}\right\}, t\right) V(t). \tag{4.19}$$

In this expression, the quantity, G, is the *conductance*

$$G\left(\{\vec{R}\}, \left\{\frac{d\vec{R}}{dt}\right\}, t\right) =$$

$$-\sum_{\nu\mu} \int d\mathbf{r} \int d\mathbf{r}' \int_{t_0}^t dt' \, u_\mu \sigma_{\mu\nu}\left(\mathbf{r}, \mathbf{r}'; t, t'; \{\vec{R}\}, \left\{\frac{d\vec{R}}{dt}\right\}\right) u_\nu, \tag{4.20}$$

where u_ν ($\nu = x, y, z$) is the component of a unit vector $\vec{u}$.

Let's now call $\{x_1\} \equiv \{\vec{R}\}$ and $\{x_2\} \equiv \{d\vec{R}/dt\}$. We can then write the total current as

$$I(t) = G(\{x_1\}, \{x_2\}, t)V(t), \tag{4.21}$$

and Newton's equations (4.16) as

$$\{\dot{x}_2\} = \left\{\frac{\vec{F}}{M}\right\} \equiv f(\{x_1\}, \{x_2\}, t),$$
$$\{\dot{x}_1\} = \{x_2\}, \tag{4.22}$$

with f some vector function of $\{x_1\}$ and $\{x_2\}$, the set of *internal* (microscopic) state variables (degrees of freedom) of the system (in this particular case, the positions and velocities of all ions in the material).

If we further write $\{x\} = \{x_1, x_2\}$ the ensemble of *all* state variables, then eqn (4.21) can be written as

$$I(t) = G(\{x\}, t)V(t). \tag{4.23}$$

The previous derivation was done by assuming a *linear* relation between current density and electric field. But this limitation is not necessary.

In fact, by including all orders in the response function we would find that the conductivity depends also on the electric field (and hence potential V) making the conductance, G, in eqn (4.23), *non-linear* (Kubo, 1957).

This means that the relation between total current and electric potential is generally

$$I(t) = G(\{x\}, V, t)V(t), \tag{4.24}$$

which needs to be solved together with eqns (4.22).

Circuit elements with memory

Equation (4.24) tells us that, under appropriate experimental conditions, the current that flows in a conductor is, in general, non-linearly dependent on voltage itself, and may depend on *memory* (internal) degrees of freedom that provide *time non-locality* to the current. This means that the current at a given time, typically depends on the *history* of the states at *previous* times. The resistor then shows memory features.

However, the derivation I have provided regarding resistors with memory can be extended to all the other circuit elements, such as capacitors and inductors (Di Ventra and Pershin, 2013a). In fact, it does not need to be limited to electrical circuit properties: it can be extended to *any* type of response functions (Kubo, 1957).

Let me then introduce the general definition of *memelements*. Let's label with $u(t)$ the quantity of interest (output), say the current, $I(t)$, in eqn (4.24), and with $v(t)$, its complementary constitutive (input) variable (the time-dependent perturbation), e.g., the voltage, $V(t)$, in eqn (4.24).

The *generalized response function* of the system, g (e.g., the conductance, G, in eqn (4.24)), can then always be written in the compact form (Di Ventra *et al.*, 2009)

$$u(t) = g(\tilde{x}, v, t)v(t), \tag{4.25}$$
$$\dot{\tilde{x}}(t) = f(\tilde{x}, v, t), \tag{4.26}$$

where $\tilde{x}$ denotes a *set* of state (memory) variables describing the internal state of the system, and f a continuous vector function.

The internal memory variables could be, e.g., the positions of the ions in the system, or the spin degree of freedom, or any other internal state variable(s) (Pershin and Di Ventra, 2011a). The important thing

Memelement

I call *memelement* any physical system such that one of its properties, call it $u(t)$, when subject to a perturbation $v(t)$, showcases a *memory-dependent dynamics* of the type (4.25)−(4.26) as due to some *memory* (internal) *degrees of freedom*, $\{\tilde{x}\}$.

is that eqns (4.25) and (4.26) represent a system with memory, because if the variation of the perturbation $v(t)$ is such that the internal state variables, $\tilde{x}$, cannot easily follow the variation of such perturbation, then the output, $u(t)$, would, at any given time t, depend on the *past dynamics* of the system.

We call systems described by eqns (4.25) and (4.26) *memelements*, short for 'memory elements' (Di Ventra *et al.*, 2009), irrespective of what type of physical response or memory properties we consider.

Of course, the amount of such memory depends on many factors, such as the form and amplitude of the perturbation $v(t)$, and the type of physical internal state variables. These have their own response times, hence a perturbation that is slow for some system, may be too fast for another.

Take as an example a memelement whose memory originates from the motion of atomic defects in the system. If we apply an external field to this system and such perturbation varies, say, as fast as a few picoseconds, then atomic defects would barely 'feel' such a variation.

On the other hand, if the internal state variable is the spin of the atomic nuclei or of the electrons, picoseconds variations of an external field that couples to spins could be more than enough to make the spin memory visible (Pershin and Di Ventra, 2011*a*).

To illustrate this point better, I plot in Fig. 4.6 the schematic of a current-voltage curve (the voltage is the perturbation $v(t)$, the current is the complementary constitutive variable, $u(t)$, hence the response, g, is the conductance—inverse of the resistance) for a sinusoidal voltage input $v(t) = \sin(\omega t)$ for different frequencies, ω.

In this case, the internal state variable provides the memory, as evidenced in the opening of the *hysteresis* curve of the current-voltage characteristics, with its maximum hysteresis at a particular frequency of the input $v(t)$ that corresponds to the inverse time scale (relaxation time) of the internal state variable.

At much higher, or much lower frequencies, the internal state variable, either cannot follow the external perturbation, or it can, respectively. Either way, the magnitude of the hysteresis (the amount of memory) reduces considerably.

We will call a resistor that behaves like the one in Fig. 4.6 a *resistive memory* or *memristive element*.

We call a capacitor with memory ($v(t)$ is the voltage, $u(t)$ is the charge, g is the capacitance) a *memcapacitive element*, while an inductor with memory ($v(t)$ is the current, $u(t)$ is the flux, g is the inductance), a *meminductive element* (Di Ventra *et al.*, 2009).

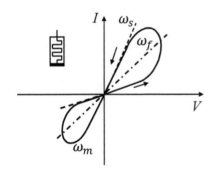

Fig. 4.6 Schematic of the current-voltage response of a resistive memory for three different frequencies, ω, of the external sinusoidal voltage $v(t) = \sin(\omega t)$. The *hysteresis* is a measure of the amount of memory in the system at each frequency. The maximum hysteresis is at a characteristic frequency ω_m, while the hysteresis is minimal at both high frequency, ω_f, and low frequency ω_s, with $\omega_f \gg \omega_m \gg \omega_s$. The inset shows the symbol we will use for this element.

An important remark

Note that there is a subtle, but very important difference between a memristive element and an *ideal memristor*. (Similarly, between memcapacitive and meminductive elements vs. *ideal memcapacitor* and *meminductor*, respectively (Di Ventra *et al.*, 2009).) The latter means that its resistance depends *only* on the charge that flows through it or on the history of the voltage across it (Chua, 1971).

This is not a minor point because the *ideal* version of this device suffers from severe *physical* limitations that make it doubtful it could be realized in practice (Di Ventra and Pershin, 2013*a*; Kim *et al.*, 2020). (The closest physical realization of this concept is the phase-dependent conductance of a Josephson junction, which is, however, just a component of the junction's total current (Peotta and Di Ventra, 2014).)

Irrespective, this difference is not particularly important for us here. However, to avoid any confusion, in this book I will use either the term 'resistive memory' or 'memristive element' to indicate *physically realizable* devices.

Mathematical statements vs. physical statements

At this stage, it is worth stressing once more (as I have done in Section 1.1), that there is a *substantial* difference between mathematical statements and statements we make about *experimentally realizable systems* (physical reality). There is no need for mathematical statements to correspond to any physically realizable quantity, so long as those statements are logically consistent within a set of foundational axioms (see Section 1.4 for the meaning of 'axiom'). Natural Sciences, such as Physics, instead, deal with *material* objects and their interactions, and rely on *experiments* as the ultimate judge of the validity of statements we make on natural phenomena. Therefore, from a physical point of view, it is not enough to 'define' a device or a computing machine mathematically. We also need to provide an experimental procedure to actually build it in hardware (in reality).

By looking at Fig. 4.6, it is clear that an element with memory may be used also as an *information processing* device. For instance, from Fig. 4.6, one sees that for the curve with maximum hysteresis one easily distinguishes two currents for the same voltage, and their difference is large enough to sustain even the unavoidable noise of any physical device.

One current, say the lowest, may be used to define the logical 0, while the largest one, the logical 1. In other words, a resistive memory can operate, under certain conditions, as a *digital* device.

Of course, if the memory is long-term, meaning that, by switching off the external perturbation, the output of the system varies very little within the time scale of a *reading* device, we could also *store* the result of such a computation in the internal state of the memelement. In other words, if we can read the result fast enough before it degrades, then for all practical purposes we have stored it (see also the definition of 'information storage' in Section 1.6)

We can then summarize these points with a general statement:

Processing and storing of information

Time non-locality may be used to *store* information as well as to *process* information within the *same* physical device.

This means that the *same* physical units can accomplish the two tasks that in our modern computers are executed in two physically distinct

Energy recycling during computation
Memcapacitive and meminductive elements (Di Ventra *et al.*, 2009), namely capacitors and inductors that show memory effects, can *store* energy (unlike resistors) in addition to providing information-processing capabilities. For instance, the energy that is expended during charging of a memcapacitive element can be recovered during its discharging. This opens up the possibility of *recycling* part of the energy that is used during computation (Traversa *et al.*, 2014).

[22]Which is, additionally, different than what is known as 'in-memory computing'; see Section 4.7.

Emulators of memelements
The response of resistive memories (as well as memcapacitive and meminductive elements) can be *emulated* with active elements such as transistors. In other words, the memory features of memelements can be reproduced by simply using an appropriate combination of transistors and other circuit elements (see, e.g., (Pershin and Di Ventra, 2010) for an example of emulator of resistive memories).

[23]For this reason, we can also call it *terminal-agnostic logic*.

locations (CPU distinct from memory). This is one of the criteria of MemComputing I have anticipated in Section 1.6 (criterion I of MemComputing).

Although this is not strictly necessary to build MemComputing machines, in this book I will mostly consider resistive memories (and, of course, other traditional circuit elements, such as standard resistors and capacitors). Indeed, one could equally design (albeit with different architectures) circuits with memcapacitive elements (Traversa, Bonani, Pershin and Di Ventra, 2014) and/or meminductive ones (Di Ventra and Pershin, 2013*b*).

The advantage of the latter two memelements is that in addition to information processing, they *store energy* and, therefore, unlike resistive memories, they may be used to *recycle* part of it during computation.

4.5 MemComputing is *not* 'memristive computing'

At this stage, I want to stress once more that MemComputing is a much more *general concept* than just 'computing with resistive memories' (sometimes called 'computing in-memory'[22]) or any other memelement.

For one, resistive memories (and other memelements as well) could easily be replaced by *emulators* made of standard transistors plus standard (namely without memory) resistors and/or capacitors (Pershin and Di Ventra, 2010).

This is not a minor point, if one wants to build MemComputing machines in *hardware*. This is because the resistive memories that are presently studied, although simple two-terminal devices, are mostly made of oxides (or some other materials) sandwiched between metals (Pershin and Di Ventra, 2011*a*).

And these types of material structures do not lend themselves easily to *integration* with modern complementary metal-oxide semiconductor (CMOS) technology that is presently used to build our modern computers. The transistor-based emulators, instead, integrate easily with such a technology. This simplifies enormously the hardware fabrication of the MemComputing machines I will discuss in this book and their hardware scalability.

4.5.1 The importance of active elements

Finally, in Chapter 6, I will show how to use memory to create a new type of Boolean logic, one that is *self-organizing* (Traversa and Di Ventra, 2017). This is a type of logic that does not distinguish between input and output terminals, so that it can operate *in reverse*.[23] It is this last property that allows *digital* MemComputing machines (DMMs) to compute hard combinatorial optimization problems efficiently.

However, in order to do that, *memelements are not enough!*

In fact, in order for any self-organizing gate of a DMM to *always* satisfy its logical proposition at any given time, it needs *feedback* (cf. Section 7.2). This feedback can only be provided by *active elements* (such as transistors), namely elements that inject energy into the system as needed.

A resistive memory, instead, is a *passive* device: it always dissipates energy (in the form of heat), and that energy cannot be recovered by the circuit. Therefore, let's anticipate this important fact:

> **The need of active elements**
> MemComputing machines need *active elements* to solve hard combinatorial optimization problems efficiently. Passive elements alone, such as resistive memories, are not enough!

Passive vs. active elements
A *passive* physical element is one that always dissipates energy, and that energy cannot be recovered. A resistor is a typical example. An *active* physical element is one that can inject energy into a system via some source. A voltage generator or battery is one such example. A transistor (see Fig. 2.9) is another example of active element.

4.6 Memprocessors

Having discussed what memelements are, and the need of active elements to build MemComputing machines, let's now define their general, basic units, we call *memprocessors*. These will be a combination of memelements and active elements performing processing and storing of information in the same physical location.

Let me then provide a formal definition of this concept that represents the link between the *mathematical* definition of a MemComputing machine (as in Chapter 5) and its real, *physical* implementation (Chapter 7).

Following (Traversa and Di Ventra, 2015) we define a memprocessor as the four-tuple (x, y, z, σ) where x is the state of the memprocessor, y is the array of internal (memory) variables, z the array of variables that connect from one memprocessor to other memprocessors (e.g., voltages of an actual circuit), and σ an *operator* that defines the *time evolution* of the memprocessor:

$$\sigma[x, y, z] = (x', y'), \quad \textit{memprocessor evolution operator}, \quad (4.27)$$

where x' and y' are the *output* state and memory variables, respectively.

To provide the reader with a better understanding of what this operator is, let's first define explicitly memprocessors made only of memelements, and then memprocessors given by arbitrary electronic units (including active elements).

Memprocessors
The fundamental units of MemComputing machines built out of memelements and active elements to perform processing and storing of information in the same physical location are called *memprocessors* (Traversa and Di Ventra, 2015).

Evolution operator
Any operator that, when acting on a given state of a physical system, describes its dynamics is called *evolution operator*. Although this concept is typically defined for *quantum* systems, it also pertains to *non-quantum* dynamical systems, with or without noise (see also Section 11.3).

4.6.1 Memprocessors made of memelements

As I have discussed in Section 4.4.2, general memelements are defined by eqns (4.25) and (4.26). According to our definition of memprocessors, eqn (4.27), we can then identify the connection variables z with both $u(t)$ and $v(t)$, namely, $z = [u(t), v(t)]$. The state x is instead the internal state array y, which is simply $\tilde{x}(t)$ in eqn (4.26).

Now, by using eqn (4.26), we can write the evolution if the internal state in a (small) interval of time Δt as

$$\tilde{x}(t + \Delta t) - \tilde{x}(t) = \int_t^{t+\Delta t} f(\tilde{x}(\tau), v(\tilde{x}(\tau), u(\tau)), \tau) d\tau, \qquad (4.28)$$

where from eqn (4.25) I have explicitly written the input variable, $v(t)$, as a function of the output, $u(t)$, and memory variables, $\tilde{x}(t)$.

By defining $x = \tilde{x}(t)$ and $x' = \tilde{x}(t + \Delta t)$ we then have from eqn (4.27)

$$\sigma[x, y, z] \equiv \sigma[x, z] = \tilde{x}(t) + \int_t^{t+\Delta t} f(\tilde{x}(\tau), v(\tilde{x}(\tau), u(\tau)), \tau) d\tau. \quad (4.29)$$

We see that eqn (4.29) has the same form as the action of an appropriate flow field (semi-group), $T(t)$, like the one I have defined in Section 3.2, eqns (3.3). This shows that memprocessors can also be described within the formalism of dynamical systems I have introduced in Chapter 3.

4.6.2 Memprocessors made of generic electronic devices

A similar analysis can be done for the memprocessors of interest to us, namely those that include any type of electronic device, including active ones. In general, such devices are described by a *differential-algebraic system of equations* (DAEs), namely a set of equations that, in addition to the ordinary differential equations (3.1), contain also *algebraic* (or *polynomial*) equations of the type $P = Q$, where P and Q are polynomials of some degrees (see, e.g., (Horowitz, 2015)).

However, this does not pose any issue, because a general DAE system can be written in the form (3.1) (see, e.g., (Bonani *et al.*, 2014)). Therefore, a general electronic device can be described by an equation of the type

$$\frac{d}{dt} q(x, y, z) = f(x, y, z, t), \qquad (4.30)$$

where the set $\{x, y, z\}$ represents all the state variables of the circuit, with q and f some functions of the arguments.

Differential algebraic equations (DAEs)

These are systems of equations (both differential and algebraic) that involve an unknown vector-valued function and its derivatives. For instance, if the function is $x(t) = (x_1(t), \ldots, x_n(t))$, and t is in a subset of $\mathbb{R}$, first-order DAEs are of the type:

$$F(t, x, dx/dt) = 0,$$

where F is a *vector function* with n components, some of which are *algebraic*. These equations appear in various fields of Science, including circuit theory (see, e.g., (Horowitz, 2015)).

Without going into much detail, from circuit theory and modified nodal analysis (Bonani *et al.*, 2014), one can show that there always exists a choice of x and y so that the Jacobian matrix

$$J_{x,y}(x,y,z) = \left[\begin{array}{cc} \dfrac{\partial q(x,y,z)}{\partial x} & \dfrac{\partial q(x,y,z)}{\partial y} \end{array} \right] \qquad (4.31)$$

is a square matrix, though not necessarily invertible. This means that one can eliminate some variables by including constraints as in the previous Section 4.6.1. This generates a reduced $J_{x,y}$ which is invertible.

Taking into account that (*chain rule*)

$$\frac{dq}{dt} = \frac{\partial q}{\partial x}\dot{x} + \frac{\partial q}{\partial y}\dot{y} + \frac{\partial q}{\partial z}\dot{z}, \qquad (4.32)$$

and the eqns (4.30) and (4.31), we then have

$$\left[\begin{array}{c} \dot{x} \\ \dot{y} \end{array} \right] = J_{x,y}^{-1}(x,y,z)\left(f(x,y,z,t) - \frac{\partial q(x,y,z)}{\partial z}\dot{z} \right). \qquad (4.33)$$

The *memprocessor evolution operator* in a time Δt is then

$$\sigma[x,y,z] = \left[\begin{array}{c} x(t+\Delta t) \\ y(t+\Delta t) \end{array} \right] = \left[\begin{array}{c} x(t) \\ y(t) \end{array} \right] +$$

$$+ \int_t^{t+\Delta t} J_{x,y}^{-1}(x,y,z)\left(f(x,y,z,\tau) - \frac{\partial q(x,y,z)}{\partial z}\dot{z} \right) d\tau. \qquad (4.34)$$

If we define the vector field

$$F \equiv J_{x,y}^{-1}(x,y,z)\left(f(x,y,z,\tau) - \frac{\partial q(x,y,z)}{\partial z}\dot{z} \right), \qquad (4.35)$$

we then see that eqns (4.34) is again equivalent to the action of the semi-group, $T(t)$, eqns (3.3). This means that memprocessors, defined with *any* type of element, whether passive, active, or both, are dynamical systems of the type discussed in Chapter 3.

4.7 MemComputing vs. 'in-memory computing'

Let me conclude this Chapter with an important clarification I anticipated in the Preface of this book. In modern computer architectures, there is a trend to employ what is called 'in-memory computing' (Plattner and Zeier, 2011). This term *has nothing to do* with MemComputing! Let me explain.

Traditionally, data are stored in hard disks or solid-state memories. Then processing of such data requires access to such media.

Near-memory architecture
In some computer architectures, the CPU chip may include one or more memory *caches* where data can be temporarily stored. These *near-memory* architectures (which are not the same thing as *edge computing*; see next margin note) allow a faster access of some data by the CPU (Horowitz, 2015). However, they have nothing to do with MemComputing!

Cloud computing
Cloud computing is a term that indicates the possibility of using computing resources (e.g., servers) remotely (via the Internet), and not directly managed by the users. In other words, the users have access to computational power and storage that are physically located far from them, without the need to maintain them (as their personal computing devices).

Edge computing
Edge computing typically means that computation and data storage are 'closer' to the user's location, namely at the 'edge' of the network. This improves both response times and reduces latency. This is very important for real-time processing of data.

Internet of Things
The *Internet of Things* (IoT) is the term used to describe the network of physical devices capable of exchanging data with each other over the Internet.

Logic in memory
Logical operations that are executed directly in memory units are often called *logic in memory*. See, e.g., (Hennessy and Patterson, 2006) for an extensive discussion of all these concepts.

'In-memory computing' (sometimes also called 'computing in memory'), instead, refers to the processing of some data that are stored *directly* in random access memory (RAM), sometimes in a distributed way, across different RAMs of a parallel computer or server.

This necessarily speeds up the processing of information considerably compared to the traditional way that requires access of data from, say, a hard disk.

However, the fundamental computing paradigm underlying this 'in-memory' access of information is still the same as the one in which a processing unit (CPU) is physically *distinct* from a memory unit. It thus suffers from the same *limitations* of the traditional von Neumann architecture (see Section 5.2) and *does not* fundamentally change the computing paradigm employed.

As I have amply discussed, and will expand even more in the following Chapters, MemComputing, instead, employs *time non-locality* to compute on the same physical location where information can be stored. Such a feature is *not* exploited by 'in-memory computing' *at all*.

Therefore, MemComputing is *neither* 'computing with memristive elements' *nor* 'in-memory computing' (nor 'computing in memory'), the latter one as implemented in some modern computer architectures.

4.8   Taking the 'with memory' out of MemComputing

However, it is important to note that the term 'in-memory computing' is also used in some literature to indicate that some *logic* operations are performed directly *in* memory, or for some analog applications also realized in memory devices (Ielmini and Wong, 2018). This type of computing is more in line with MemComputing, and, indeed, performing 'logic *in* memory' was one of the very first applications in the field (Traversa *et al.*, 2014).

Even so, as I have explained in the previous Section 4.7, 'in-memory computing' is a term that has already a well-defined meaning in some modern computing architectures (Plattner and Zeier, 2011). Therefore, it should not be used to indicate 'logic in memory' or 'analog computation in memory'.

In addition, as I have stressed repeatedly, 'memory' truly refers to the *time non-local* property of a physical system to remember its past dynamics and employ this feature to compute. That is why MemComputing does not refer simply to computation *in* memory, but also *with* memory.

Although we could take away the 'with memory' component of this computing paradigm, leaving only the 'in memory' one, by doing so we would severely limit its range of possibilities. In fact, as I will discuss in the rest of the book, it is precisely *time non-locality*, and the degrees of freedom associated to it, that allow us to solve combinatorial optimization problems efficiently.

4.9   Chapter summary

- I have discussed that *time non-locality* (the 'mem' in MemComputing) is common to *all* physical systems subject to external perturbations: no physical system can respond instantaneously to such perturbations. I have also discussed that time non-locality may induce spatial non-locality, even if the interactions among constituents are local.

- In Physics we do have an example of a machine that uses a type of spatial non-locality, known as entanglement, to compute a particularly difficult problem (factorization) efficiently: a quantum computer. Entanglement realizes some type of long-range order, in the sense that different parts of the system are intrinsically dependent on each other (correlated). However, a classical type of long-range correlations appears in many non-quantum phenomena, such as at the onset of a continuous phase transition. This is the type of Physics we want to exploit to compute efficiently.

- I have also discussed that an element with memory (memelement) may *process* information, in addition to possibly *store* it. Therefore, a machine that employs time non-locality offers advantages that are not easily achieved with standard computers.

- The machines we are after are essentially networks of *memprocessors*, processors with memory. However, as I will show in the following Chapters, *active* elements are necessary to realize specific memprocessors (*self-organizing gates*) that we can use to compute hard problems efficiently.

- Electrical circuits are ideal candidates to realize these machines in practice, because their hardware implementation is easily scalable. In fact, with the help of CMOS emulators of memelements, MemComputing machines can be fabricated with present CMOS technology without introducing new structures or materials.

- However, there may be physical systems other than electrical circuits that could be built to perform similar computations. These systems may be mechanical, optical and/or superconducting, and thus may offer some advantages for specific applications, such as those requiring reduced spatial footprint and/or lower energy consumption.

- Finally, I have stressed that MemComputing is *neither* 'computing with memristive elements' *nor* 'in-memory computing', as the latter is currently implemented in some modern architectures.

- In fact, MemComputing is even much broader than the 'in-memory computing' that is sometimes referred in the literature to indicate 'logic in memory', and a few other analog functionalities. It is the 'with memory' ('with time non-locality') part of MemComputing that gives it a much wider scope, and allows its efficient application to the problems I will discuss in Chapters 8, 9, and 10.

Part II

The MemComputing paradigm

The answers you get depend upon the questions you ask.
Thomas S. Kuhn (1922–1996)

Ideas that require people to reorganize their picture of the world provoke hostility.
James Gleick (1954–)

... they wondered originally at the obvious difficulties, then advanced little by little and stated difficulties about the greater matters, e.g., about the phenomena of the moon and those of the sun and the stars, and about the genesis of the Universe. ...

In this part of the book, I will formalize the notion of *Universal Mem-Computing Machine* (UMM), and then focus on its digital version: *Digital MemComputing Machine* (DMM). I will also introduce the novel idea of *self-organizing gates*, namely gates that are *terminal agnostic* and can solve their Boolean (or algebraic) relations *irrespective* of whether signals are fed to them from the traditional input terminals or the traditional output terminals. These gates are thus able to work *in reverse*. I will then report the mathematical properties of the *self-organizing circuits* built out of these gates. I will finally discuss that, due to their *physical* and *logical* topology, these circuits can solve combinatorial optimization problems efficiently if built in hardware.

- In Chapter 5, I will introduce the notion of Universal MemComputing Machine.
- The concept of DMMs is introduced in Chapter 6.
- Chapter 7 discusses the idea of self-organizing gates, a possible physical (hardware) implementation of such gates, and the mathematical properties circuits built out of them have.

5

Universal MemComputing machine

If you tell me precisely what it is a machine cannot do, then I can always make a machine which will do just that.
John von Neumann (1903−1957)

A mathematical machine is set up, and without asserting or believing that it is the same as Nature's machine, we put in data at one end and take out the results at the other. As long as these results tally with those of Nature, we regard the machine as a satisfying theory. But so soon as a result is discovered not reproduced by the machine, we proceed to modify the machine until it produces the new result as well.
John Lennard-Jones (1894−1954)

Takeaways from this Chapter

- Definition of universal MemComputing machine (UMM).
- A UMM is Turing-complete.
- The notion of *information overhead.*
- The importance of the *physical topology* in computing.
- A UMM can solve NP-complete problems efficiently.
- The meaning of *functional polymorphism* of a UMM.

After having introduced the notion of memprocessors in Chapter 3, I am now ready to define rigorously what we mean by MemComputing machines, along the lines of what we have done in Section 2.2.1 for the universal Turing machine (UTM). This formal definition will allow me to discuss their computational power in general terms.

5.1   Computational memory

When two or more memprocessors are connected, we have a network of memprocessors, which we call *computational memory* (this together with the *control unit* will make up the MemComputing machine, see the next Section 5.2). Following the definition of memprocessors in eqn (4.27), we define the vector $\mathbf{x}$ as the state of the network (i.e., the array of all the states x_i of each memprocessor), $\mathbf{y} = \cup_i y_i$ the array of all internal (memory) variables of the network (each memprocessor with an array y_i), and $\mathbf{z} = \cup_i z_i$ the array of all connecting variables, with z_i the connecting array of variables of the i-th memprocessor.

We say that two memprocessors are connected, if the vectors z_i and z_j of connecting variables of the memprocessors i and j, respectively, are such that $z_i \cap z_j \neq \emptyset$. Instead, a memprocessor is not connected to any other memprocessor (it is *isolated*) when $z = z(x, y)$ (namely, z is fully determined by x and y) and

$$\sigma[x, y, z(x, y)] = (x, y). \tag{5.1}$$

This means the memprocessor does not evolve in time.

The time evolution of the connecting variables $\mathbf{z}$ in a network of memprocessors is given by an evolution operator Ξ, defined as

$$\Xi[\mathbf{x}, \mathbf{y}, \mathbf{z}, \mathbf{s}] = \mathbf{z}', \tag{5.2}$$

where $\mathbf{z}'$ is the new set of connecting variables. The array $\mathbf{s}$, instead, defines the external signals that can be applied to a subset of connections to provide appropriate stimuli to the network.

The evolution of the *entire* network (assuming there are n memprocessors in the network) is then defined by

$$\begin{cases} \sigma[x_1, y_1, z_1] &= (x_1', y_1'), \\ \quad \vdots \\ \sigma[x_n, y_n, z_n] &= (x_n', y_n'), \\ \Xi[\mathbf{x}, \mathbf{y}, \mathbf{z}, \mathbf{s}] &= \mathbf{z}'. \end{cases} \tag{5.3}$$

If the evolution operators σ and Ξ are *discrete evolution operators*, namely they are operators in *discrete time* (see the difference with *continuous time* in Section 2.2.1), then their definition includes also the one of *artificial neural networks* (Haykin, 1998), when, additionally, one considers memprocessors without active elements.

In other words, formally, a collection of these memprocessors assembled into a network may also realize models of neural networks.[1] Therefore, neural networks can be viewed as specific cases (a subset) of MemComputing machines (Traversa and Di Ventra, 2015), or, equivalently,

Computational memory
A network (collection) of memprocessors is called a *computational memory* (Traversa and Di Ventra, 2015).

Artificial neural networks
They are computing models inspired by the operation of the biological neural networks in animal brains. They learn features from a set of data without a rule-based programming (see, e.g., (Haykin, 1998), and Chapter 10).

Neuromorphic computing
The application of artificial neural networks, whether in software or hardware, to brain-inspired computing tasks is often called *neuromorphic computing*.

[1] As such, MemComputing also encompasses *neuromorphic computing*, even though the latter typically refers to brain-inspired computational tasks, such as those I will discuss in Chapter 10.

MemComputing and neural networks
MemComputing machines *generalize* the notion of *artificial* neural networks.

I will not consider these types of operators in this book (even when I will discuss how to efficiently train neural networks with digital MemComputing machines in Chapter 10). Instead, I will consider only *continuous operators*, namely those representing general dynamical systems in *continuous time*.[2] In addition, if we consider an electric circuit realization of MemComputing machines, as I do in this book, the operator Ξ is defined by the Kirchhoff's laws (see Section 4.4.1) and external (current or voltage) generators that provide the external signals **s**.

[2]However, when integrated numerically, these operators give way to *discrete maps* (see Chapter 8).

5.2   Universal MemComputing machine

After the formal definition of 'computational memory' I am now ready to introduce the general (and *ideal*) concept of a *universal MemComputing machine* (UMM).

In words, I define:

Universal MemComputing machine
An *ideal* machine that performs computation *in* and *with* memory.

Here again, 'memory' is synonymous of *time non-locality* not necessarily 'storage'.

I can further refine this notion with the use of the memprocessors I introduced in Section 4.6 (see Fig. 5.1, bottom panel):

Universal MemComputing machine
A network of interconnected memprocessors (*computational memory*) that can perform either *digital* (logic) or *analog* (functional) operations (or both) controlled by a *control unit*.

Control unit
It is that unit of a computing machine that, according to the instructions received, informs the processor which program to execute and how. In modern computers it is part of the CPU. (Hennessy and Patterson, 2006).

Of course, digital machines (in the sense I have discussed in Section 2.8) are the *scalable* ones in terms of resources. These will be the focus of the remaining Chapters of this book. (An example of an *analog* MemComputing machine can be found in (Traversa *et al.*, 2015).)

However, since the concept of UMM is not limited to digital machines, here I consider it in its generality.

The computation *in* and *with* memory is *fundamentally* different from the one performed by a Turing machine (Section 2.2.1), as practically implemented in the most common *von Neumann architecture*.

The latter is attributed to mathematician John von Neumann (see Fig. 5.2). It is one possible *hardware* realization of a Turing machine (von Neumann, 1993).

Fig. 5.2 John von Neumann (1903–1957). Sketch by the author.

von Neumann architecture

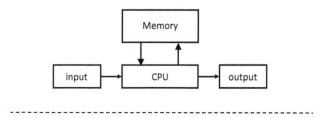

MemComputing architecture

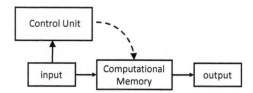

Fig. 5.1 Schematic illustration of the traditional von Neumann architecture (**top panel**) vs. the MemComputing one (**bottom panel**). Data flow and signal transmission are represented by straight and curved dashed arrows, respectively. In the MemComputing architecture the input data can be fed either directly into the network of memprocessors (*computational memory*), or into the *control unit*, which then sends a signal to the computational memory.

Following the rationale of the Turing machine, in the von Neumann architecture the tape head is replaced by an *active* device, the CPU (which contains a *control unit* and an *arithmetic/logic unit*), and the infinite tape by a *finite* physical medium that stores programs and data.[3] Importantly, mirroring the mathematical structure of a Turing machine, in the von Neumann architecture memory is *physically* distinct from the CPU (processing) (see Fig. 5.1, top panel).[4]

This separation of tasks leads to a latency and bandwidth limitations in the transfer of (syntactic) information between the CPU and the memory unit. In turn, this creates a bottleneck (the so-called *von Neumann's bottleneck*) in the actual execution speed, requiring large amounts of *energy* to move data.

The MemComputing computation, instead, can be sketched in the following way. When the memprocessors are connected, through a signal sent by the *control unit*, they change their internal states according to *both* their initial states *and* the signal supplied. They then evolve in time according to some equation of the type (3.4), which as we have discussed in Section 3.6, supports *intrinsic parallelism*. With this understanding in mind, let's now define this idea formally.

5.2.1   Formal definition

We define the UMM as the eight-tuple (Traversa and Di Ventra, 2015)

$$UMM = (M, \Delta, \mathcal{P}, S, \Sigma, p_0, s_0, F), \qquad (5.4)$$

where M is the set of possible states of a single memprocessor.

[3]There are other types of computer architectures, where program instructions and data don't share the same memory and signal pathways (Hennessy and Patterson, 2006). This distinction, however, is not very important for the discussion here, which focuses on the fact that processing and storing of information are done on physically distinct locations. Also, the von Neumann architecture is somewhat similar to the *analytical engine* proposed by Charles Babbage (1791 − 1871) in the 19th century. See, e.g., (Hamming, 1997) for a short history of computers.

[4]Note that here the term 'memory' is really intended as 'storage', namely *long-term* memory.

von Neumann's bottleneck
It is meant to indicate the latency and bandwidth limitations in the transfer of data between the CPU and the memory unit (Hennessy and Patterson, 2006).

As I have anticipated in Section 2.8, the distinction between *digital* and *analog* UMMs then rests on the finiteness, or otherwise, of the set of *input* and *output* states of the memprocessors.

If that set is *finite* (call it M_d) then it defines a *digital* machine (see Chapter 6 for an in-depth discussion of digital MemComputing machines). If it is a *continuum*, or an *infinite* discrete set of states (call it M_a) then we say that the machine operates in the *analog* regime. In general, the set M may be the direct sum of the previous two types of sets, $M = M_d \bigoplus M_a$, namely we could also define a *mixed digital-analog* UMM.

The set $\mathcal{P}$ contains the arrays of *pointers*, p_α, that select the memprocessors called by the *transition function* δ_α, and S is the set of indexes α. The set of the initial states written by the input device on the computational memory is Σ. Finally, $p_0 \in \mathcal{P}$ is the initial array of pointers, s_0 is the initial index α, and $F \subseteq M$ is the set of final states.

Δ is the set of *transition functions* (cf. with the one for a Turing machine, eqn (2.2))

$$\delta_\alpha : M^{m_\alpha} \backslash F \times \mathcal{P} \to M^{m'_\alpha} \times \mathcal{P}^2 \times S \,. \tag{5.5}$$

Here, the number of *input* memprocessors, read by the transition function δ_α are indicated by $m_\alpha < \infty$. Instead, the number of *output* memprocessors written by the same transition function are $m'_\alpha < \infty$.

The definition of a UMM seems somewhat similar to the one I have introduced for a universal Turing machine (UTM) (Section 2.2.1). However, this not quite the case, since a UMM has features *not* available to a UTM. As I will show in Section 5.5, these are the features that make the UMM a very powerful computing model. Let's first identify these features separately.

5.3   Major differences between a UTM and a UMM

5.3.1   Intrinsic parallelism

Recall that a UMM is made of a network of memprocessors and a control unit. As I have shown in Section 4.6, memprocessors are nothing other than dynamical systems. Therefore, they must share with dynamical systems the property of *intrinsic parallelism* (see Section 3.6). Intrinsic parallelism is definitely *not* a property of the UTM.

To make it clear that the transition functions δ_α of a UMM, eqn (5.5), do support the intrinsic parallelism I have discussed in Section 3.6, let me write them explicitly as

$$\delta_\alpha[\mathbf{x}(p_\alpha)] = (\mathbf{x}'(p'_\alpha), \beta, p_\beta) \,, \tag{5.6}$$

where $p_\alpha, p'_\alpha, p_\beta \in \mathcal{P}$ are the arrays $p_\alpha = \{i_1, ..., i_{m_\alpha}\}$, $p'_\alpha = \{j_1, ..., j_{m'_\alpha}\}$ and $p_\beta = \{k_1, ..., k_{m_\beta}\}$, with $\beta \in S$. The quantities $\mathbf{x}(p_\alpha) \in M^{m_\alpha}$ are the *input* memprocessor states that are fed into the transition function, while $\mathbf{x}'(p'_\alpha) \in M^{m'_\alpha}$ are the *output* states of the transition function.

Equation (5.6) is then interpreted as follows: δ_α *reads* the initial states $\mathbf{x}(p_\alpha)$ and *writes* the new states $\mathbf{x}'(p'_\alpha)$. At the same time, it prepares the new pointer p_β for the next function δ_β with input $\mathbf{x}'(p_\beta) \in M^{m_\beta}$. Therefore, the UMM—unlike the UTM—is such that *any* transition function δ_α *simultaneously* acts on a set of memprocessors, which is precisely the property of intrinsic parallelism I have discussed in Section 3.6.

Note also, from eqn (5.6), that when the control unit sends the signals $\mathbf{s}$ to $\mathbf{x}(p_\alpha)$, the memprocessor network changes *collectively* the states of $\mathbf{x}(p'_\alpha)$ into $\mathbf{x}'(p'_\alpha)$. Therefore, even if the control unit acts on *only one* memprocessor ($\dim(p_\alpha) = 1$), the UMM supports intrinsic parallelism. In other words, the change of state of a single memprocessor influences the collective state of all the other memprocessors in the machine. This is one of the desired criteria I have anticipated in Section 1.6 (criterion IV of MemComputing).

An important remark

In addition to the intrinsic parallelism that is lacking in a UTM, let me point out another substantial difference between a UTM and the UMM as defined in eqn (5.4): a UMM does *not* distinguish between states of the machine and symbols written on the tape. Instead, this information is *fully encoded* in the states of the memprocessors. This is indeed a very important feature for a machine, like a UMM, that is able to perform computation *and* storing of data on the same physical platform.

Moreover, a UTM has only a *finite*, discrete number of states and an unlimited amount of tape storage (see definition in Section 2.2.1). A UMM, on the other hand, may also access, in principle, an *infinite* number of continuous states, even if the number of memprocessors is finite. This is true in the analog and mixed digital-analog regime where each memprocessor is a continuous-time device with a continuous set of state values.

Physical topology
Given a network of units (e.g., memprocessors), its *physical topology* (sometimes called *architecture*) defines how these units are *physically* linked together. In other words, it is the topological structure of the network.

Logical topology
The *logical topology* of a network specifies how *data flow* within the network (see also Chapter 7).

Connectome
It is the full map of the 'wiring' (connections) of neurons in the animal brain.

5.3.2 Information overhead

There is yet another very important property of a UMM that is not available to a UTM. Since a UMM is defined as a network of memprocessors acting collectively, the *physical topology* of the network is of fundamental importance.

The physical topology of the network enters in the very definition of the transition functions in eqn (5.5), via the pointers, and the array of all connecting variables, $\mathbf{z} = \cup_i z_i$, of the memprocessors (Section 5.1), which take advantage of how the memprocessors are connected together to solve a particular problem.

This type of property is similar to the one discussed in biological neural networks (Kohonen, 1989), where information is thought to be *distributed* in the network of neurons, in a way that cannot be obtained from the simple union of disconnected neurons (see Section 12.10).[5]

In fact, this is where additional *physical* information enters into MemComputing: the 'connectivity' of the network—the *connectome*, to borrow a term from Neuroscience—is not just some 'ideal' mathematical concept. It originates from the actual physical coupling among memprocessors.

For instance, in the case of electrical circuits, this coupling could be realized by actual metallic wires that couple the different circuit elements in the system. In general, it is some form of physical coupling (interaction) among the different units of the machine.

Since the information that is provided by the physical topology of the network is an 'additional' type of information and substantially *different* from the *Shannon's self-information*, we call it *information overhead* (Traversa and Di Ventra, 2017).

Therefore, the information overhead, unlike Shannon's information, is a type of information that is *embedded* in the machine but *not stored* by it. It is a *global* property that is defined at the outset and *does not change* during the operation of the machine. In other words, unlike the states of a MemComputing machine that can change during computation, its information overhead does not.

To clarify this important point, and what information overhead brings to the table, let's look at the following example.

[5]This is again a property of *complex systems* (cf. Section 1.5).

Shannon's self-information
It it the (syntactic) information *stored* in a message m. It is quantified as $I(m) = -\log_2 p(m)$, where $p(m)$ is the probability that the message m is chosen from all possible choices in the message space M (Cover and Thomas, 2006). This type of information is stored in long-term memory and, for the same instance problem, it is the *same* for both Turing and MemComputing machines.

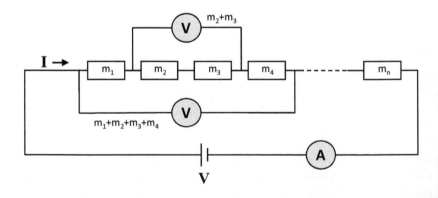

Fig. 5.3 System of n linearly connected memprocessors, m_i $(i = 1, \ldots, n)$, made of resistive memories, each storing a bit of information. The far ends of the circuit are biased by a voltage V. By using a voltmeter (circle with the letter 'V' inside) that reads the voltage across any pair of resistive memories, and an ammeter (circle with the letter 'A' inside) that measures the global current flowing in the circuit, one can directly measure all $n(n-1)/2$ sums of the intermediate memprocessors bits.

Example 5.1

Quadratic information overhead

Let's consider a simple MemComputing machine made of a set of n memprocessors connected *linearly* in series, as depicted in Fig. 5.3. To make this example even more concrete, we can think of each memprocessor as a resistive memory, connected to their nearest neighbors via some (dissipation-less) wires, so that the whole machine is a set of resistors in series obeying Kirchhoff's laws (Section 4.4.1) .

At the far ends of the series of resistors we apply a bias, of an appropriate magnitude such that, in each one of the n memprocessors, we encode one bit (one message, m_i, with $i = 1, \ldots, n$).

However, suppose now you are given a device (a voltmeter) that reads the voltage drop across *any* combination of resistive memories, say between the first two resistive memories, the first three, and so on (see Fig. 5.3). Since the current is the same across any of the resistive memories, by measuring that current with an ammeter, and reading the voltage drops with the voltmeter, we can infer the *sum* of all possible resistances (bits or messages) encoded by the single memprocessors. There are $n(n-1)/2$ possible sums, which, if given the two measuring devices (ammeter and voltmeter), allow us to access these different messages.

Therefore, we see that even a simple *linear* physical topology, as that depicted in Fig. 5.3, provides a *quadratic information overhead*, namely one that grows quadratically with the number of memprocessors. In other words, the topology of Fig. 5.3 has *embedded* ('compressed') in it extra (quadratic number of) messages the user may have access to.

We can then define:

Information overhead
It is *data compression* that is already present in the *physical topology* of the network of interconnected memprocessors.

Of course, in the previous example one would need a voltmeter (and an ammeter) to access ('decompress') each one of those sums (messages) embedded in the machine. That information, therefore, does not seem particularly useful for computations.

However, one could envision situations in which the machine may take advantage of other types of physical topologies for computing hard problems. This is particularly true if the topology of the machine supports an *exponential information overhead*, namely one that grows exponentially with a *polynomially* increasing number of memprocessors. See an explicit example of this case in (Traversa and Di Ventra, 2015).

[6]Note that this does *not* prove that NP = P (Section 2.3.2): a UMM is *not* a UTM (see also Section 5.4).

If that is the case, then one can show that a UMM can solve NP-complete problems with polynomial resources (see Section 5.5).[6] This also anticipates the fact that one needs *specific* physical topologies to tackle hard problems efficiently, as I will show explicitly in Chapters 6 and 9.

5.3.3 Functional polymorphism

Although I will not make use of it in the rest of the book, I point out here yet another property—which is also shared by the brain (Kohonen, 1989); see Section 12.10—that distinguishes a UMM from a UTM:

> **Functional polymorphism**
> The ability of a UMM to compute *different functions* without modifying the physical topology of the machine network, by simply applying the appropriate input signals.

In other words, there is no need to change the physical topology of the machine to obtain different transition functions. It is enough to change the input signals.

This property is clearly embedded in the definition of the set Δ of transition functions δ_α, eqn (5.5). In fact, the UMM, unlike the UTM, can have more than one transition function δ_α at any given time (see definition 5.4).

[7]The realization of such a functionality, could also be implemented in hardware by using, e.g., *Field Programmable Gate Arrays* (FPGAs). These are integrated circuits that form blocks of logic gates connected via interconnects that are programmable by the user for specific applications (Hennessy and Patterson, 2006).

For the solution of each combinatorial optimization problem, this property is not necessary. However, it may come in handy if one wants to use the same machine, in *hardware*, to solve a variety of different problems, with the same type of memprocessors. The interested reader may look at a practical example of functional polymorphism for the digital MemComputing machine proposed in (Traversa *et al.*, 2014), whose architecture is very similar to the one employed in modern flash memories, and employs *memcapacitive elements* (Section 4.4.2) and standard transistors as memprocessors.[7]

> **Turing completeness**
> A machine is said to be *Turing-complete* if it can solve *any* problem that a UTM can solve. It does not imply the reverse.

> **Turing equivalence**
> A machine is said to be *Turing-equivalent* if and only if it can solve the *same* problems a UTM does.

5.4 Universality of MemComputing machines

I can now address the reason for the name 'universal' used for the MemComputing machines defined in eqns (5.4) and (5.5). The term 'universal' means that *any* problem that a universal Turing machine (as defined in Section 2.2.1) can solve, can be solved also by a UMM. We would then say that a UMM is *Turing-complete* (Arora and Boaz, 2009).

> **An important remark**
> Note that, as of this writing, the reverse has not been demonstrated, namely, that any problem solved by a UMM is also solvable by a UTM. Such a demonstration would amount to stating that a UMM is *Turing-equivalent* (Fig. 5.4).

$$\text{UMM} \rightleftharpoons_{\text{?}} \text{UTM}$$

Fig. 5.4 A universal Memcomputing machine is *Turing-complete* (UMM $\Longrightarrow$ UTM). However, it is still an open question if it is also *Turing-equivalent*, namely the reverse (UMM $\Longleftarrow$ UTM) has not been shown yet.

In order to demonstrate the universality of a UMM, I follow the strategy outlined in (Traversa and Di Ventra, 2015). This consists in proving that a UMM can *simulate* any UTM. This is a sufficient condition for universality. In (Pei, Traversa and Di Ventra, 2019) the same universality was demonstrated using set theory and cardinality arguments, and it was further shown that a UMM can simulate quantum machines.[8]

Let's then refer to the definition (2.1) of UTM and consider a UMM with the memprocessor states in the set $M = Q \cup \Gamma$, where we recall that Q is the set of possible states that the UTM can be in, and Γ is the set of symbols that can be written on the tape. One memory cell is located by the pointer j_s, while the (*infinite*) remaining cells are located by the pointer $j = ..., -k, ..., -1, 0, 1, ..., k, ...$[9]

Let's also introduce the array of pointers $p \in \mathcal{P}$ defined by $p = \{j_s, j\}$. This way I can use the cell j_s to encode the state $q \in Q$, while the symbols in Γ can be encoded in the other cells. In this case, the set Δ of the UMM contains only one function: $\bar{\delta}[\mathbf{x}(p)] = (\mathbf{x}'(p), p')$. Note that we do not need the output index β that appears in eqn (5.6) because now we have only one function.

The new states $\mathbf{x}'$ written by $\bar{\delta}$ are written according to the transition function δ of eqn (2.2). In particular, $\mathbf{x}'(j_s)$ contains the new state of the UTM, while $\mathbf{x}'(j)$ contains the symbol the UTM would write on the tape.

The new pointer p' is then given by $p' = \{j_s, j'\}$, where $j' = j$, if the transition function δ of the UTM requires a 'No' shift. It is $j' = j + 1$ if 'Right' shift, and finally $j' = j - 1$ if 'Left' shift.

We then conclude that, according to the previous procedure, writing on $\mathbf{x}(j_s)$ the initial state q_0, and the initial symbols Σ, the UMM with $\Delta = \bar{\delta}$ simulates the UTM with transition function δ. This then demonstrates the Turing-completeness of the UMM, hence its universality.[10]

[8] Note that these proofs do not tell us anything about the *resources* required by a UMM to simulate another machine. They only pertain to the *type of problems* they can solve.

[9] Recall that *infinite memory* is part of the definition of Turing machines (Section 2.2.1).

> **The halting problem**
> Given an arbitrary computer program and an input, the *halting problem* consists in determining whether the program with that input will halt or continue to run forever (Arora and Boaz, 2009).

[10] It is worth pointing out that, if we could prove that a UMM is *not* Turing-equivalent, some (Turing) *undecidable* problems (cf. Section 2.3.1), such as the halting problem, may find solution within the MemComputing paradigm, thus contradicting the original Church-Turing hypothesis, see Section 2.2.2 and (Turing, 2004).

5.5 A UMM can solve NP-complete problems in polynomial time

The previous discussion on the properties of the UMM and its differences with respect to a UTM should make it clear that the MemComputing approach to computation is *fundamentally* different from the Turing one. In fact, as I have explained in Section 5.3.1, the physical coupling among memprocessors is such that, when the control unit sends an input signal to one or a group of memprocessors, the coupling induces dynamics to a larger group of memprocessors, up to the entire network. This is a feature that *does not exist* in a Turing machine, even a 'parallel one' (see the discussion in Section 3.6).

In addition, we have seen that by appropriately employing the physical coupling between the memprocessors, we may even *encode* (compress) information that grows exponentially with the number of memprocessors, and such information is *global*, in the sense that it is *independent* of the instantaneous state of the machine.

Therefore, by virtue of its intrinsic parallelism, a UMM can take advantage of such information *all at once*, in some sense similar to what a quantum computer does when employing entanglement (and state superposition) to solve prime factorization; see Section 4.3.1 and (Nielsen and Chuang, 2010).

Fig. 5.5 Left panel: an exponentially growing solution tree for a typical NP-complete problem, indicating two successive algorithmic steps, i and $i + 1$. **Right panel:** sketch of a network of memprocessors forming a UMM that attempts to solve it with transition function $\delta_{\alpha=i}$. The control unit sends the input signal to a group of memprocessors (a subset of the whole network, shaded area). The signal induces computation, and for each input signal sent by the control unit the UMM computes, *all at once*, an entire iteration of the tree via its transition function $\delta_{\alpha=i}$, changing the state of the memprocessors (different shaded area).

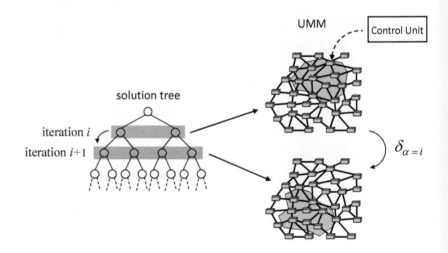

[11] Namely those problems that are classified within the Turing paradigm as NP-complete. If the UMM turns out to be *not* equivalent to a UTM, these same problems should be classified *differently* within the MemComputing paradigm of computation (cf. Section 13.8.6).

[12] This demonstration follows closely the one in (Traversa and Di Ventra, 2015).

Let's then show here that a UMM endowed with an *exponential* information overhead, and thanks to its intrinsic parallelism, can solve NP-complete problems[11] in *polynomial* time using only a *polynomially* growing number of memprocessors.[12]

An important remark

The previous statement *does not* prove NP = P! The famous NP = P question pertains *only* to Turing machines (see the exact formulation of this statement in Section 2.3.2). As I already mentioned, a

UMM is *not* a Turing machine—there is not even a proof that they are equivalent in terms of the problems they can solve. Therefore, the resolution of this question is still open (see also Chapter 13).

In Section 2.3.1 we have seen that, to solve an NP-complete problem, the known fastest algorithm, running on a *deterministic* Turing machine, requires (in the worst case) *exponential* time (steps), with respect to the dimension n of the input. This is because the solution tree grows exponentially with n (see discussion in Section 2.3.1 and Fig. 5.5).

Let's then consider a *polynomial* function of n, $P(n)$,[13] and assume that, at every iteration i, each branch of the solution tree splits into M_i new branches. The complete solution tree has then a total number of nodes, $N_{nodes} \leq \sum_{k=1}^{N_s} \prod_{i=1}^{k} M_i$, where $N_s = P(n)$ is the number of iterations. The case '$<$' occurs when some branches end with no solution before the $P(n)$-th iteration.

Now, from the definition of a UMM, eqn (5.4), let's consider a set Δ of transition functions δ_α with $\alpha = 1, ..., P(n)$. According to the definition (5.6) we take the transition functions to be

$$\delta_\alpha[\mathbf{x}(p_\alpha)] = (\mathbf{x}'(p_{\alpha+1}), \alpha + 1, p_{\alpha+1}). \tag{5.7}$$

Here, $\mathbf{x}(p_\alpha)$ is a vector encoding all the nodes belonging to the iteration $\alpha = i$ of the solution tree into at most $\dim(p_\alpha) \leq \prod_{i=1}^{\alpha} M_i$ memprocessors. On the other hand, the vector $\mathbf{x}'(p_{\alpha+1})$ encodes all the nodes belonging to the iteration $\alpha + 1 = i + 1$ of the solution tree into at most $\dim(p_{\alpha+1}) \leq \prod_{i=1}^{\alpha+1} M_i$ memprocessors.

The transition from the iteration i to the iteration $i+1$ of the solution tree (Fig. 5.5) is computed by the function $\delta_\alpha = \delta_i$ acting on the group of memprocessors encoding the iteration i to give the iteration $i + 1$ *all at once* (intrinsic parallelism).

Now, let's assume the UMM has an *exponential information overhead*. This means that, at each iteration k, the data encoded into $N_{nodes_k} \leq \prod_{i=1}^{k} M_i$ can be encoded in a number of memprocessors $\dim(p_k) \leq N_{nodes_k}$ (see Section 5.3.2).

In addition, since each transition function, δ_α, corresponds to only *one computation step* for the UMM, the latter will take a time proportional to $P(n)$ to find the solution of the original problem. Therefore, given an NP-complete problem, in principle, a UMM (which is an *ideal* machine) can solve it in a time $O(P(n))$ (polynomial time).

As I have discussed in Section 2.3.1, another hypothetical machine that can also accomplish this feat is a *non-deterministic* Turing machine (NTM). In this Section, I have then shown that, a UMM, that supports exponential overhead in its physical topology, has the same computational power (in terms of resources) of an NTM.

[13]The degree of the polynomial is not important, so long as it is a polynomial.

However, an NTM (in its 'exploring-all-possible-paths' interpretation I have discussed in Section 2.3.1) requires, in principle, *both* a number of processing units *and* a number of tapes growing *exponentially*, in order to explore all possible solution paths (Garey and Johnson, 1990).

A UMM with an exponential overhead, instead, can process an *entire* iteration of the solution tree in just one step, with a polynomially growing number of memprocessors.

This gives us great hope that an actual *physical* realization of Mem-Computing machines may be found. I will indeed discuss a practical realization in Chapter 7.

5.6   Chapter summary

– I have introduced the notion of *universal MemComputing machine*. This machine has several noteworthy properties. In addition to *intrinsic parallelism* of any dynamical system (Section 3.6), it can solve different problems without changing its physical topology (architecture), by simply providing the appropriate input signals. This *functional polymorphism* is a brain-like feature with important practical ramifications.

– Another important feature of a UMM, which is also brain-inspired, is its *information overhead*. This information is *embedded* (compressed) in the *physical topology* of the network of interconnected memprocessors, but *not* stored in memory. Therefore, this information arises from the *physical* interaction among memprocessors and does not change during time evolution. However, it can be exploited during computation, rendering the latter more efficient.

– In fact, I have shown that a UMM, with a physical topology supporting an *exponential* information overhead, has the same computational power of non-deterministic Turing machines, meaning that such an *ideal* machine can solve NP-complete problems in polynomial time and with a polynomially growing number of memprocessors. This statement however does *not* prove NP = P: a UMM is not a Turing machine.

– We proved that a UMM is Turing-complete but could not prove (yet) that it is equivalent to a universal Turing machine.

Digital MemComputing machines

6

You can't live in the digital and die in the analog.
Dean Cavanagh (1966−)

Analog is more beautiful than digital, really, but we go for comfort.
Anton Corbijn (1955−)

Takeaways from this Chapter

- The mathematical definition of digital MemComputing machines (DMMs).
- The dynamical system picture of DMMs.
- What features make such machines compute efficiently?

So far, I have considered the concept of MemComputing in its generality. This has led us to the introduction of the *mathematical* notion of a universal MemComputing machine (UMM). I have then analyzed its strengths and differences with respect to Turing machines.

However, recall from Section 2.8 that, in order to be easily *scalable* when built in hardware, MemComputing machines[1] need to map a *finite* string of symbols (such as the integers 0s and 1s) into a *finite* string of symbols: the machines have to be *digital*. As we have seen in Section 5.2.1, this is indeed one possible type of UMM, a subclass of the most general concept, I call *digital MemComputing machines* (DMMs) (Traversa and Di Ventra, 2017; Di Ventra and Traversa, 2018*b*).

In this Chapter, I will first provide their formal definition and then make a connection with the theory of dynamical systems I have introduced in Chapter 3. This connection is in anticipation of their actual hardware implementation that I will discuss in Chapter 7.

I will then show in Chapters 8, 9, and 10 the numerical *simulations* of

[1] As any other type of machine.

their equations of motion on a variety of combinatorial optimization, Machine Learning, and quantum mechanical problems. These simulations already show great advantages compared to traditional state-of-the-art algorithms,[2] even without an actual hardware realization.

6.1 Formal definition

Following the definition of a UMM as provided in Section 5.2.1, we can mathematically define DMMs as the eight-tuple (Traversa and Di Ventra, 2017)

$$DMM = (\mathbb{Z}_2, \Delta, \mathcal{P}, S, \Sigma, p_0, s_0, F), \tag{6.1}$$

where (although not strictly necessary) we consider the space of states $\mathbb{Z}_2 = \{0, 1\}$. Generalization to any finite number of states is, of course, trivial. Δ is the set of *transition functions*

$$\delta_\alpha : \mathbb{Z}_2^{m_\alpha} \backslash F \times \mathcal{P} \to \mathbb{Z}_2^{m'_\alpha} \times \mathcal{P}^2 \times S, \tag{6.2}$$

where the number of memprocessors used as *input* of the function δ_α is $m_\alpha < \infty$, while the number of memprocessors used as *output* is $m'_\alpha < \infty$. $\mathcal{P}$ is the set of the arrays of pointers p_α that select the memprocessors called by δ_α, and S is the set of indexes α. Σ is the set of initial states, $p_0 \in \mathcal{P}$ is the initial array of pointers, s_0 is the initial index α, and the set of final states is $F \subseteq \mathbb{Z}_2^{m_f}$ for some $m_f \in \mathbb{N}$.

> **An important remark**
>
> It is worth stressing once more the following: the *digital* (hence *scalable*) character of a machine—as defined previously—rests on the *finiteness* of the set of *input* and *output* states in eqn (6.1), *not* on the transition function, eqn (6.2). The transition function can be chosen at will, so long as it maps a finite string of symbols to a finite string of symbols (cf. discussion in Section 2.8). It is this freedom that allows us to use dynamical systems with memory as representations of the transition function (see Section 6.3).

[3] In fact, the discussion in Section 6.2 is not limited to DMMs. It could equally be applied to any type of MemComputing machine.

Having defined DMMs formally, I can now make a connection with the theory of dynamical systems of Chapter 3. However, before doing that let me say a few (very important) words on how we can operate such machines.[3] This discussion will be key to understanding the transition from traditional Boolean gates and circuits to *self-organizing* gates and circuits, namely gates and circuits that do *not distinguish* between signals originating from the traditional input or the traditional output.

6.2 Solution mode vs. test mode

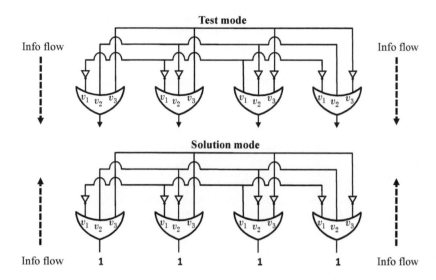

Fig. 6.1 Sketch of the two modes a DMM can operate to solve a given Boolean problem, like the 3-SAT problem of Fig. 2.8. The **top panel** shows the *test mode* for the verification of a given solution, while the **bottom panel** the *solution mode* for the *inverse protocol* implementation. In the test mode one inputs the variables assignment to verify the solution of the problem. The information flows from the 'input' gate terminals to the 'output' gate terminals. In the solution mode, one provides the variables assignment ('1' or 'true') to the 'output' gate terminals to find the correct variables assignment of the 'input' gate terminals. The flow of information is *reversed* with respect to the test mode. The *logical topology* of DMMs is then *bi-directional*.

Let's first point out that, due to its *digital* input and output structure, the problems a DMM is specifically designed to solve are *Boolean problems*,[4] and more generally, *combinatorial optimization problems* (Section 2.5). I will focus here only on the Boolean class, and mention the *algebraic* problems in Chapter 7.[5]

Consider then a set of Boolean variables $\mathbf{y} \in \mathbb{Z}_2^n$ and $\mathbf{b} \in \mathbb{Z}_2^{n_o}$, with $n, n_o \in \mathbb{N}$. The problem defined by the system of Boolean statements, $f(\mathbf{y}) : \mathbb{Z}_2^n \to \mathbb{Z}_2^{n_o}$, is solved if the machine returns a possible input $\mathbf{y}$ of f, if it exists, that satisfies $f(\mathbf{y}) = \mathbf{b}$. There are then two modes a given DMM can operate: *test* and *solution modes* (see Fig. 6.1).

---

[4]I have defined this particular class of problems in Section 2.4.1.

[5]See also Section 9.2.5.

Test mode

In the test mode (see top panel of Fig. 6.1) the *control unit*[6] feeds the appropriate elements of the machine with a signal encoding the input $\mathbf{y}$. The machine then starts its computation by formally applying the transition function (6.2).[7] This way, we obtain the output $f(\mathbf{y})$, to be compared against $\mathbf{b}$, to determine whether or not $\mathbf{y}$ is a solution of the Boolean problem. In simple words, we can check if $\mathbf{y}$ is a solution of the problem by simply inputting it in f and see if the result is $\mathbf{b}$.

Solution mode

Conversely, in the solution mode (see bottom panel of Fig. 6.1) the

[6]That unit that tells the machine which program to execute and how (see Section 5.2).

[7]The transition function could be an ordered combination of successive transition functions, namely it could be written as the *composition* $(\delta_\zeta \circ \cdots \circ \delta_\alpha)(\mathbf{y})$, where $\circ$ is the symbol of composition of the different transition functions, δ_j, in that particular order.

control unit feeds the appropriate elements of the machine with a signal encoding **b**.

Let's now consider the *inverse* of the transition function, which we call δ^{-1}. To this function we can now feed the vector **b**, which is the vector resulting from the test mode. The inverse of the transition function, δ^{-1}, then receives **b** and produces the 'output' **y**, which was the 'input' in the test-mode operation.[8]

In simple terms, the (syntactic) information flows from the traditional output to the traditional input!

[8] Or, if there is more than one transition function, the composition of all transitions functions $(\delta_\zeta^{-1} \circ \cdots \circ \delta_\alpha^{-1})(\mathbf{b})$ provides the output.

Note that this type of operation is quite unusual for the Boolean gates and circuits we are typically familiar with (like the ones in Fig. 6.1), and is *not* a feature of Turing machines. In fact, it will require us to rethink the very Boolean logic itself, and will lead us to the idea of *self-organizing gates* (see Chapter 7 and (Traversa and Di Ventra, 2017; Di Ventra and Traversa, 2018b)).

Their *logical topology* (see definition in Section 5.3.1) is *bi-directional*, although not necessarily one-to-one (bijective). We can then conclude this Section by saying that

Bijective function

Consider a function, f, between two sets $\mathcal{A}$ and $\mathcal{B}$: $f : \mathcal{A} \to \mathcal{B}$. We say that f is *bijective*, or *one-to-one* if each element of $\mathcal{A}$ is paired with only one element of $\mathcal{B}$, and vice versa (Arfken, Weber and Harris, 1989).

Logical topology of DMMs

The information flow (*logical topology*) of DMMs is *bi-directional*, namely they can work in both *test* and *solution* mode, without distinction between traditional 'input' and traditional 'output'.

Let's now try to put these ideas into practical, *physically realizable* systems that can be built in hardware.

6.3   Toward a *physical* realization of DMMs

As I have discussed in Section 2.4.1, if we are given a specific Boolean problem, we can represent it as a Boolean circuit. We then map this circuit into the *physical topology* of the network of memprocessors of a DMM, and employ the latter to solve the original problem.

In view of the properties of a UMM, which transfer also to the subclass of DMMs, we may then expect an actual physical (*hardware*) realization of DMMs to be able to solve hard problems efficiently (see Section 5.5).

In fact, in line with my discussion in Section 2.8, this physical realization does *not* preclude their scalability because the input and output, namely what we write in and read out, respectively, are *digital*, hence they can be represented with finite precision, or with an error that does *not* scale with the size of the problem. I now want to make these statements even more precise.

6.3.1   From mathematical concept to physical system

First of all, we know, from their definition (6.1), that DMMs are a collection of memprocessors (guided by a control unit).

And, as I have shown in Section 4.6, memprocessors are simply *non-quantum dynamical systems* whose general dynamics are described by ordinary differential equations of the type (3.1).

Let's then assume that a given DMM is represented *physically* (in hardware) by a dynamical system described by the general eqns (3.1).[9] Following the discussion in Section 3.11, we also choose *equilibrium points* of eqns (3.1) to represent the *solutions* of a given problem.

If possible, the dynamics should also be *integrable* (no chaos), no (quasi-)periodic orbits should exist, if equilibria are present, and no other minima in the phase space, X,—other than the solutions of the problem—should be present. However, we will see later (Section 6.6) that we can lift, to a certain degree, these restrictions as well.

Let's also recall from Chapter 3 that the system dynamics are embedded in an D-dimensional phase space, X, in which $\mathbf{x}(t)$ is a trajectory of eqns (3.1).[10] Each component $x_i(t)$ of $\mathbf{x}(t)$ has initial conditions $x_i(0) \in X$. We also assume that the space X is a *metric space*, namely we can define on it the concept of a distance.

We now want to know which features allow such a *physical* system to represent a *digital* machine that computes *efficiently*, namely with resources that grow at most *polynomially* with input size.[11]

6.3.2   What do we need to compute efficiently?

Let me first enumerate these properties and I will discuss them later. We will then require eqns (3.1) to have the following properties, to be able to solve efficiently (namely with polynomial resources) a given Boolean problem (Traversa and Di Ventra, 2017):

Physical properties a DMM needs to compute efficiently:

(i) The stable equilibria $\mathbf{x}_s \in X$ of eqns (3.1) are associated with the solutions of the given Boolean problem the machine needs to solve. In other words, if the given Boolean problem has solutions, they must be represented by the equilibrium points of eqn (3.1).

(ii) To consistently define our machine as a *digital* machine, its 'input' and 'output' (e.g., $\mathbf{b}$ and $\mathbf{y}$, respectively, in solution mode) must be mapped into a set of physical parameters $\mathbf{p}$ (input) and equilibria $\mathbf{x}_s$ (output) such that $\exists \hat{p}_0, \hat{p}_1, \hat{x}_0, \hat{x}_1 \in$

---

[9] For now, let's not worry about unavoidable noise. I will discuss it in Chapters 11 and 12, where I will show that DMMs employ topological quantities to compute. As such, they are robust against noise and perturbations.

[10] Note that in this Chapter the dimension, D, of the phase space, X, is equal to the dimension, n_o, of the 'input', $\mathbf{b}$, of the Boolean problem a DMM solves in solution mode: $D = n_o = \dim(\mathbf{b})$. In Chapter 7, it will be larger than that.

[11] The analysis Section 6.3.2 follows closely the one in (Traversa and Di Ventra, 2017).

Metric space

It is an ordered pair (X, d), where X is a set and d is a *metric* on X: $d : X \times X \to \mathbb{R}$, with the following properties $(x, y, z \in X)$:

$d(x, y) = 0$, if and only if $x = y$,

$d(x, y) = d(y, x)$,

$d(x, z) \leq d(x, y) + d(y, z)$.

In this book, I will use the symbol $| \cdot |$ to indicate d, e.g., $d(x, y) \equiv |x - y|$. An example of metric space is the Euclidean space endowed with the Euclidean norm (Arfken *et al.*, 1989).

$\mathbb{R}$ that encode the logical 0s and 1s. Then the inputs p_j can be within a distance ϵ_p from $\hat{p}_0$ or $\hat{p}_1$, and the outputs x_{s_j} can be within a distance ϵ_x from $\hat{x}_0$ or $\hat{x}_1$, and $|\hat{p}_0 - \hat{p}_1| = c_p$ and $|\hat{x}_0 - \hat{x}_1| = c_x$ for some $\epsilon_p, \epsilon_x, c_p, c_x > 0$, *independent* of $D \equiv n_o = \dim(\mathbf{b})$ (the dimension of the 'input') and $n = \dim(\mathbf{y})$ (the dimension of the output). Moreover, if we indicate a polynomial function of n of maximum degree γ with $\mathfrak{p}_\gamma(n)$, then $\dim(\mathbf{p}) = \mathfrak{p}_{\gamma_p}(n)$ and $\dim(\mathbf{x}_s) = \mathfrak{p}_{\gamma_x}(n_o)$ in *test mode*; or $\dim(\mathbf{p}) = \mathfrak{p}_{\gamma_p}(n_o)$ and $\dim(\mathbf{x}_s) = \mathfrak{p}_{\gamma_x}(n)$ in *solution mode*, with γ_x and γ_p *independent* of n and n_o.

(iii) Other stable equilibria, (quasi-)periodic orbits or strange attractors (chaos) that we generally indicate with $\mathbf{x}_w(t) \in X$, not associated with the solution(s) of the problem (the 'w' stands for 'wrong solutions'), may exist, but their presence is either irrelevant, or can be accounted for with appropriate initial conditions (see Section 6.6).

(iv) The system has a *compact* (see definition in Section 3.10) *global asymptotically stable attractor* (Hale, 2010), which means that there exists a compact subset $J \subset X$ that attracts the *whole* space X.

(v) For a given problem size, the system converges to equilibrium points *exponentially fast* starting from a region of the phase space whose measure is *not* zero, and which can decrease at most *polynomially* with the size of the problem. Moreover, the convergence time (in fact, all resources) can increase at most *polynomially* with the input size.

Global attractor

A compact subset, J, of the phase space, X, is a *global attractor* under the dynamics (3.1), if *all* trajectories, $\mathbf{x}(t)$, in the phase space are asymptotically attracted by J.

Asymptotically stable set

The set J is *asymptotically stable* under the dynamics (3.1), if it is stable and attracts points locally in the phase space. See (Hale, 2010) for a more rigorous definition of all these concepts.

Set of measure zero

For the purposes of this book, a set, $\mathcal{B} \subset \mathbb{R}^n$, is of *measure zero* if it is *countable* (cf. Section 2.2) or *finite*.

Note that it is property (ii) that renders our machine *digital*, like our modern computers (cf. discussion in Section 2.8). It says that one can find physical values of some quantities (e.g., voltages) that represent a logical 1 and a logical 0 that are sufficiently distinct from each other in magnitude, and that all input parameters, at the beginning of the computation, and output trajectories, at the end of it, are within a certain distance from those values, with that distance being *independent* of the size of the problem.

If that is the case, the input and output states of the machine are sufficiently *distinct* to be able to sustain reasonable levels of noise and perturbations. Again, this is similar to how we represent 0s and 1s with our physical transistors. We assign a logical 0 and a logical 1 to a low and high current, respectively, when the value of that current passes a certain threshold, with that threshold being *independent* of the number of transistors we put together (cf. Fig. 2.9).

The transition function, is instead replaced by the equations of motion (3.1) (see also Section 3.6) so that, unlike a Turing machine (Section 2.2.1), a DMM is:[12]

Digital MemComputing machine (DMM)

A *dynamical* mapping from a *finite* string of symbols to a *finite* string of symbols:

$$\{\textit{finite string of symbols}\} \xrightarrow{\dot{\mathbf{x}}(t)=F(\mathbf{x}(t))} \{\textit{finite string of symbols}\},$$

in *continuous (physical) time*.

---

[12]So long as the dynamics expressed by eqns (3.1) representing the transition function are *robust* against noise and perturbations, then the dynamics will maintain the digital structure of the input and output. We will see in Chapters 11 and 12, that the DMMs we have introduced do satisfy this important property.

Taken all together, these physical and mathematical properties, (i)−(v), simply state that if the digital machine described by eqns (3.1) has to solve a Boolean problem with at most polynomially growing resources, it must identify the long-time limit of its dynamics with the equilibrium points of such equations, and these equilibrium points have to be reached exponentially fast from some region of the phase space.

We also note that all these properties are *mutually consistent*. This means that we should be able to (and indeed we do) find physical systems that satisfy all of them at once (see Chapter 7).

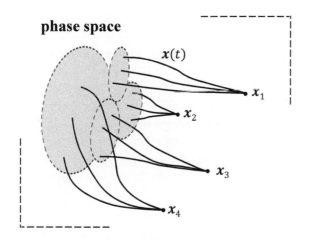

phase space

$x(t)$

x_1

x_2

x_3

x_4

Fig. 6.2 Schematic of the phase space, X, that an ideal physical (digital) machine needs to have to compute efficiently (with polynomial resources). It consists of equilibrium points (four in the figure) corresponding to the solutions of the problem at hand. The equilibrium points have their own basins of attraction (grey areas). The basins of attraction of (quasi-)periodic orbits, chaos, or other 'wrong' equilibria, if they exist, must satisfy condition (iii) in the text.

If all these requirements are satisfied, the phase space, X, is completely clustered into regions that are the attraction basins of the equilibrium points and, possibly, other stable equilibria, (quasi-)periodic orbits, and strange attractors, if they exist (see Fig. 6.2).

An important remark

I anticipate here that a class of systems that *always* has a *global attractor* is the one of *point-dissipative* dynamical systems (Hale, 2010), and indeed these are the non-quantum dynamical systems we are after. In addition, this property has allowed us to engineer certain dynamical systems representing DMMs so that they do not have any stable equilibria, other than the solutions of the given problem, no (quasi-)periodic orbits, and no chaos (Traversa and Di Ventra, 2017; Di Ventra and Traversa, 2017a; Di Ventra and Traversa, 2017b).

I will give a more detailed description of these physical systems in Chapter 7, but I point out here that by 'dissipative' I do not necessarily mean 'passive' in the physical sense I discussed in Section 4.5: active elements (e.g., those that provide energy to a circuit) can be dissipative as well in the mathematical (functional analysis) sense (Hale, 2010).

In fact, as I have anticipated in Chapter 4, and will show explicitly in Chapter 7, the practical realization of DMMs *does* require active elements.

Having discussed the general properties equations of the type (3.1) need to satisfy to represent a digital machine, let's make another important point.

6.4   Linear or non-linear systems?

Equations (3.1) can be either *linear* or *non-linear*. Linearity means that the flow vector field F is a linear function of the state $\mathbf{x}$.

The advantage of a linear differential equation is that if we find two or more of its solutions, a *general* solution of such an equation can be written as a *linear combination* of those solutions. More precisely, a linear system is one that satisfies the following properties

Linear system

(i) *Homogeneity*: $F(\alpha\mathbf{x}) = \alpha F(\mathbf{x}), \quad \forall \alpha \in \mathbb{R}$

(ii) *Superposition*: $F(\mathbf{x} + \mathbf{y}) = F(\mathbf{x}) + F(\mathbf{y}), \quad \forall \mathbf{x}, \mathbf{y} \in X$.

A prominent example in Physics is the Schrödinger equation for the state vector of a quantum-mechanical system (Messiah, 1958).

On the other hand, a *non-linear* differential equation does *not* have these features. The flow vector field of such an equation is a *non-linear function* of the state $\mathbf{x}$. In fact, we could use as definition of a non-linear system the following (Hilborn, 2001):

Non-linear system
A system whose equations of motion are *non-linear functions* of the
dynamical variables **x**.

A consequence of this property is that if we found several solutions of
such an equation we could not sum them in a linear combination, and
claim that such a sum is a solution of the original equation. Equations
describing fluid dynamics (Landau and Lifshitz, 1975) are examples of
this type. In fact, many interesting phenomena occur due to the non-
linearity of the underlying equations of motion.

What about computing?
Should we look for linear or non-linear systems to solve complex problems efficiently? Of course, any of these systems have to satisfy the
properties I have discussed in Section 6.3.2. However, this is not enough.

As an example, let's consider a simple *linear* dynamical system de-
scribed by $\dot{\mathbf{x}} = -(\mathbf{x} - \mathbf{x}_s)$, where $\mathbf{x}_s$ is the solution (equilibrium point)
of the problem we want to solve. This simple system *does* satisfy all
requirements I have identified previously. However, in order to build it
we would need to know *a priori* the solution of the problem, which defies
the purpose!

6.5   Guiding the system toward the solution

Instead, we want a system that does satisfy the requirements discussed
in Section 6.3.2, and is able to 'select' the correct solution path, in
the vast phase space, *on its own*. This is the property (v) I discussed
in Section 3.11: the system has to be *constrained* to always reach the
equilibrium points. It should not have any other choice.

In other words, we would like to have a machine that, at any given
time of the computation, is able to always 'guess' the correct path to
follow in order to solve the problem.

The attentive reader would recognize this requirement as the main
feature of a non-deterministic Turing machine (see Section 2.3). Indeed,
this is what we want of our machines: to be as powerful as a non-
deterministic Turing machine, while, at the same time, being physically
realizable.

As I will demonstrate with several numerical examples in Chapters 8, 9,
and 10, and explain in more detail in Chapters 11 and 12, *non-linear* dynamical systems hold considerable computational advantages compared
to linear ones, as they may support *long-range correlations* (as discussed

**Quantum computers and lin-
ear machines**
Apart from the measurement process, and possible coupling with
the environment (which leads to
decoherence), quantum computers are fundamentally *linear machines*. This means they manipulate state vectors in a lin-
ear/vector space (a Hilbert space;
cf. Section 3.1) using *linear* operators (quantum gates) on that
space. This may be the reason
why quantum computers have not
been shown to be able to solve
NP-complete problems efficiently,
not even in principle (Nielsen and
Chuang, 2010).

in Section 4.3.2), namely the ability of the different parts of the machine to *correlate* (both spatially and temporally) during dynamics, with those correlations decaying as a *power law*, not exponentially, with system size.

This feature will be key to tackle hard combinatorial optimization problems efficiently. For this reason, *non-linear* dynamical systems will be the *only* ones we will be considering from now on.

An important remark

The presence of memory (time non-locality) alone does *not* guarantee non-linearity. In fact, by referring to eqns (3.1), one could consider a flow vector field, F, that is linear in *all* degrees of freedom, including those providing memory. This would correspond to a linear system with time non-locality.

However, note that, for all the physical cases I will discuss in this book, the flow vector field has non-linear components, including those provided by the memory variables.

6.6   The role of initial conditions

Before concluding this Chapter let me make a further important point regarding the initial conditions of eqns (3.1), and how they affect our ability to find the solution(s) of a given problem, in particular if there are also periodic orbits, chaos, and/or other 'spurious' minima.

All these would be the 'wrong solutions' of the given problem. This pertains to the condition (iii) in Section 6.3.2.

Let's first define $V = \text{vol}(X)$ the volume of phase space X defined by some *Lebesgue measure* on X. Let's then consider $J_s \subseteq J$, the compact subset of the global attractor J, that only contains all equilibria $\mathbf{x}_s$—solutions of the given problem.

In addition, I introduce $J_w \subseteq J$, the compact subset of the global attractor J, containing all other 'wrong' long-time dynamical states, $\mathbf{x}_w(t)$. Therefore, we have $J = J_s \cup J_w$ and $J_s \cap J_w = \emptyset$.

Let's also define the sub-volume $V_s = \text{vol}(X_s)$ where $X_s \subseteq X$ is the subset of X attracted by J_s, and the sub-volume $V_w = \text{vol}(X_w)$ where $X_w \subseteq X$ is the subset of X attracted by J_w.

By definition of attractors (see Section 3.9) applied to J_s and J_w, we have that $\text{vol}(X_s \cap X_w) = 0$ and since we require of J to be a *global* attractor, we must have

$$V = V_s + V_w. \tag{6.3}$$

Let's now use these quantities to define the probabilities that, starting from a *random* initial condition, our machine finds a solution of a

Lebesgue measure

It is the most common (although not the unique) way of assigning a *measure* to the subsets of Euclidean space. For instance, in one dimension it is simply the length of the subset (Arfken, Weber and Harris, 1989).

Boolean problem, or that it fails to find it:[13]

$$P_s = \frac{V_s}{V}, \qquad P_w = \frac{V_w}{V}, \qquad (6.4)$$

respectively, with $P_s + P_w = 1$.

Then it is easy to see that, in order for the machine to find the solution(s) of the given problem with *polynomial* resources, the condition $P_s > \mathfrak{p}_\gamma^{-1}(n)$ must be satisfied, for some polynomial, $\mathfrak{p}_\gamma(n)$, of some degree $\gamma \in \mathbb{N}$ for a given size, n, of the problem. This is because, even if there are 'wrong' attractors in the phase space, by operating the machine with a polynomially growing number ($\mathfrak{p}_\gamma(n)$) of different initial conditions, then we are guaranteed (with high probability) to obtain a solution. This is the formal way of writing condition (iii) in Section 6.3.2.

An important remark

Note that for all the combinatorial optimization problems I will discuss in this book (Chapters 8, 9, and 10) the corresponding equations of motion *do not* support chaos and other minima in the phase space, other than those corresponding to the solutions of the given problem. In addition, for satisfiable problems, no (quasi-)periodic orbits exist if equilibria are present (Traversa and Di Ventra, 2017; Di Ventra and Traversa, 2017a; Di Ventra and Traversa, 2017b).

6.7 Chapter summary

- I have introduced the formal definition of DMMs, the digital and *scalable* subclass of a UMM.

- I have also discussed the features a dynamical system representing them needs to satisfy in order to compute efficiently (with *polynomially growing* resources).

- In particular, we need a system whose dynamics always end up in a restricted region of the phase space. This way, irrespective of the initial conditions, the orbits of the system are bounded and will eventually enter such a region, and stay there.

- The dynamical systems that satisfy this mathematical property are called *point dissipative*. We cannot exclude, however, that other types of mathematical properties may be sufficient to accomplish the same goal.

[13] For simplicity, I am assuming the machine finds the solution $\mathbf{y}$ of a Boolean problem in only one step. Generalization to multiple computational steps is trivial and can be found in (Traversa and Di Ventra, 2017).

7

Self-organizing gates and circuits

A self-organizing system is intrinsically adaptive: it maintains its basic organization in spite of continuing changes in its environment. Perturbations may even make the system more robust, by helping it to discover a more stable organization.

Francis Paul Heylighen (1960–)

Takeaways from this Chapter

- Time non-locality (memory) allows gates to be 'inverted'.

- *Self-organizing gates* (SOGs) are those *terminal-agnostic* gates that self-organize to *always* satisfy their *logical* or *algebraic* proposition.

- SOGs can be fabricated with memelements and *active* elements: they are memprocessors ideally suited to solve combinatorial optimization problems.

- A collection of SOGs forms a *self-organizing circuit*.

The discussion in Chapters 5 and 6 has set the conceptual and mathematical foundations of MemComputing, even its digital, hence *scalable* version. However, we have yet to answer the question of how we realize these machines in hardware, with *actual* electronic devices? The goal of the present Chapter is precisely to answer this question.[1]

[1]And, from now on, I will focus only on the *digital* machines.

7.1 The need for terminal-agnostic gates

We have seen in Section 6.2 that when we want to solve a given Boolean problem with a digital MemComputing machine (DMM), we first write it as a Boolean circuit. Such a circuit is a collection of logic gates.

In fact, this circuit representation is not unique, because one can choose different logic gates as a basis of Boolean logic. For instance,

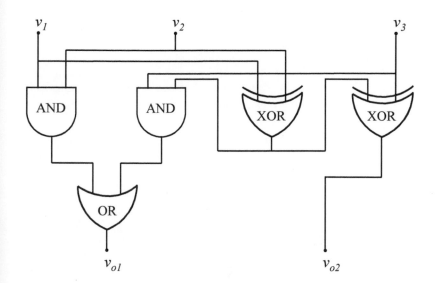

Fig. 7.1 Sketch of a possible Boolean circuit that performs the sum of 3 bits, indicated with v_1, v_2, and v_3. Their sum is represented in the bits v_{o1} (sum) and v_{o2} (carry). In this circuit, one OR gate, two AND gates, and two XOR gates are used.

the pairs (AND, NOT) or (OR, NOT) are two sets of Boolean gates that can be used to represent *any* Boolean gate.

The gates NOR (not-OR) or NAND (not-AND) represent another *complete* basis set (Rautenberg, 2009). In this case, since the set has only one element, we call it a *singleton*.

Irrespective of the chosen Boolean representation, the first step in solving any combinatorial optimization problem will be done by writing it as a Boolean circuit. Or a combination of Boolean circuits and *algebraic* relations, if we have, e.g., inequalities between variables.

Let me provide a couple of concrete examples.

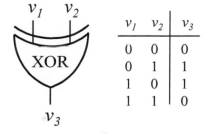

v_1	v_2	v_3
0	0	0
0	1	1
1	0	1
1	1	0

Fig. 7.2 **Left:** symbol of a XOR gate. **Right:** its truth table in a binary representation.

Example 7.1

Full adder

Take, for instance, the simple Boolean circuit that performs the addition of three bits: a *full adder* (see Fig. 7.1). It consists of one OR gate, two AND gates, and two XOR (exclusive-OR) gates. I reported the truth tables of the AND and OR gates in Fig. 2.7. The truth table of the XOR gate is in Fig. 7.2.

By using these truth tables, it is easy to verify that given all possible binary values of the input (v_1, v_2, and v_3 in Fig. 7.1) one obtains the correct sum (v_{o1}) and its carry (v_{o2}).

Complete set of gates
A set of gates is said to be *complete* if the truth table of any other gate can be obtained by the appropriate combination of the elements of that set (Rautenberg, 2009).

The next example has important applications in many fields, notably modern *cryptography* (Goldreich, 2001).

Example 7.2

Prime factorization

Consider an integer $n \in \mathbb{N}$. Question: what are the prime numbers that multiplied together give n? The answer to this question is the famous problem of *prime factorization*.

Due to the *fundamental theorem of Arithmetic* (Hardy and Wright, 1979), this set of prime factors is unique (except for the order of the factors).

Let's then consider an integer n that, for simplicity, we assume to be factored in only two prime numbers, say p and q, $n = p \times q$.

All these numbers can be expressed in *binary* representation as

$$n = \sum_{j=0}^{N} n_j 2^j, \tag{7.1}$$

$$p = \sum_{j=0}^{N-1} p_j 2^j, \tag{7.2}$$

$$q = \sum_{j=0}^{N-1} p_j 2^j, \tag{7.3}$$

> **Prime number**
> It is a natural number ($n \in \mathbb{N}$) greater than 1 that is not a product of two smaller natural numbers.
>
> **Prime factorization**
> It is the problem of finding the primes that multiplied together give the integer $n \in \mathbb{N}$.
>
> **Fundamental theorem of Arithmetic**
> It states that every positive integer larger than 1, is either a prime, or can be represented in exactly one way, apart from rearrangements of the factors, as a product of two or more primes (Hardy and Wright, 1979).

where, n_j, p_j, and q_j are the 0s and 1s representing the respective integers; $N = \lfloor \log_2 n \rfloor$, where the symbol $\lfloor \cdot \rfloor$ is the *floor function* that rounds the elements of $\log_2 n$ to the nearest integer toward 0.[2]

In this problem, we have made the assumption that p and q are primes. Therefore, we can choose $p_0, q_0 \neq 0$. This choice guarantees that p and q are not divisible by 2. This also means that $n_0 = 1$, and by setting $n_N = 1$, we are guaranteed that n has the shortest binary representation.

Suppose now you have solved the problem, and found the primes p and q. Following standard (binary) arithmetic, by multiplying p and q to obtain n, we would obtain the following remainders, r_j, at each step of the multiplication process:

[2]Note that there are two possible solutions to this problem: either p times q, or q times p. One can eliminate this degeneracy by choosing $N_p = N - 1$ and $N_q = \lfloor N/2 \rfloor$ (or the reverse), where N_p and N_q are the numbers of bits representing the factors p and q, respectively. This choice also guarantees that, if p and q are primes, the solution is unique, and the trivial solution $n = n \times 1$ is forbidden.

$$r_0 = p_0 q_0 - n_0 = 1 \times 1 - 1 = 0; \tag{7.4}$$

$$r_j = \sum_{k=0}^{j} p_{j-k} q_k + \frac{r_{j-1}}{2} - n_j, \quad j = 1, ..., \lceil N/2 \rceil - 1; \tag{7.5}$$

$$r_j = \sum_{k=0}^{\lceil N/2 \rceil - 1} p_{j-k} q_k + \frac{r_{j-1}}{2} - n_j, \quad j = \lceil N/2 \rceil, ..., N - 1; \tag{7.6}$$

$$r_N = \sum_{k=0}^{\lceil N/2 \rceil - 1} p_{N-1-k} q_{k+1} + \frac{r_{N-1}}{2} - 1. \tag{7.7}$$

These remainders must satisfy the following properties: $r_j \geq 0$ and $r_j = 0$, mod 2, for $j = 1, ..., N-1$ (where 'mod 2' indicates the 'modulo' operation which gives the remainder of the division of a given number by 2), while $r_N = 0$.

In the previous expressions, the symbol $\lceil \cdot \rceil$ is the *ceiling function* that rounds the elements of $N/2$ to the nearest integer toward ∞.

The interesting thing about the remainder expressions, eqns (7.4)-(7.7), is that they can be readily written in Boolean form. To see this, let's consider again the truth tables of the AND and XOR gates (see Figs. 2.7 and 7.2, respectively).

From these tables it is easy to recognize the following mapping between the arithmetic operations of sum, $\sum$, and product $\times$, with the corresponding logic gates:

$$\sum \rightarrow \text{XOR, AND,} \tag{7.8}$$

$$\times \rightarrow \text{AND,} \tag{7.9}$$

where the logical AND in (7.8) accounts for the *carry bit* in the sum (see Fig. 7.3 for the Boolean circuit of the 2-bit sum, also called *half adder*).

This means that the entire arithmetic operation of factoring an integer can be implemented as a Boolean problem. For example, in Fig. 7.4 I show one possible circuit that multiplies p and q to give $n = 35$, which in binary representation is $(100011)_2$.

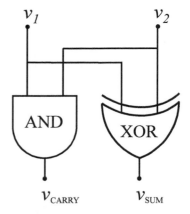

v_1	v_2	v_{SUM}	v_{CARRY}
0	0	0	0
0	1	1	0
1	0	1	0
1	1	0	1

Fig. 7.3 Top: Boolean circuit for a 2-bit sum (*half adder*) composed of a XOR and an AND gate. **Bottom:** its truth table in a binary representation for both the sum and its carry bit.

RSA numbers
A set of large numbers, with exactly two prime factors, provided to the scientific community to test factorization algorithms. 'RSA' stands for 'Rivest', 'Shamir', and 'Adleman', who invented one of the first public-key cryptography algorithms (Goldreich, 2001).

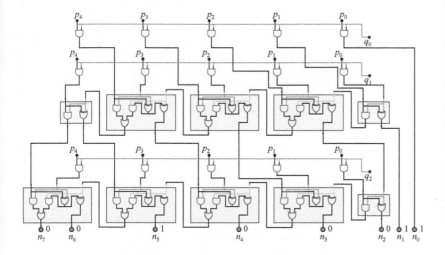

Fig. 7.4 A possible Boolean circuit that multiplies two integers p and q to give $n = 35 = (100011)_2$. Reproduced with permission from (Di Ventra and Traversa, 2018a).

If we knew such primes it would then be easy to check that their product is the given number n (whether we write it in Boolean form or not). However, to answer the prime factorization problem we need to solve it 'in reverse': given n, what are the primes p and q that factor such a number?

[3]As an example, the largest RSA number factored so far (February 2020) contains 829 bits. It took the equivalent of about 2700 years of core-time (CPU time of a single processor) to complete, and was done on tens of thousands of machines over the course of a few months (see, e.g., Wikipedia under RSA-250).

Logical irreversibility

If the logical topology of a computing machine is not bijective we say that the computation is *logically irreversible* (Landauer, 1991). This is the case with our modern computers and all the DMMs I will consider in this book.

Landauer principle

Logical irreversibility implies *loss of information* ('erasure' of information), hence an increase of *entropy* (*physical irreversibility*), with concomitant *minimal heat generation* of $k_B T \ln 2$ per logical step, where $k_B = 1.38 \times 10^{23} \text{J/K}$ is the *Boltzmann constant*, and T is the external temperature. Modern computers erase information at almost every step of the computation.

Reversible computation

An ideal computation that *preserves information* at every step is called *reversible* (Bennett, 1973). It requires storing *all* intermediate results during computation, namely its entire history, but it does not have, in principle, any limit on minimal heat generation.

In fact, this problem is thought to be in the NP class of problems (although it is not NP-complete) (Arora and Boaz, 2009). This means that as soon as the number of bits of n is large (say in the hundreds), the time to solution can be incredibly long.[3]

7.1.1   What does it mean to solve 'in reverse'?

Solving 'in reverse' the Boolean circuit of factorization in Fig. 7.4 is similar to asking, for the full adder of Fig. 7.1, which binary values the variables v_1, v_2, and v_3 have to acquire in order to be *consistent* with a given sum v_{o1} and carry v_{o2}. In other words, if we knew the values of sum and carry, which variables assignment would produce them?

Equivalently, in reference to Fig. 6.1 of Section 6.2, it is like asking what literals need to be assigned to the three 'top' terminals (in-terminals) of each 3-variable OR gate of that circuit, once the logical value of 1 is assigned to the 'bottom' terminals (out-terminal) of each gate (solution mode of Fig. 6.1).

An important remark

Note that there may be *several* variable assignments that give the *same* output. This, however, does not pose any problem. Here, I am not asking for a one-to-one correspondence between input and output. I am only looking for a *logically consistent* relation between input and output. Namely, as in Section 6.2, I want the flow of information (*logical topology*) to be *bi-directional*, but not necessarily bijective.

Since *any* combinatorial optimization problem can be written in CNF format (see Section 2.4), then solving such problems is tantamount to operating *each* logic gate *in reverse*. In other words, if we had logic gates that, in addition to working as usual from in-terminals to out-terminals, would also 'adjust' to provide the correct logical proposition when a logical value is fed at the out-terminals, those gates would solve the original problem, when operating all together in a given circuit.

As we have seen in Section 6.2, DMMs can operate in both 'solution mode' and 'test mode'. In the solution mode they indeed operate 'in reverse', because, given a certain 'input' (e.g., the number n to factor), their state will change to provide the answer to a given problem (the primes p and q in the case of factorization). Since DMMs are a collection of memprocessors arranged according to a specific *physical topology*, our next goal is then to find the appropriate fundamental units (memprocessors) that, once interconnected, would solve the original problem. The easiest way to proceed is to realize that the physical topology of the network of memprocessors can be simply chosen as the physical topology of the Boolean circuit corresponding to the problem that needs to be solved. With this choice, the memprocessors must then represent the logic gates of that Boolean circuit.

However, standard gates are *unidirectional* (or *sequential*), namely they can *only* operate by providing an input to the in-terminals and outputting a value to the out-terminals. Therefore, the best traditional Boolean logic can do is working in *test mode*.

This means we need to come up with a *new* class of gates that are *bi-directional* or *terminal-agnostic*, namely they will *always* satisfy their logical proposition, *irrespective* of whether the truth value is assigned at the traditional in-terminals or the traditional out-terminals. This way, the machine would also operate in *solution mode*. Time non-locality (memory) allows us to do just that.

> **Terminal-agnostic gates**
> A gate, whether Boolean or algebraic, whose *logical topology* (flow of syntactic information) is *bi-directional* (but not necessarily bijective), is called *terminal agnostic* or *self organizing* (Traversa and Di Ventra, 2017; Traversa and Di Ventra, 2018). It means that the gate can accept signals from its traditional input and output terminals, and *always* satisfy its Boolean or algebraic relations. Our traditional gates are *not* terminal agnostic. They are *unidirectional*: information flows *only* from input to output terminals.

7.2 Self-organizing gates (SOGs)

Before proceeding further let me make an important distinction. Although all of the problems I have discussed so far, and most of those that I will cover in Chapters 9 and 10, pertain to *logic* (Boolean) propositions, there are other (equally important) ones, like the Integer Linear Programming (see Section 9.2.5), that require finding an assignment of variables satisfying *algebraic* relations (e.g., inequalities).

One can define terminal-agnostic gates in *both* cases, and even though their practical realization may be different, the overall strategy is the same. Therefore, in the following I will discuss in detail Boolean gates only.

7.2.1 Self-organizing *logic* gates (SOLGs)

To make the discussion more concrete, let's consider an AND gate (recall that its symbol and truth table are shown in Fig. 2.7). This gate has only four *logically consistent states*, according to the truth values assigned to the in-terminals: 0 AND 0 = 0; 0 AND 1 = 0; 1 AND 0 = 0; 1 AND 1 = 1.

From logic to dynamics
Let's now *design* a physical system, following some appropriate equation of motion of the type (3.1), which has *only* these four states as its *equilibrium points*, and no other attractor, namely no periodic orbits or

chaos.

This physical system will then dynamically *self-organize* (SO) into any one of these four states, according to the *initial conditions* (cf. discussion in Section 6.3.2). Namely, its phase space clusters into four basins of attraction, each one corresponding to only one equilibrium point representing one of the four logically consistent states of an SO-AND gate (Fig. 7.5).

SO-AND phase space

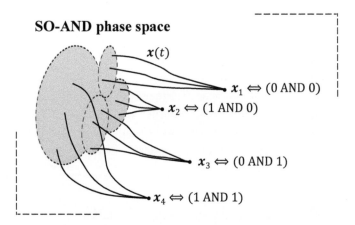

Fig. 7.5 The phase space of a self-organizing AND (SO-AND) gate with only four equilibria, each one corresponding to a *logically consistent state* of an AND gate. In the absence of any other attractor, the phase space of this gate clusters into four basins of attraction (grey areas).

If we could accomplish this program, we would then obtain a completely new type of gate, I call *self-organizing logic gate* (SOLG), in which only the final states are important, not how such states are reached during dynamics (Traversa and Di Ventra, 2017).

Note also that although this gate can operate *in reverse*, it is *not* generally invertible in a one-to-one sense. For instance, in the case of an AND gate there are two in-terminals and only one out-terminal, so an AND gate does not define a bijective function (it is *logically irreversible*, hence *physically irreversible*; see discussion in the preceding Section 7.1.1).

The need of extra (memory) degrees of freedom

Since the original two in-terminals and one out-terminal of the AND gate can only acquire digital values (0 or 1), we realize that during dynamics, the three terminals of this new SO-AND gate are not necessarily digital, because the *transition function* they represent is the dynamical set of eqns (3.1) (see also Section 6.3.2). We have then relaxed the digital condition at the terminals of the gate and allowed them to *adapt* to any *bounded* value they may support during dynamics.[4]

Therefore, when the system is at equilibrium (in one of the four logically consistent states), *no dynamics* should occur. On the other hand, away from the logically consistent states, the system will change its state, *always* attempting to converge to one of the equilibrium points, while

---

[4] I have highlighted the word 'bounded' because it is important that during dynamics the orbits in the phase space are *bounded*, which is equivalent to saying that *no* physical (*observable*) quantity can diverge during dynamics (whether voltages, currents, energy, etc.).

always keeping any physical property finite (see Fig. 7.6 for a sketch of the possible dynamics of a SO-AND gate).[5]

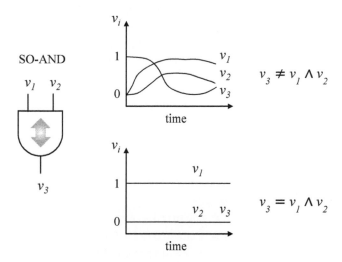

[5] Although we do not encounter this issue in the actual implementation of DMMs, it is worth noting that only the *observable* quantities, such as voltages, currents, and energy need to be finite. Derived quantities, such as resistances and capacitances, which are *response functions* (Section 4.1), may indeed diverge. This would not create any physical issue because response functions are *not* observables. They are *derived* quantities (Di Ventra and Pershin, 2013*a*).

Fig. 7.6 The SO-AND gate is in an *unstable* configuration if its logical relations are *unsatisfied*. It is in a *stable* configuration if one of its logical relations is *satisfied*.

In fact, at any given time during dynamics (after the initial condition has been set, but before the output is reached), the state of the system could be in any *non-linear combination* of 'input' and 'output' states. This behavior can be realized if we introduce *extra degrees of freedom*, namely the dimension of our (phase) space of dynamical variables is not three (or whatever is the number of gate terminals), but larger.

An important remark

For concreteness, and since this is the practical (*hardware*) realization of DMMs I envision in this book, from now on I will map the *variables* of the original Boolean gate into *voltages* of an actual physical circuit (see also Section 4.4).

Instead, I will represent the extra (*memory*) degrees of freedom with the symbol $\tilde{x}$, or simply x, if no confusion arises. They could be due to any physical mechanism that induces *time non-locality* (memory) in the system (Pershin and Di Ventra, 2011*a*).

That said, we can accomplish the task of transforming the original gate into an SO-gate in many ways, and I will now provide explicit schemes to do just that. The important thing to remember is that we need to attain two goals:

From original gate to SO-gate

(i) Provide dynamics to the literals at the terminals, $v_1 \to v_1(t)$, $v_2 \to v_3(t)$, $v_3 \to v_3(t)$, ..., for as many terminals as there are in the gate.

(ii) Add as many extra dynamical (*memory*) degrees of freedom, $\tilde{x}_1(t)$, $\tilde{x}_2(t)$, ..., as necessary to let the system evolve to the *only* logically consistent equilibrium states of the gate.

An important remark

Let's consider first the step (i) of the procedure to transform an original gate into an SO-gate.

By providing dynamics to the literals of the terminals of the original gate we go from a set of *discrete* states (the logically consistent solutions of the gate) to a dynamical system of the voltage variables only. The resulting phase space may have several types of *critical points*, in addition to the equilibrium points which correspond to the correct logical proposition of the gate. In particular, this (*reduced*) phase space of the voltages may contain local minima that could trap the system dynamics.

This is where the step (ii) of the procedure comes in handy.

The memory variables, if appropriately introduced, *expand* the reduced phase space of the voltages, transforming *any* possible local minimum into *saddle points* (see Section 3.7 for their exact definition). This leaves the equilibrium points representing the logically consistent solutions as the only equilibrium points, with possibly center manifolds (see Fig. 7.7).

Feedback

For the purpose of this book, I define *feedback* as any *self-regulating* mechanism by which the system adjusts its dynamics according to its instantaneous value, and how 'far' it is from a desired outcome. It requires *active* elements (Section 4.5.1).

The condition (ii) suggests that the extra degrees of freedom could be provided by the memelements I have defined in Section 4.4.2. However, as I have anticipated in Section 4.5, *memelements alone are not enough!* We need to add *active* elements as well. The reason is because the dynamical system has to be able to *always* end up in the correct equilibrium points, which represent the logical states of the gate. In other words, we need to *guide* it toward the solution (cf. discussion in Section 3.11). To accomplish this, the system needs *feedback* so that if it 'strays away' from the correct path in the phase space, it will immediately 'correct' it.

Before providing explicit examples of electrical circuits that can be used to build SOLGs in hardware, and their equations, let me summarize their properties:

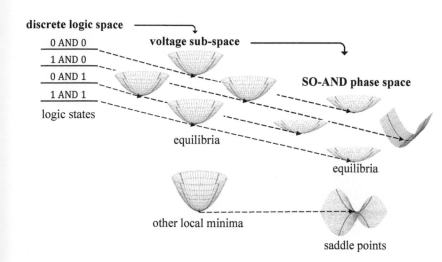

Fig. 7.7 The original logic space consists of a *discrete* number of states (e.g., four for an AND gate). The first step is to transform the logic states into continuous (voltage) variables (*linear relaxation* of the discrete variables) with the logic states corresponding to equilibria. However, this may introduce additional local minima (and possibly saddle points, which are not of concern; see Section 11) in the (*reduced*) voltage phase space. By introducing extra memory degrees of freedom, the reduced voltage phase space is enlarged into the SO-gate phase space. Thanks to the memory variables, the additional local minima are transformed into saddle points, while maintaining the one-to-one correspondence between logic states and equilibria. Note, however, that the structure of these new equilibria may change, e.g., they may also have center manifolds (see Section 3.7).

Self-organizing logic gates

A novel type of logic gates that can use *any* terminal *simultaneously* as 'input' or 'output'. SOLGs are *terminal-agnostic*. Signals can go in and out at the same time at *any* terminal, resulting in a non-linear combination of input and output signals. In addition to time non-locality (memory), SOLGs require *feedback* to operate as desired, hence the need of *active* elements.

Linear relaxation of discrete variables

The step consisting in tranforming *discrete* variables into *continuous* ones can be called *linear relaxation* of the original variables within some compact (namely closed and bounded) manifold (cf. definition in Section 3.10).

The symbols I will use for the SO-AND, SO-OR, and SO-XOR gates are shown in Fig. 7.8. The double arrow inside the traditional logic symbol indicates precisely that the signals can come in and out from *any* terminal, without distinction between the traditional input and output terminals. Symbols for all other gates can be similarly constructed.

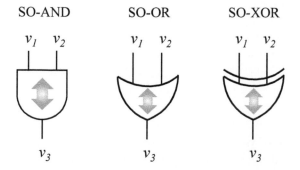

Fig. 7.8 Symbols I will use for the SO-AND, SO-OR, and SO-XOR logic gates. *Any* of the terminals v_1, v_2, and v_3 can function as *both* input *and* output.

7.2.2   Self-organizing *algebraic* gates (SOAGs)

Although I will mention them only once in this book (in Section 9.2.5), I note here that the strategy for building SOLGs can be employed also in the design of *self-organizing algebraic gates* (SOAGs) (Traversa and Di Ventra, 2018).

These gates are similar to the SOLGs in their mode of operation. However, instead of satisfying a Boolean relation, they satisfy an *algebraic* relation, e.g., an inequality relation between variables:

> **Self-organizing algebraic gates**
> A novel type of gates that can use *any* terminal *simultaneously* as 'input' or 'output' to satisfy an *algebraic* relation. SOAGs are *terminal-agnostic*. In addition to time non-locality (memory), SOAGs require *feedback* to operate as desired, hence the need of *active* elements.

Such gates can be used in tackling, e.g., *Integer Linear Programming* (ILP), an optimization problem in which one needs to find a variable assignment that satisfies also algebraic relations (see Section 9.2.5 for the explicit definition of this problem). The interested reader should consult (Traversa and Di Ventra, 2018) for the application of SOAGs to ILP.

7.3   Electrical circuit realization of SOLGs

After these preliminaries I can now suggest an actual *physical* realization of SOGs. As anticipated, the electrical circuit realization of SOLGs is the easiest to accomplish in *hardware*, and the only one considered in this book. However, the concept is quite general, and one could envision other realizations as well, by employing, e.g., superconducting or optical devices.

Although the specific gate realization may be different, the general idea is the same for all types of Boolean gates. As an illustrative example, let's then consider an SO-AND, as shown in Fig. 7.9, and choose a *reference voltage v_c*.

Referring to Fig. 7.9, let's also assume the top-left terminal is fixed at voltage +1V (logical 1), while the initial voltage at the top-right terminal is, say, at −1V (logical 0), and the initial voltage of the bottom terminal is +1V (logical 1). Of course, this is just one of the many possibilities that may occur: any intermediate (finite) voltage at those terminals is possible during dynamics, and the top-left terminal does not need to be fixed. The general operation of the SOLG would be the same.

This choice of values for the terminal voltages means that the SOLG is in an initial *unstable* configuration, and will attempt to dynamically

Reference voltage v_c

As already mentioned, we can choose to represent with $v_c = +1V$ the logical 1, and with $-v_c = -1V$ the logical 0. Or, more simply, we can represent a logical 1 whenever the voltages are positive, or a logical 0 when they are negative (cf. Section 8.2.4). Either choice is, however, arbitrary. In actual hardware realizations of these gates, different choices for the reference voltage may be more advantageous.

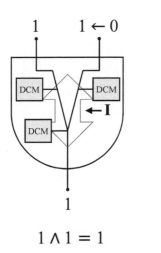 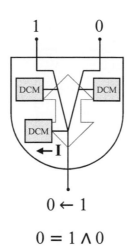

Fig. 7.9 Principle of operation of a self-organizing AND gate in which the top-left terminal is fixed at voltage +1V (logical 1), while the initial voltage at the top-right terminal is −1V (logical 0), and the initial voltage of the bottom terminal is +1V (logical 1). **Left panel:** the dynamic correction module (DCM) attached to the top-right terminal injects a current *into* that terminal (left-pointing arrow), so that it attempts to bring that terminal to +1V, thus satisfying the logical proposition (1 AND 1 = 1). **Right panel:** at the same time, the DCM attached to the bottom terminal, instead, takes current *out* of that terminal, so that its voltage will become −1V (logical 0). This way, the system tries to satisfy the logical proposition (0 = 1 AND 0).

change its state to reach one of the four possible equilibrium states compatible with an AND gate (see Fig. 7.6).

7.3.1 Dynamic correction modules

To accomplish this, we envision a set of *active* devices, we call *dynamic correction modules* (DCMs), attached to each terminal, which read the voltages of the other terminals as well: they provide the necessary *feedback* to the system we have discussed in Section 7.2.1 (Traversa and Di Ventra, 2017; Di Ventra and Traversa, 2018*b*).

The DCM attached to the top-right terminal then reads the voltages at the other two terminals, and 'realizes' that the gate is in the 'wrong' configuration. Therefore, it *injects* a current to that terminal. This attempts to bring that terminal to +1V, thus satisfying the logical proposition (1 AND 1 = 1); left panel of Fig. 7.9.

Simultaneously, the DCM attached to the bottom terminal, by reading the voltages at the other two terminals, 'realizes' that the gate is not logically satisfied. It therefore takes current *out* of that terminal, so that its voltage may tend to −1V (logical 0). This way, the system tries to satisfy the 'inverted' logical proposition (0 = 1 AND 0); right panel of Fig. 7.9.

These two DCMs will then compete to force the system to go either to the logical configuration (1 AND 1 = 1) or the logical configuration (0 = 1 AND 0). Since the SOLG is a dynamical system, which one will be satisfied depends on the initial conditions imposed on the system.

Note also that the SOLG operates *simultaneously* in reverse as well

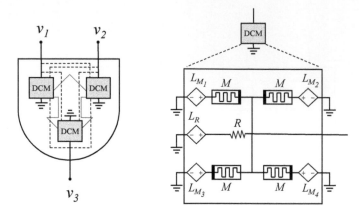

Fig. 7.10 Self-organizing (SO) AND gate, **left panel**, formed by dynamic correction modules (DCMs), **right panel**. The resistive memories M have minimum and maximum resistances R_{on} and R_{off}, respectively, and $R = R_{off}$. The linear functions L_{VCVG} drive the voltage-controlled voltage generators (VCVG) which depend only on v_1, v_2, and v_3 (eqn (7.10)). They are reported in Table 7.1.

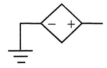

Fig. 7.11 Symbol of a voltage-controlled voltage generator. Voltage generators are very important *active* elements in electrical circuits. They are *voltage sources* that generate a voltage output that is independent of the load current. When the voltage-source magnitude depends on the voltage input, then they are called *voltage-controlled* (Horowitz, 2015).

as a traditional gate: it is *agnostic* to the direction of the (syntactic) information flow at its terminals.

An example of a SOLG that can function as a logical AND is schematically shown in Fig. 7.10, where the DCMs are made of resistive memories (Section 4.4.2) and *voltage-controlled voltage generators* (VCVGs), which are *active* devices (Fig. 7.11).

For the resistive memories, one can choose *any* physically valid device, according to eqns (4.25) and (4.26) (Traversa and Di Ventra, 2017; Di Ventra and Traversa, 2018*b*). The important point is that the resistive memories have a minimum resistance, R_{on}, and a maximum resistance, R_{off}.

In addition, in actual physical circuits one would necessarily have *parasitic capacitive effects*, which, incidentally, (for our purpose) have the beneficial effect of limiting the maximum voltages attainable. These capacitive effects can easily be introduced, e.g., as a small capacitance in parallel with the resistive memories (Traversa and Di Ventra, 2017).

The VCVGs are *linear voltage generators* piloted by the voltages v_1, v_2, and v_3 at the terminals of the SOLG. The output voltage of the VCVG is given by

$$v_{VCVG} = a_1 v_1 + a_2 v_2 + a_3 v_3 + dc. \qquad (7.10)$$

The parameters a_1, a_2, a_3 and dc are determined to satisfy a set of constraints characteristic of the gate, namely constraints that will *guide* the physical system to *always* satisfy the gate logic.

Therefore, these parameters are different for an AND, OR, XOR, or any other gate (Traversa and Di Ventra, 2017).

These constrains can be summarized in the following scheme:[6]

---

[6] The choice of v_c/R_{on} as current magnitude to inject is not universal and may be changed appropriately. This, however, would also change the parameters of the various L_{VCVG} in Table 7.1.

Table 7.1 Values of the parameters for the voltage-controlled voltage generators, L_i, as defined in eqn (7.10) for a SO-AND gate. From (Traversa and Di Ventra, 2017).

	Terminal 1				Terminal 2				Terminal 3			
	a_1	a_2	a_3	dc	a_1	a_2	a_3	dc	a_1	a_2	a_3	dc
L_{M_1}	0	-1	1	v_c	-1	0	1	v_c	1	0	0	0
L_{M_2}	1	0	0	0	0	1	0	0	0	1	0	0
L_{M_3}	0	0	1	0	0	0	1	0	0	0	1	0
L_{M_4}	1	0	0	0	0	1	0	0	2	2	-1	$-2v_c$
L_R	4	1	-3	$-v_c$	1	4	-3	$-v_c$	-4	-4	7	$2v_c$

SO-gate constraints

- If the gate is connected to a network and the gate configuration is correct, no current flows from any terminal: the gate is in *stable equilibrium* (steady state).

- Otherwise, a current of the order of v_c/R_{on} flows with sign *opposite* to the sign of the voltage at the terminal.

These are the criteria that accomplish what I have schematically shown in Fig. 7.9 (cf. also discussion in Section 7.2.1). A possible set of parameters that satisfies these conditions is given in Table 7.1.

Other SOLGs (e.g., SO-OR or SO-XOR) can be equally designed. The interested reader can look in (Traversa and Di Ventra, 2017) for a detailed description of these additional gates, and Chapter 8 for a dynamical system representing a 3-terminal OR gate.[7]

Parasitic capacitance

It is any capacitance that exists due to the simple proximity of conductors at different voltages (Horowitz, 2015). It exists in any actual (experimentally realized) circuit element.

[7] The explicit construction of an SO-NOT is typically not necessary, since NOT gates appear as simple negations of variables in the connection lines between gates (see, e.g., Fig. 2.8 in Chapter 2). This can be easily done by satisfying Kirchhoff's laws at the terminals of an electronic circuit.

An important remark

Note that one could choose a *different* set of parameters (or even a different design altogether for SOLGs) that satisfies the *same* conditions, or simply the general principles of operation of SOLGs.

Example 7.3

A different SO-AND

As an example of a different design of an SO-AND, consider the one in Fig. 7.12.

We choose for the equation of motion for the internal memory variables, $\tilde{x}_j$, the following (Traversa and Di Ventra, 2017):

$$\frac{d}{dt}\tilde{x}_j = -\alpha h(\tilde{x}_j, v_{M_j})g(\tilde{x}_j)v_{M_j}, \tag{7.11}$$

SO-AND

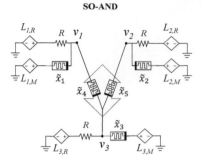

Fig. 7.12 Circuit diagram of a possible realization of a self-organizing AND (SO-AND) gate. The voltages v_1, v_2, and v_3 are the variables of the gate. The resistive memories are represented by a rectangle with a square waveform inside, and have internal memory variables $\tilde{x}_j$, $j = 1, \ldots, 5$, and an orientation, denoted by the bar on one side (see Section 4.4.2). They also contain a *parasitic capacitor* in parallel (not shown). All resistors have the same resistance value R. The functional form of the VCVGs is given in Table 7.2. Adapted with permission from (Bearden *et al.*, 2018)

[8]The functions $h(\tilde{x}_j, v_M)$ serve precisely to bound the dynamics of the memory variables. Ideally, they should then be represented by *step functions* (Traversa and Di Ventra, 2017). However, in hardware realizations of these devices, and in numerical simulations, the step functions would be replaced by appropriate differentiable functions. See, e.g., (Traversa and Di Ventra, 2017) for an explicit form of these functions.

where $\tilde{x}_j$ is the state variable for the j-th resistive memory element. The function h serves to bound the dynamics of the memory variable to certain regimes. The conductance, $g(x)$, will depend on the physical system used as memory element (Pershin and Di Ventra, 2011a).

For simplicity, let's consider the following function $g(x) = ((R_{off} - R_{on})x + R_{on})^{-1}$, where we set $R_{off} = 1\ \Omega$ and $R_{on} = 0.01\ \Omega$. The ratio between these two extreme resistances, determines the 'degree of memory' of this memory element, as it affects its hysteresis properties (cf. Section 4.4.2).

The voltage drop, v_M, is measured based on the orientation of the resistive memory: $v_M = v_a - v_b$, where v_b is measured from the thick-bar side of the electronic symbol for the resistive memory. The coefficient α is a rate of change of the memory, and following (Bearden, Manukian, Traversa and Di Ventra, 2018) we choose $\alpha = 60$. Finally, for convenience, we choose all memory variables to be bounded $\tilde{x}_j \in [0,1]$, $\forall j = 1, \ldots, 5$.[8]

Now, using circuit theory (Horowitz, 2015), we can easily derive the equations of motion for the currents and voltages in the circuit.

To be specific, let's then try to reproduce the behavior of an SO-AND represented in Fig. 7.9. As in that figure, let's then set the voltage v_1 to 1 V. Since this is an AND gate, we then expect the system to evolve to either $v_2 = v_3 = 1$ V or $v_2 = v_3 = -1$ V, which are both logically consistent. The final state will, of course, depend on the initial conditions, namely the initial values of all voltages and internal memory variables.

If we apply Kirchhoff's current law (see Section 4.4.1) to terminals 2 and 3 of Fig. 7.12 we then get (recall, from Fig 7.12, that we have a capacitance—not shown—in parallel with each resistive memory)

$$C(-\frac{d}{dt}v_1 - 2\frac{d}{dt}v_2 + 2\frac{d}{dt}v_3) = -I_2 + (v_2 - v_3)g(\tilde{x}_5) + $$
$$(-v_{2,M} + v_2)g(\tilde{x}_2) + \tag{7.12}$$
$$\frac{-v_{2,R} + v_2}{R},$$

$$C(-3\frac{d}{dt}v_1 - 3\frac{d}{dt}v_2 + 4\frac{d}{dt}v_3) = -I_3 + (v_1 - v_3)g(\tilde{x}_4) + $$
$$\frac{v_{3,R} - v_3}{R} + (v_2 - v_3)g(\tilde{x}_5) + \tag{7.13}$$
$$(-v_3 + v_{3,M})g(\tilde{x}_3),$$

where I_2 and I_3 are the currents at nodes 2 and 3, respectively, C is the capacitance, and R is the resistance that appear in the circuit of Fig. 7.12.[9]

However, since terminals 2 and 3 are 'floating' (namely they are free to change their values), the corresponding currents must be zero: $I_2 = I_3 = 0$. In addition, according to our initial choice, terminal 1 is attached to a voltage generator that holds its value constant. Therefore, $dv_1/dt = 0$.

The VCVGs generate a voltage $v_{i,j}$ (with $i = 1, 2, 3$ and $j = R, M$), according to the relation (7.10). The parameters for this particular realization of an SO-AND are given in Table 7.2. As I have explained previously, the VCVGs inject a large current when the gate is in an inconsistent configuration, a small current otherwise.

The results of the numerical integration of eqns (7.11), (7.12), and (7.13) are reported in Fig. 7.13, for the particular initial condition $\mathbf{x}_{in} = \{v_2, v_3, x_1, x_2, x_3, x_4, x_5\} = \{0, 0, 1, 1, 0.75, 1, 1\}$.

It is clear from Fig. 7.13 that we obtain a consistent solution for the SO-AND. In fact, by starting at the logical 1 for v_1, we obtain the logical 1 at both v_2 and v_3.

It turns out that $\mathbf{x}_{in}$ is actually an *unstable* (saddle) critical point in the phase space, which can be seen by slightly perturbing, say, the voltage v_2 (there are three such perturbations shown in Fig. 7.13).

This can be confirmed by performing the *linear stability analysis* I have discussed in Section 3.7; see eqn (3.8). In fact, by diagonalizing the Jacobian, eqn (3.9), of eqns (7.11), (7.12), and (7.13), one obtains the signs of the real part of its spectrum: $\{\text{sign}(\text{Re } \lambda_i)\} = \{-, -, +, 0, 0, 0, 0\}$ (Bearden *et al.*, 2018), which means there are 2 stable directions, 1 unstable direction, and 4 center manifolds (see Section 3.7 for a refresh of these features of critical points).

The final point of the dynamics is also a critical point with spectrum: $\{\text{sign}(\text{Re } \lambda_i)\} = \{-, -, 0, 0, 0, 0, 0\}$, namely with 2 stable directions and 5 center manifolds. (Note that in Fig. 7.13, only one flat direction is indicated at the final critical point, but bear in mind that there are 4 center manifolds (Re $\lambda_i = 0$) in the initial state, and 5 center manifolds in the final state.)

Instantons (in dissipative systems)

In Fig. 7.13 we also see that for these particular initial conditions, only the memory variables x_2 and x_4 evolve in time. In addition, we see from that figure a peak in the voltage vs. time dynamics of v_2 and v_3.

As I will discuss in Chapter 11, this trajectory is an *instanton* that connects the initial-time critical point with one unstable direction, to the final-state critical point with all stable directions and center manifolds.

[9]The values $C = 10^{-5}$ F and $R = 1$ Ω have been used in (Bearden *et al.*, 2018) to obtain Fig. 7.13.

Table 7.2 Coefficients for the VCVGs's voltage relations given by $v_{i,j} = a_1 v_1 + a_2 v_2 + a_3 v_3 + dc$, where $i = 1, 2, 3$ and $j = R, M$. All voltages are in Volts. From (Bearden *et al.*, 2018).

	a_1	a_2	a_3	dc
$L_{1,M}$	0	-1	1	1
$L_{1,R}$	3	1	-2	-1
$L_{2,M}$	-1	0	1	1
$L_{2,R}$	1	3	-2	-1
$L_{3,M}$	2	2	-1	-2
$L_{3,R}$	-3	-3	5	2

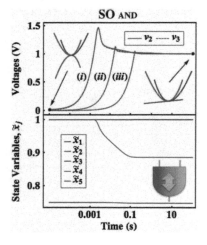

Fig. 7.13 Time evolution of the voltages (**top panel**) and memory variables (**bottom panel**) of the SO-AND of Fig. 7.12. The initial condition is perturbed by (*i*) $\delta v_2 = 10^{-2}V$, (*ii*) $\delta v_2 = 10^{-3}V$, (*iii*) $\delta v_2 = 10^{-4}V$. Adapted with permission from (Bearden *et al.*, 2018).

[10] In Chapter 11, I will also discuss the role of initial conditions in the instantonic trajectories.

Importantly, slightly different initial conditions (three in Fig. 7.13, obtained by perturbing the unstable initial state with three different variations of v_2), generate *different* trajectories (instantons), but the *same* logically consistent solution. As I will discuss at length in Chapter 11, this is not surprising, precisely because the initial critical point has an unstable direction.[10]

Indeed, unstable directions resemble *locally* (in phase space) something that is characteristic of chaotic dynamics (see Section 3.9). However, as already mentioned, DMMs are *not* chaotic, namely they do *not* support the *global* non-integrability of their dynamics (cf. Section 3.10 and also Section 12.9).

7.4   Fast and slow degrees of freedom

Time scale and characteristic frequency

The typical time over which a physical property varies is called the *time scale* of that particular observable. Its inverse is called the *characteristic frequency*.

The previous example presents us with the opportunity to discuss a very important difference between the *time scales* of the various *degrees of freedom* in the equations of motion of these SOGs as well as their circuits.

From Fig. 7.13, we see that the voltage degrees of freedom (which ultimately represent the solution(s) to the given problem) have a *faster* dynamics than the memory degrees of freedom (whose role is to eliminate local minima from the phase space, see Fig. 7.7).[11] This difference in time scales is *not* accidental: it was chosen on purpose! The reason is simple.

[11] We could also say that the characteristic frequency of the voltages is much higher than the characteristic frequency of the memory variables.

If the memory degrees of freedom were to vary too fast, e.g., as fast as the voltage degrees of freedom, the latter ones would *continuously* experience large variations in their dynamics, not just the 'instantonic' ones. These large variations would not allow them to easily reach equilibrium points, and, in fact, could easily lead to *chaotic dynamics*.[12]

[12] As an analogy, consider the pinball Example 1.3 or the waterfall Example 1.4. If we shake the pinball machine or the ground too violently, the ball in the first case, and the water in the second, would have a very hard time to reach the exit point, and their dynamics would be essentially chaotic.

Therefore, SOGs (as well as their circuits) should preferably be built with two distinct *time scales* in mind:[13]

[13] Of course, within these two different sets of time scales, there could be some additional differences. For instance, in Chapter 8 and 9, I will discuss 'short-term' and 'long-term' memories. Their time scales are different, but still *slower* than the time scales of the voltages.

Fast and slow degrees of freedom

- *Fast degrees of freedom*: those associated with the variables of the problem to solve. These are the voltages in an electrical circuit realization of MemComputing machines.
- *Slow degrees of freedom*: those related to the memory dynamics in the system.

Of course, the *actual* difference between the time scales of the fast and slow degrees of freedom will depend on the specific realization of the self-organizing gates and circuits. As a rule of thumb, though, an order of magnitude difference seems to be a reasonable starting point (Traversa and Di Ventra, 2017).

7.5   The magnitude of time non-locality

After the discussion on time scales, I can further employ Example 7.3 as a test bed to understand what role the *magnitude* (or strength) of time non-locality (memory)[14] plays in SOGs.

In fact, we expect that if the memory is *negligible* then the system will have no degrees of freedom that help in the adaption (self-organization) of the voltage degrees of freedom toward the logically consistent truth table of the gate.

This is indeed the case, as it is evident from Fig. 7.14, where the time interval to achieve equilibrium for the SO-AND gate of the previous Example 7.3 is plotted versus the ratio R_{on}/R_{off}, which, for the particular choice of memory devices I have made in that example, is a measure of the degree of time non-locality in the system.

When $R_{on}/R_{off} \to 1$, then time non-locality tends to zero, and, as expected, this causes the dynamics to slow to a halt: the gate cannot satisfy its logical proposition.

Interestingly, though, the same trend also occurs (although less dramatically) when $R_{on}/R_{off} \to 0$, which implies an approach to *infinite* memory. This can be interpreted as due to the system possessing too many trajectories to 'explore' in the phase space, or, equivalently, having the strength of the unstable directions of critical points (with that strength depending on the ratio R_{on}/R_{off}) so large that the system dynamics are *repelled* too strongly away from equilibrium (see also discussion in Chapter 11).

For the particular SO-AND discussed here, we then see from Fig. 7.14 that there is an *optimum* degree of time non-locality for speeding up the dynamics, which will strongly depend on the particular physical systems used to implement these gates.

[14]Namely, how far in the past the system remembers its state.

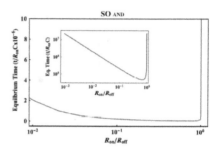

Fig. 7.14 Main panel: dependence of the equilibrium time on R_{on}/R_{off} (memory content for the SO-AND of Example 7.3). Equilibrium is defined as the time necessary for voltages to be within 1% of their steady-state values for a given ratio R_{on}/R_{off} ($R_{on} = 0.01 \ \Omega$ held fixed). The **inset** illustrates the dependence on a log-log scale. Reprinted with permission from (Bearden *et al.*, 2018).

7.6   From thermodynamic equilibrium to steady state

Finally, let me stress an additional point on the operation of these SOGs. This point applies equally to their circuits (Section 7.7).

In the absence of any voltage (bias) applied to their terminals, SO gates (whether logic or algebraic) are in *thermodynamic equilibrium* (cf. Section 3.9). This means that no current flows in or out of those terminals, *irrespective* of the value of the internal (memory) degrees of freedom. The latter ones are, at thermodynamic equilibrium, *decoupled* from the voltage degrees of freedom.[15]

As soon as we switch on the voltages at some terminals of the gate, dynamics ensue, and the system will *self-organize* toward an appropriate equilibrium point, which is a *steady state* of the dynamics. This is shown, e.g., in Fig. 7.13.

In fact, as soon as the voltages at some terminals are switched on, the *coupling* between the (internal) memory variables and the voltage variables switches on as well and persists till a steady state is reached. Once such a state is reached, the memory variables *decouple* from the voltage variables, creating *center manifolds* (Fig. 7.7).[16]

This steady state persists *so long as* the input voltages are on. As soon as those voltages are switched off, the system will return to its *thermodynamic equilibrium*.

The timescale for such a decay back to thermodynamic equilibrium is again on the order of the RC time of the circuit, with R the typical resistance, and C the typical capacitance. It then strongly depends on the actual materials/devices used to build these machines.

I will come back to these issues in more detail in Chapters 11 and 12. For now, let's move on to the next stage of how to use these gates to solve combinatorial optimization problems, the latter defined in Section 2.5.

7.7 Self-organizing (SO) circuits

Now that we have seen how to construct individual SOLGs, it is straightforward to assemble them to obtain *self-organizing logic circuits* (SOLCs). The construction of these SOLCs is indeed quite easy to do.

The procedure can be summarized as follows:

Physical topology and information overhead
The *physical topology* (or architecture) of a SO circuit defines its *information overhead* (cf. Section 5.3.2).

Building SOLCs:

(i) Transform a given problem into Boolean format.[a]

(ii) Design one Boolean circuit (out of possibly many) that represents such a problem. This sets its *physical topology* (architecture).

(iii) Replace the standard logic gates of this Boolean circuit with SOLGs.

(iv) Apply the Boolean values that represent the input of the problem to the appropriate terminals of this circuit.

(v) Let the circuit self-organize to the solution.

(vi) Read the solution at the appropriate terminals.

[a]This transformation is exact for a combinatorial optimization problem. Oth-

erwise, it is an approximation.

Note that step (ii) is precisely the one that defines the *information overhead* I have introduced in Section 5.3.2.

It is this type of 'extra' information, which is embedded in the *physical topology* of the circuit, that is important to solve hard combinatorial optimization problems efficiently (see also Chapters 8, 9, and 10).

In fact, a *random* assembly of SOLGs *cannot* solve these problems, as it cannot provide any means to 'guide' the machine toward the solution.

Note also that this type of information is *not* available to Turing machines, since *no* physical information about the problem to solve is ever introduced in the definition of such machines.

An important remark

We can follow a similar strategy to assemble SOAGs into *self-organizing algebraic circuits* (SOACs) to solve *algebraic* problems, like the Integer Linear Programming (ILP) one (Traversa and Di Ventra, 2018).

I will now provide an explicit example of how to construct a SOLC by following the described procedure. This example will also show that one needs to proceed with particular care when assembling the individual SOLGs into a circuit.

In fact, the practical realization of SOLGs I have introduced in the preceding Section 7.3, when assembled together to represent a given problem, still requires extra circuitry to guarantee absence of equilibria other than the logically consistent solutions of a given problem (Traversa and Di Ventra, 2017). I will discuss a different strategy in Chapter 8 as to how to accomplish this goal, which is particularly useful for numerical simulations of these machines.

However, the general idea is always the same: one must guarantee that the only equilibria in the phase space are the solutions of the problem to be solved (Di Ventra and Traversa, 2018*a*). Or at very least, property (iii) I introduced in Section 6.3.2 has to be satisfied.

Example 7.4

3-bit sum SOLC

As an example of the previous procedure we can take again the circuit that performs the 3-bit sum shown in Fig. 7.1.

In that case, if we want to operate the circuit *in reverse*—namely given the sum v_{o1} and carry v_{o2}, determine the three bits v_1, v_2, and v_3 that produce such a result—we can simply take that circuit, and replace the corresponding standard OR, AND, and XOR gates with the SO-OR, SO-AND, and SO-XOR gates. This is shown in Fig. 7.15.

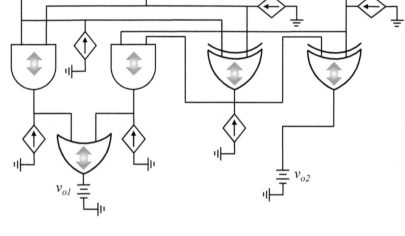

Fig. 7.15 Sketch of a possible 3-bit sum SOLC such that, given the bits v_{o1} and v_{o2}, outputs the consistent bits v_1, v_2, and v_3. Note that in addition to SO-OR, SO-AND, and SO-XOR gates, the circuit employs VCD-CGs (diamonds with arrows inside) at each gate terminal, except at the terminals where the input is supplied. These VCDCGs eliminate, as possible equilibrium points, the zero-voltage state of the terminals.

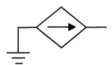

Fig. 7.16 Symbol of a voltage-controlled differential current generator. Current generators are *active* elements that are *sources* of current. When the current source depends on the voltage input, then they are called *voltage-controlled*. When the current-source magnitude varies in time according to voltage differences, it is called a voltage-controlled *differential* current source (Horowitz, 2015).

Note, however, that there is an extra circuit element in that figure: a *voltage-controlled differential current generator* (VCDCG), which is also an active element (Fig. 7.16). This device is connected to all the terminals of the SOLGs, except the ones at which we feed the input signals.

The reason for these VCDCGs is because, for the *particular* choice of parameters in Table 7.1 and in (Traversa and Di Ventra, 2017), SOLGs may also admit 'spurious solutions' other than the logically consistent equilibrium points (for instance, by setting $v_3 = -1$V we could also obtain $v_1 = v_2 = 0$V, which is *not* one of the four correct equilibrium points).

Although these extra minima satisfy property (iii) of Section 6.3.2, we can eliminate them altogether.

In fact, one can design the VCDCGs to admit as unique stable solution either $v = +1$V or $v = -1$V, essentially making the SOLG solution with voltage $v = 0$V *unstable*.

There are several ways one could design these VCDCGs, and I refer the reader to (Traversa and Di Ventra, 2017), for an explicit example.

Here, I just give its general principle of operation by considering a simplified equation that governs the VCDCG dynamics

$$\frac{dI}{dt} = f_{DCG}(v), \qquad (7.14)$$

where I is the current generated by the VCDCG, and the function f_{DCG} is sketched in Fig. 7.17.

In reference to that figure, if we consider voltages v around 0, eqn (7.14) gives $\frac{dI}{dt} = \frac{\partial f_{DCG}}{\partial v}|_{v=0} v < 0$. This relation is equivalent to the one of a *negative* inductor, and it renders the solution 0 unstable.

On the other hand, around $v = \pm v_c$ we obtain $\frac{dI}{dt} = \frac{\partial f_{DCG}}{\partial v}|_{v=\pm v_c} (v \mp v_c) > 0$. This is equivalent to an inductor in series with a DC voltage generator of magnitude $\pm v_c$.

This means that any other voltage v induces an increase or decrease of the current and, therefore, there are no other possible stable points, other than $v = \pm v_c$.

(Incidentally, it is due to the presence of these 'inductor-like' elements that the voltages at the gate terminals may exceed the boundary values of $+1$V and -1V; see, e.g., Fig. 7.18.)

However, a different circuit could be designed to limit their values between those bounds; see Section 8.2.1. In addition, other types of elements may be devised that accomplish the same tasks I have outlined previously.

This leads me to the following *very important* remark:[17]

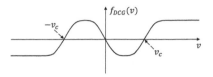

Fig. 7.17 Sketch of the function $f_{DCG}(v)$ of eqn (7.14).

[17]This remark is relevant to both the *hardware* implementation of these machines, as well their *software* simulation (see Chapter 8).

An important remark

It is worth stressing once more that the previous choice of circuitry is just *one possible realization*.

The important point to remember is that either we choose SOLG's parameters and design so that *no solution* other than the logically consistent one is present, or if that is difficult to do, when we connect different SOLGs together to form a SOLC, we may connect a VCDCG at each terminal but the ones at which we send the inputs, as it is shown in Fig. 7.15.

This guarantees absence of any equilibria other than the logically consistent relations of the gates.

Example 7.5

Prime factorization SOLC

For another example of SOLC, let's look at the one that solves the *prime factorization* problem, as I have described it in Example 7.2.[18]

We could use the same circuit as in Fig. 7.4, appropriately modified according to the bit size of the integer to factor, with the addition of the VCDCGs at each gate terminal, except at the terminals where the input is supplied.

[18]In Section 9.1.2, I will show another SOLC that solves the same problem, and which is more conducive to numerical simulations.

[19]Assuming that the number, n, to factor is composed of only two primes. If it is not, that circuit will result in two numbers, q and p, (whose value depends on the chosen initial conditions) which are not primes. However, by repeating this process with each of these numbers, eventually all the primes of the original number n can be obtained.

By replacing the traditional logic gates of that circuit with the same SOLGs as the ones in Section 7.3 (see also (Traversa and Di Ventra, 2017) for more details), we then obtain a SOLC that solves the prime factorization problem.[19]

In Fig. 7.18, I plot examples of voltage dynamics at the nodes of SOLCs for prime factorization of numbers of different bit size.

(The numerical simulations have been performed using an *implicit* method of integration—see Section 8.4.1 for the difference between 'implicit' and 'explicit' methods of integration.)

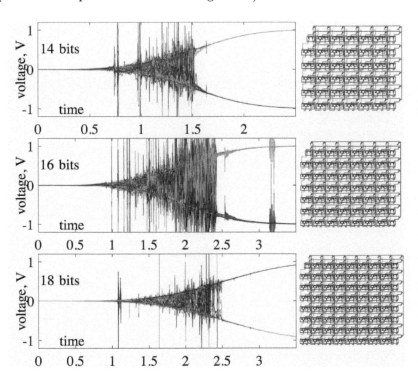

Fig. 7.18 Voltages at the nodes of the SOLCs for the prime factorization of numbers of different bits as a function of time (in arbitrary units). A schematic of the circuit is reported on the right of each plot. These circuits have the same architecture as the one in Fig. 7.4. The solution is found when all voltages are either +1V or −1V (logical 1 and 0, respectively). Reprinted with permission from (Traversa and Di Ventra, 2017).

Hardware scalability for factorization	
Quantum Computing *probabilistic*	MemComputing *deterministic*
$O(N^2 \log_2 N \log_2^2 N)$ on a <u>quantum</u> computer + $O(\text{poly}(N))$ on a <u>classical</u> computer	$O(N^2)$

Fig. 7.19 *Hardware* scalability comparison for the factorization of a number with N bits using a *decoherence-free* quantum computer implementing Shor's algorithm (Shor, 1997) vs. a DMM with the architecture of Fig. 7.4 (Traversa and Di Ventra, 2017).

It is clear from Fig. 7.18 that the system starts from *thermodynamic equilibrium* (all voltages and currents in the circuit are zero; cf. Section 7.6), and when the voltages that represent the number to factor are switched on, the system undergoes a non-trivial dynamics inter-spaced by sudden changes of voltages, before it settles into a *steady state* (equilibrium), which corresponds to the correct factorization of the number. The sudden changes in voltages are again what I called *instantons* in the Example 7.3 (cf. Fig. 7.13). The memory degrees of freedom, instead have a much slower dynamics, similar to the one I showed in Fig. 7.13.

In Chapters 11 and 12, I will discuss more in depth instantons and the approach to equilibrium of SOLCs. For now, it is enough to realize that we can indeed build these circuits in hardware to, e.g., factor numbers.

> **An important remark**
> If this particular DMM for factorization were built in *hardware* it would scale as $O(N^2)$ in space (i.e., with the number of self-organizing logic gates employed) and $O(N^2)$ in convergence time with input size (number of bits) N, and since all voltages and memory variables are bounded, the energy expenditure would also grow polynomially with size (Traversa and Di Ventra, 2017); see Fig. 7.19.

Absence of solutions/equilibria

Now, suppose you want to factor, with the same type of SOLCs, a number n which is *already* a prime, e.g., the number 47. The way in which we have built the circuit for factorization, the trivial solution of that number times 1 is forbidden (see discussion in Example 7.2).

Therefore, the dynamical system has *no* equilibrium points. Rather, it settles into a (quasi-)periodic orbit from which it will never exit.[20] This is illustrated in Fig. 7.20 for the prime number 47 (a 6-bit number).

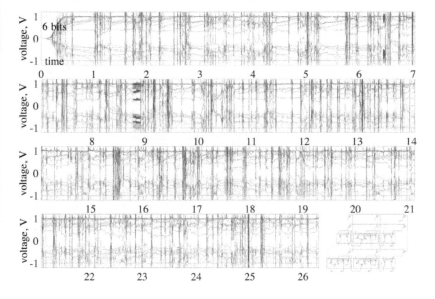

After a short transient, which is very similar to the one observed in Fig. 7.18, the system is incapable of reaching a steady state, and keeps on experiencing (instantonic) jumps in the voltages forever.[21]

Shor's algorithm and field-sieve method for factorization

The *Shor's algorithm* is a probabilistic algorithm that, if implemented on a hypothetical (*decoherence-free*) quantum computer, would factor numbers of bit size N into primes with $O(N^2 \log_2 N \log_2^2 N)$ operations on the quantum computer, together with a polynomial (in N) post-processing on a classical computer (Shor, 1997). The *general number field-sieve method* is the best known prime factorization algorithm (for numbers larger than 10^{100}) that, if implemented in software, (heuristically) scales exponentially as a function of input size, $O(\exp((c + o(1))N^{1/3}(\ln N)^{2/3}))$, with $c = (64/9)^{1/3}$ (Goldreich, 2001).

[20] Absence of chaotic behavior for these SOLCs, even in the absence of equilibria, has been shown in (Di Ventra and Traversa, 2017a).

Fig. 7.20 Voltages at the nodes of the SOLCs for the prime factorization of the prime 47 (a 6-bit number) as a function of time (in arbitrary units). Schematic of the circuit simulated is reported on the far bottom right of the plot. Reprinted with permission from (Traversa and Di Ventra, 2017).

[21] Until, of course, we switch off the input signals, after which the system returns to thermodynamic equilibrium (cf. Section 7.6).

> **An important remark**
>
> Although, this is *not* an analytical (logical) proof that a solution to the problem does not exist, by estimating the time it takes the system to find the solution[a] of an *ensemble* of similar (in bit size) factorization problems, one may infer (with a certain level of confidence) that a solution to the given problem does not exist.[b]
>
> ---
> [a]Either the median time or some other measure of time to solution (cf. Chapter 8).
> [b]In the language of algorithms, we would say that this procedure represents an *incomplete* algorithm (see Section 2.6).

After discussing a practical realization of SOLGs using electrical circuits, let me conclude this Chapter by enumerating the mathematical properties that the equations of motion of these SOLGs satisfy, which makes them a powerful computing model to tackle hard combinatorial optimization problems.

7.8   Mathematical properties of SOLCs

So far, I have discussed the general ideas behind the design of self-organizing gates, some of their possible circuit realizations, and how to assemble them to form self-organizing circuits. What needs to be answered now is a series of important mathematical questions which complement the physical properties a DMM needs to have to compute efficiently (c.f. Section 6.3.2).

After the discussion in the previous Section 7.7, and putting aside the issue of noise which I will discuss explicitly in Chapter 12, it should now be clear that SOLCs and SOAGs are simply dynamical systems whose equations of motion are of the type (3.1).

The state of these systems, $\mathbf{x}$, represents all voltages, currents and internal (memory) variables of the system. Instead, the flow vector field, $F(\mathbf{x})$, depends on the *type* of devices employed (e.g., DCMs and VCDCGs), and the *physical topology* of the circuit assembled to solve a specific problem.

Therefore, the questions that we need to address are as follows.

Are the equilibrium points of these dynamical systems in a one-to-one correspondence with the solutions of the problem? How fast are these equilibria reached from some region of the phase space? Can we guarantee that by increasing the size of the input, the rate at which DMMs reach equilibrium scales at most polynomially?

In fact, can we guarantee that all *resources* (see definition of computational resources in Section 2.2.2) scale at most polynomially with problem size? If there are equilibrium points (representing solutions to the problem at hand), can we design the dynamical systems to have no periodic orbits and chaos?

All these questions have been addressed in a series of theorems demonstrated in (Traversa and Di Ventra, 2017; Di Ventra and Traversa, 2017*a*; Di Ventra and Traversa, 2017*b*) using Functional Analysis and Topology.[22]

The detailed demonstration of some of these theorems is not trivial and requires quite an extensive background that is best left to the specialized literature. Therefore, to avoid unnecessary distractions from the main features of these machines, I will not report them here, and refer the reader to the published papers (Traversa and Di Ventra, 2017; Di Ventra and Traversa, 2017*a*; Di Ventra and Traversa, 2017*b*), where these demonstrations are provided. I will just enumerate the results of these theorems in the form of main properties of these dynamical systems.

[22]See also Chapters 8, 9, and 10 for other types of equations of motion for which these same questions have been addressed mathematically. Also, in Chapter 12, I will show—using (supersymmetric) *Topological Field Theory*—the polynomial scalability of the time to solution of the continuous dynamics (Section 12.3), and absence of chaos in these dynamical systems (Section 12.9).

An important remark

Note that some of these theorems have been demonstrated only for a *particular* class of SOLGs (and corresponding SOLCs) whose parameters are in Table 7.1 for the SO-AND, and can be found in (Traversa and Di Ventra, 2017) for other SOLGs.

Although one would need to prove the same theorems for any other type of physical realization of SOLCs/SOACs, the properties they represent are nonetheless desirable for *all* these dynamical systems.

This means that if some other physical realization of these gates is proposed, an effort should be made to address the previous questions mathematically. I will discuss in Chapters 8, 9, and 10 another realization of SOLCs for which these mathematical demonstrations have been carried out. See also Sections 12.3 and 12.9.

In addition, the theorems have been demonstrated only for the *continuous* dynamical systems representing SOLCs, i.e., for equations of the type (3.1). They have *not* been proved for their *discrete version*, which is relevant for their *numerical integration*.

This is not a minor point. I will return to it in Section 13.8.6, because after the *simulation* results I will report in Chapters 8, 9, and 10, it would be tempting to conclude that we have proved the famous question of whether NP = P or not (see Section 2.3.2 for a rigorous statement of this question). As I will explain better in Section 13.8.6, this *mathematical* question is still unanswered.

After this discussion I can now enumerate the main results of (Traversa and Di Ventra, 2017) for the particular class of SOLGs and SOLCs I have outlined in this Chapter:

SOLC's properties

(i) The dynamical systems describing SOLCs are *point dissipative*. As anticipated in Section 6.3.2, this feature allows SOLCs to reach equilibria *irrespective* of the initial conditions.

(ii) The only equilibrium points are the solutions of the given problem SOLCs represent. Any other critical point in the phase space is a saddle point.

(iii) For each size of the problem, equilibrium points have exponentially fast convergence rate in all their attraction basin.

(iv) The convergence rate scales at most *polynomially* with input size.

(v) SOLCs's *physical* resources (physical time, space, and energy) grow at most *polynomially* for problems whose solution tree grows exponentially (see Section 2.3.1).

(vi) In the presence of solutions there cannot be (quasi-)periodic orbits or chaos.

> **Point-dissipative systems**
> It is the class of dynamical systems that admit a *compact* (see definition in Section 6.3.2) subset J of the phase space X, that attracts *every* point of X under the flow vector field of eqns (3.1) (Hale, 2010).

Note, again, that the SOLCs in (Traversa and Di Ventra, 2017) (or, in general, those designed to have the enumerated features) are such that their *hardware* implementation would then solve combinatorial optimization problems efficiently, namely with resources that scale only *polynomially* with input size. No conclusion can be derived on the *numerical* solution of the corresponding equations of motion (see, however, numerical evidence in Chapters 8, 9, and 10).

> **An important remark**
> If the dynamical system representing a self-organizing circuit has been proved to be *integrable* (not chaotic), and devoid of quasi-periodic orbits, then one can infer the *polynomial* scalability of the solution search by means of (supersymmetric) *Topological Field Theory* (Di Ventra and Ovchinnikov, 2019). See Section 12.3.

Metric vs. Banach spaces

Another important property to highlight is that the phase space, X, of the dynamical systems representing SOLCs/SOACs is just a *metric* (non-linear) compact space (see definition in Section 6.3.1), not a *Banach* (vector/linear) space, which is a particular type of metric space (see the *hierarchy* of mathematical spaces in Fig. 7.21).

If eqns (3.1) represent a point dissipative system whose phase space is simply a metric space, then they do not always admit an equilibrium point (Hale, 2010). Instead, if the phase space were a Banach space (such as a Hilbert space), a point-dissipative system would always admit an equilibrium point, and the global attractor of a Banach space is always connected, which means that equilibrium point can always be reached (Hale, 2010).

This implies that, if the phase space were a Banach space, we could not design point-dissipative dynamical systems whose *only* equilibrium points are the solutions of the given problem we are trying to solve, because that would mean the chosen problem would *always* have at least one solution, which is clearly not true.

The mathematical properties of the phase space are then very important for the correct mapping between equilibrium points and solutions of a combinatorial optimization problem. It also shows that the extension of the ideas I have discussed in this Chapter to *quantum* systems is not straightforward: by construction, quantum systems have their natural state dynamics in a Hilbert space, which is a Banach space. (Of course, we could follow a different strategy altogether to generalize to the quantum case; cf. Section 13.6.)

Banach space
It is a *complete normed vector space*. 'Complete' means that any (Cauchy) sequence of vectors converges to a well-defined limit which is within the space. 'Normed' means that a 'norm' or 'vector length' is defined on it (Arfken *et al.*, 1989). An example of Banach space is the Hilbert space (which is endowed with an *inner product*; Section 3.1). Since a normed vector space is also a metric space, a Banach space is also a metric space (see definition in Section 6.3.1). The reverse, however, is not necessarily true (see Fig. 7.21).

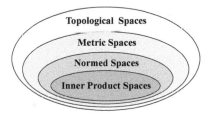

Fig. 7.21 Hierarchy of mathematical spaces.

7.9 Decision vs. optimization

At this stage it is important to clarify an important point regarding the meaning of 'solving a given problem'. If the problem is a *decision* problem, or its *search* version (see Section 2.3.4 for a refresh of these definitions), then SOLCs/SOACs appropriately designed as I have discussed previously, *do* indeed solve it, in the sense that given an initial condition, they *guarantee* the system to converge to the solution (or one of the solutions) of the given problem (if a solution exists).

On the other hand, if the problem is an *optimization* one, where one needs to find the *global* optimum out of several *local* optima, then the SOLCs/SOACs cannot guarantee that the solution found after an *assigned time* is the optimum, even if the corresponding dynamical system has reached what appears to be a steady state.

The reason has nothing to do with the gates or circuits discussed here, rather with the nature of such problems: one cannot check in polynomial time if what has been found is indeed the global optimum. That would require comparing it with all possible local optima. In other words, one cannot take whatever 'solution' has been found and easily check that it is the global optimum.

Therefore, in the case of an optimization problem, the best one can do is to assign a given time, and determine *within* such a time the 'best' assignment of the variables that satisfies a given proposition.[23] This is what we will mean by 'solving' optimization problems in Chapters 9 and 10.

[23] By 'best' assignment we mean the maximization (or minimization) of some *objective function*, subject to some constraints; see Section 2.5.

An important remark

What happens if we let a DMM for an optimization problem to run for an infinite time? The answer is that it will eventually settle into a (quasi-)periodic orbit. This is because, even if it finds the optimum after a finite time, since the optimum is not an equilibrium point of the dynamics, the DMM would keep on changing its dynamical state, and settle into the only attractor available to it, which is a (quasi-)periodic orbit, not a chaotic one (Traversa and Di Ventra, 2017).

7.10   Chapter summary

- In this Chapter I have introduced the notion of *self-organizing gates*, whether *logic* or *algebraic*.

- These gates differ fundamentally from standard gates in that they are *terminal-agnostic*, namely they can satisfy their proposition *regardless* of the terminal(s) to which a truth value is assigned.

- SOLGs/SOAGs can be realized if extra degrees of freedom (internal state variables) are introduced. However, they require *feedback* to drive the corresponding dynamical system toward the correct logical or algebraic proposition.

- This means that in addition to time non-locality (memory), which can be realized in practice with passive devices, such as resistive memories, these gates require *active* elements.

- SOLGs/SOAGs, when assembled together in architectures that represent a given combinatorial optimization problem, allow the system to solve these problems efficiently. This statement pertains to the *hardware* implementation of these self-organizing circuits, whether SOLCs or SOACs.

- In other words, SOLCs/SOACs represent a *practical* realization of the mathematical notion of DMMs.

- However, to avoid overburdening the reader with too many names, in the following I will employ the term DMM to indicate both the *mathematical* concept and its *practical* realization as a self-organizing circuit. I will distinguish the two terms only when confusion may arise.

Part III

Applications

... And a man who is puzzled and wonders, thinks himself ignorant; ...

L'esperienza è il solo insegnante in cui possiamo confidare.
(Experience is the only teacher we can trust.)
Leonardo da Vinci (1452−1519)

One can argue with a philosophy; it is not so easy to argue with a computer printout, which says to us: 'Independently of all your philosophy, here are the facts of actual performance'.
Edwin Thompson Jaynes (1922−1998)

So far, I have introduced the MemComputing paradigm with the idea that these machines need to be built in hardware. However, as I have explained in Part II, the underlying physical dynamics of digital MemComputing machines (DMMs) are *non-quantum*. Therefore, *unlike* quantum computers, the equations of motion representing DMMs *can* be simulated efficiently on our modern computers.

In this part of the book, I will discuss various *simulations* of DMMs to tackle a variety of combinatorial optimization problems, ranging from satisfiability (SAT) to maximum satisfiability (MAX-SAT) to applications in Machine Learning and Quantum Mechanics. These results will showcase the substantial advantages the simulations of DMMs can already deliver, even without building these machines in hardware.

- In Chapter 8, I will discuss different numerical methods for solving the ordinary differential equations of DMMs and compare their limitations and advantages.
- In Chapter 9, I will show simulations of DMMs applied to a variety of hard combinatorial optimization problems.
- In Chapter 10, I will demonstrate how MemComputing can be leveraged to train, in an unsupervised manner, (deep) neural networks at optimality and as a single unit. I will then show some of its applications in Quantum Mechanics.

Numerical simulations of DMMs

8

Numerical analysis is very much an experimental Science.
Peter Wynn (1931−2017)

Takeaways from this Chapter

- How to design digital MemComputing machines (DMMs) for numerical simulations.
- The difference between *implicit* and *explicit* methods of integration.
- *Explicit* integration schemes are *sufficient* for simulating DMMs.
- It is found *empirically* that the integration *time step*, Δt, needs to decay as a *power law* as a function of problem size, not exponentially, to control numerical errors.

8.1 Coarse-graining from hardware to software

In Chapter 7 I have introduced the *physical* realization of self-organizing (SO) gates and circuits. Since I have considered their electrical circuit representation, these gates and circuits can be built in *hardware* with present complementary metal-oxide semiconductor (CMOS) technology (see 'third advantage of non-quantum systems' in Section 3.3).

As I have anticipated in Section 3.2, this also affords an incredible advantage ('first advantage of non-quantum systems'): MemComputing machines can be described by *ordinary differential equations* (ODEs) of the type (3.1). And ODEs can be *simulated* efficiently on our modern computers.

This advantage begets yet another one: we are free to *simplify*, or *coarse-grain*, the ODEs of DMMs for the benefit of numerical simulations, so long as we maintain the most *relevant* physical and mathe-

---

Simplified models vs. physical reality

As I have anticipated in Chapter 1, physical systems are always (infinitely) better at carrying out whatever 'computation' they naturally perform, than any mathematical description (a *model*) of the same can ever do. This is because actual physical systems take into account *all* possible effects, phenomena, and interactions they are subject to. However, as it is well known in the Sciences, simplified models, if properly designed, can indeed capture the *most relevant* aspects (degrees of freedom) of physical systems. These are the models (inspired by actual physical devices) I will discuss in this Chapter, and in Chapters 9 and 10. They will also help us better understand the workings of DMMs and their emerging phenomena, as I will show in Chapters 11 and 12.

matical properties of the original system. This means that, even if the chosen ODEs do not represent *exactly* hardware devices, as the ones I have introduced in Chapter 7, as long as they satisfy the conditions in Section 7.8, or the ones in Section 6.3.2, they would still take advantage of the main properties of DMMs to solve combinatorial optimization problems efficiently.[1]

In fact, while a *physical* DMM with all its passive and active elements would solve a combinatorial optimization problem in 'one shot', on (physical) time scales dictated by the actual properties of its constituents, the *numerical* simulation of the DMM's ODEs has to contend with the discretization of time,[2] random access memory (RAM) limitations, speed of the processor used, etc. This brings with it its own additional issues. This is the reason why, if one is interested in *simulating* DMMs, the choice of equations that provide the most efficient numerical implementation is paramount.[3]

In this Chapter, I will provide *one* strategy to accomplish this 'transition' (*coarse-graining*) from *hardware* to *software*. In particular, I will follow closely the one outlined in (Bearden, Pei and Di Ventra, 2020). However, the reader should keep in mind that the same results can be accomplished in several other ways. I will then show the role of the different integration schemes in solving such equations. In particular, I will show *empirically* that, for the machines to be able to solve hard problems, one does not need to reduce the integration step, Δt, exponentially with increasing problem size: Δt needs to decrease at most as a *power law* to control numerical errors (Zhang and Di Ventra, 2021).[4]

In the next two Chapters, 9 and 10, I will employ these equations (or some variations of them) to solve a variety of combinatorial optimization problems, and in applications to Machine Learning and Quantum Mechanics. In Section 12.7.5, instead, I will show that the dynamics of DMMs in the presence of (numerical and physical) noise is akin to *directed percolation* of the state, $\mathbf{x}(t)$, in the phase space. In Chapters 11 and 12, I will also explain why the numerical simulations are robust against the unavoidable (and large) numerical errors that accumulate during computation. I will show that this is due to the *topological* nature of the solution search.

8.2   From physical system to simplified dynamics

As I have anticipated in Section 2.4, any combinatorial optimization problem can be written, in principle, in conjunctive normal form (CNF) (Arora and Boaz, 2009). Let's then start by assuming that the problem we want to solve is written that way. As an important practical example

---

[1]In other words, in going from hardware to software, we are keeping the most relevant degrees of freedom of the actual physical systems, while discarding the less relevant ones. It is like performing some sort of *Renormalization Group* (Goldenfeld, 1992) (*coarse-graining*) procedure on the equations of motion of the actual hardware realization of DMMs (cf. Section 13.7.5).

[2]Hence, the accumulation of numerical errors.

[3]Of course, one may also proceed in reverse. First, an appropriate set of ODEs for DMMs is chosen, and then one looks for actual physical devices that closely reproduce their behavior.

[4]This is true for all the instances studied so far, but we do not have an analytical and general proof of this statement yet.

let's consider a 3-SAT, namely a (NP-complete) satisfiability problem with exactly three literals per clause (see Section 2.4 for a refresh of these concepts). It is straightforward to generalize to any type of problem written in CNF.[5]

8.2.1 A self-organizing logic circuit for 3-SAT

Let us recall from Section 2.4 that a 3-SAT formula is a collection of disjunctions (OR gates or clauses) of *exactly* three literals (variables or their negations), with the clauses related to each other by logical ANDs. Each formula would then have N Boolean variables (v_i), M clauses, and thus $3M$ literals. Our goal is to find an assignment of the variables such that all clauses are satisfied.

As I have mentioned in Section 2.6 (cf. also Section 4.3.4), the complexity of the problem emerges precisely from the fact that variables appear in different clauses (constraints), so that, when the number of clauses and the number of variables have a certain ratio, $\alpha_r = M/N$, the problem is difficult to solve with traditional algorithms with increasing number of variables. In fact, it has been empirically shown that there is an 'easy-hard-easy' transition in 3-SAT, with increasing $\alpha_r = M/N$, with the hardest instances occurring at $\alpha_r \sim 4.27$ (Mezard and Montanari, 2009).

If we follow the strategy I have outlined in Chapter 7, we would first transform the Boolean variables of the 3-SAT into continuum ones (e.g., voltages in the electrical realization of DMMs), and then replace the traditional gates with self-organizing (SO) gates. In fact, if we refer again to Fig. 2.8 of Chapter 2, it is enough to replace the traditional OR gates with three-terminal SO-OR gates, since the AND operation is simply carried out by connecting the variables appearing at different clauses with the appropriate negation (NOT gate). We then set all the 'output' terminals of the OR gates to logical 1 (to satisfy the CNF formula), and let the SO circuit self-organize till it finds a solution. This

[5]Here, I am only interested in the general strategy. Of course, the *functional form* of the equations and the choice of parameters may be different for different problems, with some forms better suited for certain problems than others (see also Chapters 9 and 10).

Clause-to-variable ratio
An important parameter in satisfiability problems is the *clause-to-variable ratio* that I will indicate with $\alpha_r = M/N$, where M is the number of clauses, and N is the number of variables. For some problems, this parameter is a good indication of the difficulty of the instances. For example, for the 3-SAT, it has been shown empirically that the hardest instances occur at $\alpha_r \sim 4.27$ (Mezard and Montanari, 2009).

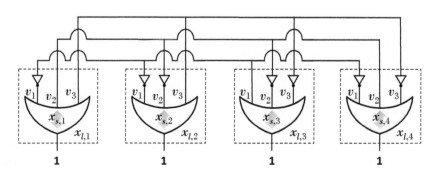

1   1   1   1

Fig. 8.1 Example of a self-organizing circuit for a 3-SAT representing the CNF formula (2.4) in Chapter 2. The output of each 3-terminal SO-OR gate is set to 1 (true), and the problem is to look for an assignment of the variables v_1, v_2 and v_3 that satisfies each clause. The triangle symbols indicate negations of the variables. For each clause, m ($m = 1, \ldots, 4$), we add a *short-term* memory variable, $x_{s,m}$, and a *long-term* memory variable, $x_{l,m}$.

is shown schematically in Fig. 8.1 for the formula (2.4) in Chapter 2. However, recall that an SO gate just needs to satisfy two properties:

(i) if the gate is not in a satisfied logic configuration, it has to modify its terminal values so as to evolve toward one of the satisfied configurations,

(ii) if it is in a satisfied configuration, it should try to stay in that configuration (see Fig 7.6).

8.2.2   Short-term memory

As we have seen in Chapter 7, property (i) can be satisfied by adding memory variables, so as to transform local minima in the (voltage) phase space to saddle points (of the extended memory-voltage phase space; Fig. 7.7). Let's then call these memory variables, x_s, *short-term memory*, because their only scope is to expand the voltage phase space, so that the individual gates always evolve toward a logically consistent solution. Once the gate is satisfied, there is no need of these variables any longer. Therefore, they just need to 'operate' only for a relatively short period of time.[6]

Here, we have the freedom to assign a short-term memory variable to each voltage variable (like in the hardware implementation, where a *dynamic correction module* is assigned to each terminal of the gate, Fig. 7.10), or one memory variable to all three of them.[7] For numerical convenience, I choose to assign one short-term memory variable to *all* three voltages of each clause, namely I label these variables as $x_{s,m}$, with m the clause number (in Fig. 8.1 there are only four clauses, so $m = 1, \ldots, 4$).

8.2.3   Long-term memory

However, the gates are also connected to each other via the sharing of variables or their negations (see Fig. 8.1). Therefore, if some gate is satisfied at a given time, its satisfaction may 'compete' with the one of another gate with which it shares variables. This could create a situation like the one I discussed in Section 7.7, in which different gates end up into an 'intermediate' state which is *not* a logically consistent state of the whole circuit.

In Section 7.7, I discussed a fix for this problem that involved adding *voltage-controlled differential current generators* (VCDCG) at each gate terminal, except at the terminals where the input is supplied (see, e.g., Fig. 7.15). This procedure, however, adds substantially to the physical topology of the circuit (hence the number of ODEs to solve), and instead can be somewhat emulated by the following one.[8] Let's add an extra *long-term memory* variable for each clause. This memory variable

---

[6]Of course, longer than the time of the individual instantonic jumps of the voltage dynamics (cf. discussion in Section 7.4).

[7]In fact, any possible combination is admissible.

[8]Of course, in hardware, VCDCGs are a valid choice to accomplish the task at hand.

varies on longer time scales than the short-term memory variable. This way, it can 'learn' the state of *un*-satisfaction of the given clause, by slowly increasing in magnitude, so long as the clause is unsatisfied, and decreasing in magnitude whenever it is satisfied. We can then use this memory variable to 'weight' the amount of current we need to inject into the terminals of such a gate for it to approach the logically consistent solution. This is indeed what VCDCGs also accomplish. However, now we have a physical interpretation of these objects as 'long-term learning' devices.[9]

[9]In Section 12.10, I will discuss the analogies of these two types of memories (short- and long-term) with those in the animal brain.

8.2.4   A clause function

After these preliminaries we can now start putting together a dynamical system with memory that solves the 3-SAT problem. First, to each m-th Boolean clause, $(l_{i,m} \vee l_{j,m} \vee l_{k,m})$ (where the ls are literals), we associate the following *clause function*:

$$C_m(v_i, v_j, v_k) = \frac{1}{2} \min[(1 - q_{i,m}v_i), (1 - q_{j,m}v_j), (1 - q_{k,m}v_k)], \quad (8.1)$$

where $q_{i,m} = 1$ if $l_{i,m} = v_i$, and $q_{i,m} = -1$ if $l_{i,m} = \bar{v}_i$ (negation of v_i). If we choose the voltages bounded between $+1$ and -1 ($v_i \subset [-1, +1]$, in appropriate units), the clause function is bounded, $C_m \in [0, 1]$.[10] We can further interpret $v_i > 0$ as the logical 1, and $v_i < 0$ as the logical 0. Since a three-terminal OR gate is satisfied whenever at least one of the three 'input' terminals is 1 (Section 2.4), then a clause is necessarily satisfied when its clause function is $C_m < 1/2$. Therefore, the 3-SAT instance is solved when $C_m < 1/2$ for *all* clauses.[11]

[10]This is an example where we bound *explicitly* the voltages between the two logically consistent states. In Chapter 7 we did not limit their value, so that they could exceed those extremes (see, e.g., Fig. 7.18). They were, nonetheless, bounded by the capacitive effects of the circuit elements.

[11]By thresholding the clause function we also avoid the ambiguity associated with $v_i = 0$.

An important remark
Note that in eqn (8.1) I could have used another form for C_m that does not involve the *minimum function*. For instance, a simple product of the arguments would have sufficed. However, the minimum function reinforces the fact that a clause can only constrain one literal, so that if such a literal satisfies the clause, the value of the other two literals is irrelevant.

We can then use this clause function to determine the dynamics of the short- and long-term memories. Since the short-term memory, $x_{s,m}$, is an indicator of the recent history of the clause, we can take it to lag the value of C_m.

The long-term memory, $x_{l,m}$, instead 'learns' how difficult it is to satisfy a clause. Therefore, it should weight more the dynamics of those clauses that are 'historically' difficult to satisfy.

Voltage variables' threshold
As I have discussed in property (ii) of Section 6.3.2, the choice of the *threshold* to identify a voltage variable as a logical 0 or a logical 1 is *arbitrary*, so long as it is *independent* of the size of the problem. In this Section 8.2.4, we are using the simplest thresholding possible (for both hardware and software implementations), namely a positive value of the voltages represents the logical 1, a negative value represents the logical 0.

8.2.5 The DMM equations for 3-SAT

As I have discussed in Chapters 6 and 7, the *form* of the ODEs describing DMMs is not unique. The important point is that they satisfy the physical *and* mathematical properties I have outlined in Sections 6.3.2 and 7.8, respectively, in particular absence (or irrelevance) of (quasi-)periodic orbits and chaos.

Also, there are several ways we could implement property (ii) of the SO-OR gates, I have discussed in Section 8.2.1. When the gate is not in a satisfied configuration, we require it to evolve toward such a state. In terms of literals/voltages, we want them to evolve toward the 'true' value, so as to satisfy their gate.

We can accomplish this by letting them follow a *gradient-like* term along their direction in the phase space, weighted by both the short- and long-term memory variables, namely according to how long the gate, to which that literal belongs, has been unsatisfied.

On the other hand, when the literal is close to a value that satisfies the gate, we would like to keep it there. This can be realized by adding a *rigidity* term, with the name indicating that the literal is held 'frozen' (rigid) in that state.

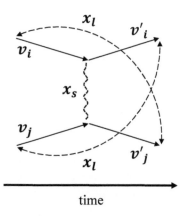

Fig. 8.2 Graphical representation of eqns (8.2), (8.3), and (8.4). The role of the short-term memory x_s (that couples voltages, v_n, in a clause) is to expand the voltage phase space, so that the individual gates always evolve toward a logically consistent solution. The role of the long-term memory, x_l, is to learn the state of un-satisfaction of the given clause. It would then weight the amount of current that needs to be injected into the terminals of such a clause for it to approach the logically consistent solution. Here, the short- and long-term memory variables are coupled *indirectly* via the voltages. See Fig. 9.5 and eqns (9.5) and (9.6) for an example of *direct* coupling between these memory variables.

By putting all this together (Fig. 8.2), we arrive at a possible dynamics for voltages and memory variables (Bearden, Pei and Di Ventra, 2020)

$$C\dot{v}_n = \overbrace{\sum_m x_{l,m} x_{s,m} G_{n,m}}^{\text{gradient-like}} + \overbrace{\sum_m (1 + \zeta x_{l,m})(1 - x_{s,m}) R_{n,m}}^{\text{rigidity}}, \tag{8.2}$$

$$\dot{x}_{s,m} = \beta(x_{s,m} + \epsilon)(C_m - \gamma), \quad x_{s,m} \in [0,1] \to \textit{short-term memory}, \tag{8.3}$$

$$\dot{x}_{l,m} = \alpha(C_m - \delta), \quad x_{l,m} \in [1, 10^4 M] \to \textit{long-term memory}, \tag{8.4}$$

where the summation is over all constraints, m, in which the voltage v_n appears.

The two terms $G_{n,m}$ and $R_{n,m}$ in eqn (8.2) are:

$$G_{n,m}(v_n, v_j, v_k) = \frac{1}{2} q_{n,m} \min[(1 - q_{j,m} v_j), (1 - q_{k,m} v_k)], \tag{8.5}$$

which is the (negative) gradient of the clause function (8.1), and

$$R_{n,m}(v_n, v_j, v_k) = \begin{cases} \frac{1}{2}(q_{n,m} - v_n), & C_m(v_n, v_j, v_k) = \frac{1}{2}(1 - q_{n,m} v_n), \\ 0, & \text{otherwise}, \end{cases} \tag{8.6}$$

where $G_{n,m}$ and $R_{n,m}$ equal 0 when the voltage/variable n does not appear in clause m.

I have explicitly introduced the capacitance C in front of the time derivative of the voltages in eqn (8.2) to highlight that their equation of motion can be derived from Kirchhoff's current law (see Section 4.4.1), with the gradient-like and rigidity terms representing the combined effect of the dynamic correction modules and the VCDCGs (cf. eqns (7.12) and (7.13) in Example 7.3). In the following, I will assume $C = 1$, in whatever physical units we choose.

Since in physical systems we cannot have divergences, we choose the memory variables to be *bounded*, with $x_{s,m} \in [0,1]$ and $x_{l,m} \in [1, 10^4 M]$.[12]

The parameters β and α determine the rate of variation for the short-term and long-term memory variables, respectively. In addition, each memory variable has a threshold parameter (γ for the short-term memory, and δ for the long-term memory) with the two parameters restricted as $\delta < \gamma < 1/2$.[13] These parameters are used for evaluating the state of the clause function C_m. In eqn (8.3) we have also added a small, positive parameter, $0 < \epsilon \ll 1$, to remove the spurious solution $x_{s,m} = 0$.[14]

Finally, the rigidity term in eqn (8.2) is also weighted by the long-term memory, $x_{l,m}$, and reduced by a parameter ζ. This parameter can be interpreted as a 'learning rate' (see also Section 10.2). This is because, the more difficult the instances, the more time we expect is needed for $x_{l,m}$ to evolve, so the phase space can be more efficiently explored.

From eqn (8.6), we see that the rigidity term, $R_{n,m}$, will try to hold one literal/voltage whenever it satisfies the clause m it belongs to. Instead, the gradient-like term, $G_{n,m}$, is maximal when its value is 1, namely when the literal v_n needs to vary the most to satisfy the clause m. It is zero when that literal is already satisfying that clause (irrespective of the value of the other two literals).

As I have anticipated, the long-term memory variable weights the gradient-like dynamics, so that it influences the most those clauses that have been harder to satisfy (we could also say that these clauses are more *frustrated*) during the solution search.

Finally, as anticipated, if we define $\mathbf{x} \equiv \{v_n, x_{s,m}, x_{l,m}\}$, the collection of all voltages and memory variables in the system, then the initial-value problem defined by eqns (8.2), (8.3), and (8.4) can be compactly written as in eqns (3.1):

$$\dot{\mathbf{x}}(t) = F(\mathbf{x}(t)); \quad \mathbf{x}(t = t_0) = \mathbf{x}_0, \tag{8.7}$$

where the flow vector field can be read from the right-hand side of eqns (8.2), (8.3), and (8.4).

With the aid of eqns (8.2), (8.3), and (8.4), it should now be even clearer what I have anticipated in Section 7.2.1 (see Fig. 7.7): the transforma-

[12] The upper limit of $x_{l,m}$ is *arbitrary*, and, if chosen appropriately, it is never reached if a solution exists (see (Bearden *et al.*, 2020) and Fig. 9.2). However, it is important to have it, so that *all* variables in the system are bounded, and there cannot be any divergences during the solution search.

[13] $\gamma < 1/2$ avoids the situation $\dot{x}_{s,m} = 0$, when $C_m = 1/2$. Assigning δ less than γ allows $x_{l,m}$ to grow, even when $\dot{x}_{s,m} = 0$ and $C_m \neq 0$.

[14] In addition, ϵ serves as a 'trapping rate' in the sense that the smaller ϵ, the more difficult it is for the system to 'flip' voltages when some C_m begins to grow larger than γ.

Physical frustration
Whenever elements of a physical system (e.g., atoms or spins) on a lattice cannot easily satisfy their lowest-energy configuration because of 'conflicting' nearest-neighbor interactions, we say the system is *frustrated*. Clauses that share variables in a combinatorial optimization problem can also be classified as examples of this phenomenon. This is because, whenever one clause is satisfied by a variable assignment, the latter may not satisfy the other clauses. In fact, the larger the 'level of frustration' the more difficult we expect the problem to be (see, e.g., (Pei *et al.*, 2020) for an example of this for the MAX-SAT).

tion of the original *discrete* problem variables into *continuous* ones (voltages) creates a *reduced* sub-space for the voltages that may contain local minima, in addition to those representing the solutions of the problem.

However, the topological landscape of this reduced sub-space, in the presence of memory, is *not* static. Rather it is *dynamically* modified ('deformed') by the memory variables, so that the local minima of the voltage sub-space are transformed into saddle points. As a consequence, the only available trajectory in the *full* phase space (of voltage and memory variables) is the one that leads to the global equilibria, which are *invariant* with respect to the action of the memory variables.

An important remark

It was shown in (Bearden, Pei and Di Ventra, 2020) that eqns (8.2), (8.3), and (8.4) satisfy the following mathematical properties:

(i) The system described by these equations is *dissipative*, namely the volume of any initial set in the phase space contracts under the flow vector field (cf. Section 3.8).

(ii) The dynamics are *bounded* by a positive invariant compact set (see Sections 3.9 and 3.10 for a definition of 'invariant compact' set).

(iii) The dynamics terminate *only* when the system has found the solution to the 3-SAT problem (if it exists). This guarantees a correspondence between the fixed points of the dynamics and the solutions of the 3-SAT problem, and *absence* of local minima.

(iv) The *basin of attraction* of the equilibrium points (solutions) contains a large hypercube in the voltage space. Once the trajectory has entered this region, the dynamics are guaranteed to converge to a solution.

(v) The dynamics do *not* support *periodic orbits* in the voltage dynamics (if a solution exists).

(vi) The flow vector field is *not topologically transitive* (cf. Section 3.9).

(vii) The previous results imply *absence of chaos* (à la Devaney) (cf. Section 3.9), namely the dynamics described by eqns (8.2), (8.3), and (8.4) are *integrable* (cf. Section 3.10).

(viii) The *continuous-time* dynamics reach the fixed points in a physical time that scales with problem size, N, as $O(N^{\theta})$ with $\theta \leq 1$. I will demonstrate this explicitly in Section 12.3.1.[a]

---

[a]Note that this result is valid for *continuous* time, not necessarily for *discrete* time, which is the relevant one for numerical solutions.

Having introduced *one* possible set of ODEs for DMMs to solve 3-SAT instances, I can now discuss how we can *numerically* solve such equations.

8.3 Stiff equations

Let me first note an 'issue' with eqns (8.2), (8.3), and (8.4). Since we have different rates for the dynamics of the short- and long-term memory variables compared to the rate of variation of the voltages, the flow vector field, F, defined by these equations, has very *different* time scales/characteristic frequencies. In fact, the short-term memory variables vary much more slowly than the voltages, and the long-term memory variables even slower.

If we look at this from the phase-space perspective, then the critical points of the flow vector field will display unstable directions with a curvature that is very different from that of the stable directions (see also discussion in Section 12.1 and Fig. 12.6).

Therefore, the corresponding ODEs are what we call *stiff* (Sauer, 2018). This is not a minor issue if we want to integrate *numerically* these equations. In fact, for stiff equations much more care needs to be put in choosing the appropriate methods of integration.

8.4 From continuous-time dynamics to discrete maps

Suppose we want to solve eqns (8.7) *numerically* with given initial conditions. We then discretize continuous time into *time steps*, t_i, with i running from 0 till the time we decide to stop the integration of the ODEs.

We can choose to divide these time steps uniformly or not, namely with a *time interval*, Δt, that is either constant over the entire simulation, or varying, at each time step, according to some procedure.[15] Irrespective, at any given time step t_i, the time derivative that appears on the left-hand side of such equations can be approximated as:

$$\dot{\mathbf{x}}(t_i) \rightarrow \frac{\mathbf{x}(t_i + \Delta t) - \mathbf{x}(t_i)}{\Delta t}. \tag{8.8}$$

Stiff ODEs
An ODE is called *stiff* when it involves two or more, quite distinct, time scales of its dynamical variables (Sauer, 2018). As a rule of thumb, stiff equations require *implicit* methods of integration to closely follow the correct solution in time. The ODEs of DMMs do not: *explicit* methods are sufficient for the solution of combinatorial optimization problems (with, of course, a 'reasonable' time step; see Section 8.6 and Chapters 9 and 10). The reason can be traced to the topological nature of the solution search (see Chapters 11 and 12).

Stiff ODEs and problem instances
It is important to stress that the stiffness of the ODEs of DMMs is an *intrinsic* property of such equations. It is *not* related to the choice of problem instance to solve. In fact, if a numerical integration method for those equations works well for a certain class of problems, it would work well for all the instances of such a class, whether 'easy' or 'hard'.

[15]This is known as an *adaptive* integration scheme (Sauer, 2018).

With this approximation, eqns (8.7) would have the following *discrete time* representation (the question mark is not a typo! See Section 8.4.1):

$$\dot{\mathbf{x}}(t) = F(\mathbf{x}(t)) \rightarrow \mathbf{x}(t_i + \Delta t) = \mathbf{x}(t_i) + F(\mathbf{?})\,\Delta t, \qquad (8.9)$$

with the same initial condition:

$$\mathbf{x}(t = 0) = \mathbf{x}(t_0). \qquad (8.10)$$

Equations (8.9) (one equation per variable), with their initial conditions (8.10), define a *discrete map*, or simply a *map*.

Maps

In Mathematics, we typically call *map* any function between two sets, A and B, that relates every element of A to exactly one element of B (cf. Section 1.5). In the context of dynamical systems, instead, a 'map' indicates any *discrete dynamical system*, namely one whose dynamics vary in discrete 'time' (steps). This is what I mean by 'maps' in this book.

Major differences between continuous dynamics and their discrete maps

It should not come as a surprise that a discrete map may introduce *substantial* differences compared to the continuous-time dynamics it is attempting to approximate. In particular, these differences affect two important aspects: the *set of critical points* and the *basin of attraction* of equilibria (see Sections 3.7 and 3.9, respectively, for the definition of these terms).

The first issue is related to the fact that a discrete map may have a *larger* set of critical points than the original, continuous-time dynamics (Stuart, 1994). I will defer the discussion of this issue to Chapter 12, Section 12.7.3, because it is directly related to the *topological* properties of the phase space.[16]

The second issue is related to the fact that by increasing Δt, the basin of attraction of the equilibrium points (assuming they are preserved by the map) *shrinks* for the discrete map (Stuart, 1994). This is not a minor issue because it could determine how small Δt has to be chosen in order for the *simulations* of the ODEs of DMMs to still solve the original problem with increasing size. I will discuss this issue in Section 8.6.

[16]Different numbers (and indexes) of critical points of the flow vector field may be related to different topologies of phase space (Fomenko, 1994).

After the discussion of the main differences between continuous dynamics and their discrete maps, let us now discuss some methods of integration. Notice first that in eqns (8.9) I have put a question mark in the argument of the flow vector field, F, on purpose. This is because we have different choices to approximate such a quantity.

8.4.1　Explicit vs. implicit methods

For the purpose of this book, these choices can be classified into two main groups that lead to what are called *explicit* and *implicit* methods of integration (Sauer, 2018). To better understand what they are, let me give you an example of each.

Example 8.1

Forward Euler method

Suppose we choose the simplest approximation to eqns (8.7):

$$\dot{\mathbf{x}}(t) = F(\mathbf{x}(t)) \rightarrow \mathbf{x}(t_i + \Delta t) = \mathbf{x}(t_i) + F(\mathbf{x}(t_i))\Delta t, \quad \textit{forward Euler.}$$
$$(8.11)$$

This approximation is known as *forward Euler*, and it is indeed the simplest method we can think of.

With this choice, we have a *closed* form for calculating the value of the state $\mathbf{x}$ at the time step $t_i + \Delta t \equiv t_{i+1}$, in the sense that all I need is information about the *previous* time step t_i. In other words, I just need to compute the flow vector field F at the previous time step, to obtain the state at the next time step.

We call a method like this *explicit*, because we have an 'explicit' knowledge of the value of the next step in the integration procedure.[17]

[17] In other words, the knowledge of the state at the next iteration time step is known in *closed form*. No additional information is required.

Instead, let us now consider a slightly different approximation.

Example 8.2

Backward Euler method

Let us write the following map:

$$\dot{\mathbf{x}}(t) = F(\mathbf{x}(t)) \rightarrow \mathbf{x}(t_i + \Delta t) = \mathbf{x}(t_i) + F(\mathbf{x}(t_i + \Delta t))\Delta t, \quad \textit{backward Euler,}$$
$$(8.12)$$

which simply replaces the evaluation of the flow vector field, F, at the *previous* time step t_i, as in the forward Euler, with its evaluation at the *next* time step, $t_i + \Delta t \equiv t_{i+1}$. This map is known as the *backward Euler method*.

This deceptively simple-looking change creates, however, a fundamental difference with respect to the forward Euler method I discussed previously. Now the value of the state at the time step t_{i+1} is *not* written in a closed form. In fact, to find it, I need to solve the following equations for the unknown $\mathbf{x}(t_{i+1})$:[18]

$$\mathbf{x}(t_{i+1}) - F(\mathbf{x}(t_{i+1}))\Delta t - \mathbf{x}(t_i) = 0. \qquad (8.13)$$

[18] Notice that eqns (8.13) constitute a *system* of equations, one for each component of the vector $\mathbf{x}$. We then need to find the *roots* of such a system of equations, given $\mathbf{x}(t_i)$.

If the flow vector field, F, in eqns (8.13) is a *linear* function of $\mathbf{x}(t_{i+1})$ (e.g., $F(\mathbf{x}(t_{i+1})) = \alpha\mathbf{x}(t_{i+1}) + \beta$, with α and β some constants), then such equations are straightforward to solve. This can be easily checked by direct substitution of this linear function into (8.13).

Unfortunately, as I have amply discussed in this book *all* ODEs of DMMs are *non-linear*. Therefore, eqns (8.13) require finding the roots of a system of (polynomial) equations in $\mathbf{x}(t_{i+1})$ (one polynomial equation per dynamical component of $\mathbf{x}$).

A method like the backward Euler is called *implicit*, since the knowledge of the state at the next time of the iteration is *not* known in closed form.

Of course, more advanced strategies/techniques of integration can be devised (see, e.g., Section 8.5, and (Sauer, 2018) for a more in-depth discussion). However, the methods of integration for the ODEs I will consider in this book can all be grouped into the two main classes I have exemplified previously, and which we classify here as:

Fixed-point iteration

Consider an equation in one variable of the type $f(x) = x$, where f is some function. The *fixed-point iteration* procedure starts from a guess x_0, and iteratively evaluates the next step, x_1, by setting $x_1 = f(x_0)$, and so on, namely $x_{i+1} = f(x_i), \forall i = 0, 1, \dots$. If the procedure converges to a number $x = r$, then this number is a *fixed point* of the equation (cf. eqn (3.14)). The same procedure can be extended to functions of several variables.

Newton-Raphson's method

Suppose you want to find the roots of the function f, namely the solutions of $f(x) = 0$. Start from a guess x_0 and compute the derivative of f at that point: $f'(x_0)$. The first-order Taylor expansion of the function at that point is then $f(x) = f(x_0) + f'(x_0)(x - x_0)$. Since we are looking for the roots of f, this expansion gives the first iteration of the *Newton-Raphson's method*: $x \equiv x_1 = x_0 - f(x_0)/f'(x_0)$. Iterating this procedure we obtain $x_{i+1} = x_i - f(x_i)/f'(x_i), \forall i = 0, 1, \dots$. The convergence of this method is typically faster than the one of the fixed-point iteration. However, for multivariate equations of the type (8.13), it requires the inversion of the Jacobian matrix. See (Sauer, 2018) or (Press *et al.*, 2007) for more details.

Explicit and implicit methods for ODEs integration

An *explicit* method to integrate the ODEs (8.7), from time step t_i to time step t_{i+1}, only requires the evaluation of the flow vector field, F, in terms of *known* quantities at the *previous* time step, t_i. The state $\mathbf{x}(t_{i+1})$ at time step t_{i+1} is fully specified by the knowledge of quantities evaluated at t_i.

An *implicit* method, instead, requires the evaluation of the flow vector field, F, in terms of *unknown* quantities at the *next* time step, t_{i+1}. The state $\mathbf{x}(t_{i+1})$ at time step t_{i+1} is *not* fully specified by the knowledge of quantities at t_i, and requires an additional procedure to determine the state at t_{i+1}. See, e.g., (Sauer, 2018) or (Press *et al.*, 2007).

Root-finding procedures

The additional procedure I am referring to could be, e.g., the *fixed-point iteration* or the *Newton-Raphson's method* to find the roots of equations of the type (8.13). The thorough description of these methods is beyond the scope of this book, and in the margin note I provide just their bare-minimum description. The interested reader can consult either (Sauer, 2018) or (Press *et al.*, 2007) for a discussion of these numerical techniques and examples of their applications.

Here, I just mention that the results in Figs. 7.18 and 7.20 in Chapter 7, those in Fig. 8.3, and those of Figs. 11.2 and 11.6 in Chapter 11 have all been generated using *implicit* methods (with the Newton-Raphson's method for root finding).

Numerical solution of the subset-sum problem

In Fig. 8.3, I report the solution of the *subset-sum problem* (that I defined in Section 2.3). Recall that this problem seeks to find a subset of N integers such that their sum equals another integer, say s.

For this NP-complete problem, the hardest cases correspond to the number of elements, N, in the set equal to the precision (in number of bits), p, required to represent each element (Arora and Boaz, 2009).

When $N = p$, no polynomial algorithm is known. In fact, in this case the standard algorithm for its solution follows the strategy I have discussed in Section 2.3, where one checks the sums of all possible subsets of the given set, until one is found that sums to the given integer s.

An implementation of this algorithm is shown in Fig. 8.3 and, as expected, it diverges exponentially for these hard cases.[19]

In the same figure, I report *simulations* of a DMM solving the same instances. The DMM circuit is shown in the inset of Fig. 8.3, and it has been designed following a strategy similar to the one in the Example 7.2 for factorization (see (Traversa and Di Ventra, 2017) for more details).

The ODEs employed are the ones of experimentally realizable SOLGs, as those I have discussed in Chapter 7, using again an implicit method and the Newton-Raphson's procedure for root finding (see (Traversa and Di Ventra, 2017) for more details of these equations).

The simulations have been done by employing a sequential MATLAB implementation of the equations of motion of the DMMs running on a state-of-the-art single core (namely no parallelization has been employed) (Di Ventra and Traversa, 2018a). For direct comparison, also the traditional algorithm has been implemented in MATLAB and run on the same core.

As I have anticipated in Section 3.2, the *memory* (RAM) overhead of integrating numerically ODEs scales *linearly* with the number of *physical* variables (voltages and internal memory variables) of the system. In turn, this number for the subset-sum problem—as formulated in (Traversa and Di Ventra, 2017)—is proportional to $Np = N^2 = p^2$.

Therefore, the *memory* requirement for the DMM simulations of the subset-sum problem (with the circuit in Fig. 8.3), scales *quadratically* with the size, N, of the problem (when $N = p$). The wall time to solution, instead, fits well with a *polynomial* of approximate degree four, up to the instance sizes tested.

An important remark

Of course, these numerical results do not prove that the numerical algorithm is polynomial for *all* sizes and *all* instances. Nonetheless, they demonstrate the great advantage of *simulating* the ODEs

---

MATLAB
MATLAB (which stands for 'matrix laboratory') is a programming language and a computing environment developed by MathWorks, Inc.

[19]Notice that the exponential could be 'reduced' to $2^{N/2}$ if we *stored* partial sums of the computation in memory. However, in that case, the memory required would scale as $2(p + \log_2 p)2^{N/2}$ (Di Ventra and Traversa, 2018a). Either way, for these hard cases, $N = p$, the computation and/or the memory requirements would still diverge exponentially with problem size.

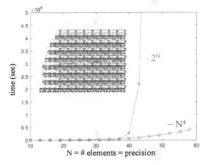

Fig. 8.3 Simulation wall time of the ODEs of DMMs (curve scaling as $\sim N^4$) solving the subset-sum problem vs. wall time of a traditional algorithm (which scales exponentially) for the same problem. The ODEs are those reported in (Traversa and Di Ventra, 2017). Both the ODEs simulations and the traditional algorithm are implemented in MATLAB and run on the same state-of-the-art single core. The instances of the problem correspond to the worst cases, namely when the number, N, of elements in the set is equal to the precision (number of bits), p, per element. The inset shows an example of a small self-organizing circuit (for $N = p = 9$) used to represent this problem. Reprinted with permission from (Di Ventra and Traversa, 2018a).

Local truncation error
Consider the eqns (8.7). The *local* truncation error of a numerical method to solve such equations is the error made on a *single step*, i, of the integration, by taking the previous solution approximation as the starting point. If $\mathbf{x}_{i+1}$ is the *approximation* at step $i+1$, while $\mathbf{x}_{\text{exact}}(t_{i+1})$ is the *exact* solution at the time step $i+1$, obtained by starting from the *approximate* initial condition $\mathbf{x}_i$, the local truncation error at step $i+1$ is:

$$e_{i+1} = |\mathbf{x}_{i+1} - \mathbf{x}_{\text{exact}}(t_{i+1})|.$$

Global truncation error
The *global* truncation error at a step i of the integration procedure is the *accumulated* error from the first i steps:

$$g_i = |\mathbf{x}_i - \mathbf{x}_{\text{exact}}(t_i)|,$$

where both approximate and the exact solutions start from the initial value $\mathbf{x}_0 = \mathbf{x}(t_0)$. The global truncation error at a given iteration *accumulates* all local truncation errors from previous iterations.

Function evaluations
It is the number of times the function F in eqns (8.7) needs to be computed in a numerical method. See, e.g., (Sauer, 2018) or (Press *et al.*, 2007) for more discussion of these notions.

[20] Mostly forward Euler, with at most an adaptive time step.

[21] Indeed, if that were the case, explicit methods would *not* be enough.

of DMMs versus traditional algorithms. If the same DMM were built in *hardware*, it would scale as $O[p(N + \log_2(N - 1))]$ in space (number of self-organizing logic gates employed), and $O((N + p)^2)$ in convergence time with size N and precision p, and since all voltages and memory variables are bounded, the energy expenditure is also polynomially growing with size (Traversa and Di Ventra, 2017).

The previous example shows that indeed we can efficiently simulate the ODEs of DMMs using *implicit* methods of integration. In fact, as I mentioned in Section 8.3, these ODEs are *stiff* (due to several dynamical time scales).

Therefore, implicit methods are supposed to be *the* methods of choice for such equations (Sauer, 2018; Press *et al.*, 2007). In other words, *explicit* methods of integration, like the forward Euler of Example 8.1, should not work properly for stiff equations.

In the remainder of this Chapter, I will show that this is not the case for DMMs. In fact, all the numerical examples of applications I will provide in Chapters 9 and 10 have been done with *explicit* methods.[20]

The reason why even explicit methods for the particular (stiff) equations of DMMs are good enough to provide the solution to the problems they are designed to solve will become clear in Chapters 11 and 12. I just anticipate here that the important point is *not* to reproduce with extreme accuracy the state trajectory $\mathbf{x}(t)$.[21] Rather, the numerical method needs to *preserve* the structure of *critical points* in the phase space (cf. Section 12.7.3).

8.5 Explicit Runge–Kutta methods

Let us refer again to the eqns (8.7). A general *explicit* numerical integration method for such equations can be formulated with the following map, also known as a *Runge-Kutta* time-step integration from time step t_i to time step t_{i+1} (Sauer, 2018):

$$\mathbf{x}_{i+1} = \mathbf{x}_i + \Delta t \sum_{j=1}^{q} \omega_j k_j, \quad \mathbf{x}_0 = \mathbf{x}(t_0), \quad \textit{explicit Runge-Kutta methods},$$

$$(8.14)$$

where

$$k_j = F(\mathbf{x}_i + \Delta t \sum_{l=1}^{j-1} \lambda_{jl} k_l).$$

$$(8.15)$$

We have several choices for the dimension q of the vector $\vec{\omega}$ (which also defines the *order* of the method), and the matrix elements λ_{jl}, leading to different explicit methods. I will now describe only three, which are the ones we have applied so far in the simulations of DMMs (in fact, we have used mostly one: forward Euler).

I will also enumerate (without proving) the main properties of each of these methods in terms of (local and global) *truncation errors*, namely the errors that originate from *approximating* the exact solution (see margin note for the definition of these errors). The derivation of these results can be found in specialized books, such as (Sauer, 2018). In the same books one can find the whole zoo of different numerical methods (both explicit and implicit) that could be equally applied.[22]

[22]In fact, some other integration method may provide an even better numerical scalability for combinatorial optimization problems.

8.5.1 Forward Euler

If we choose $q = 1$ and $\omega_1 = 1$ then we obtain eqns (8.11), which describe the *forward Euler* method we have already discussed in Section 8.1.

The *local truncation error* of this method is $O((\Delta t)^2)$, while the *global truncation error* is bounded by an exponential, namely the global error of this method at time step i is:

$$g_i = |\mathbf{x}_i - \mathbf{x}_{\text{exact}}(t_i)| \leq \frac{c_E \Delta t}{L}(e^{L(t_i - t_0)} - 1), \tag{8.16}$$

where $c_E > 0$ is a constant (related to the absolute value of the second time derivative of $\mathbf{x}$), and L is a *Lipschitz constant*.

Since in eqn (8.16) the time interval, Δt, appears linearly, we say that the forward Euler method is *first order*.

We expect that, by increasing the size of the combinatorial optimization problem to solve, on average, it should take a longer time to find a solution. Therefore, from eqn (8.16) we would expect the time interval Δt needs to decrease *exponentially* as the time of the simulation progresses (problem size increases), in order to control the exponential bound on the global error.

As I will show in Section 8.6 this is *not* the case. In fact, we find empirically (at least for all the instances considered so far) that it is enough for that time interval to decrease as a *power law* with increasing size of the problem.

Lipschitz continuity
A function $f(x)$ is said to be *Lipschitz continuous* if there exists a constant L (*Lipschitz constant*) such that

$$|f(x) - f(y)| \leq L|x - y|,$$

where $|\cdot|$ denotes a metric (possibly different) on the domain and codomain spaces of the function (cf. discussion in Section 6.3.1, and see, e.g., (Arfken *et al.*, 1989)). The ODEs of DMMs I discussed in Chapter 7 are Lipschitz continuous. Those defined in Section 8.2.5 are not. Therefore, particular care has to go into analytically deriving the properties I have outlined in the same Section (see (Bearden, Pei and Di Ventra, 2020)).

8.5.2 Trapezoid

Let us choose $q = 2$, $\vec{\omega} = (\frac{1}{2}, \frac{1}{2})$, and $\lambda_{21} = 1$. This defines the following map:

$$\mathbf{x}(t_{i+1}) = \mathbf{x}(t_i) + \frac{\Delta t}{2}[F(\mathbf{x}(t_i)) + F(\mathbf{x}(t_i) + \Delta t F(\mathbf{x}(t_i)))], \quad trapezoid, \tag{8.17}$$

which is called the *trapezoid method* of integration.

Since, from the Euler method, $\mathbf{x}(t_i) + \Delta t F(\mathbf{x}(t_i)) = \mathbf{x}_{i+1}$, we see that the trapezoid method improves on the Euler method by evaluating the *average* of the flow vector field at the time t_i and the value of the flow vector field the Euler method would have given at time t_{i+1}.

The *local* truncation error of this method is $O((\Delta t)^3)$, and the *global* truncation error at time step i is:

$$g_i = |\mathbf{x}_i - \mathbf{x}_{\text{exact}}(t_i)| \leq \frac{c_T (\Delta t)^2}{L}(e^{L(t_i - t_0)} - 1), \qquad (8.18)$$

where $c_T > 0$ is a constant.

We see from (8.18) that the trapezoid method is *second order*. This means that for the same (small) Δt it should yield a better value for the trajectory $\mathbf{x}(t)$ compared to the forward Euler method.

However, note from eqns (8.17) that the trapezoid method requires, for *each* step of integration, *two* function evaluations vs. the *single* function evaluation of the forward Euler, eqn (8.11). Irrespective, as in the forward Euler, we would expect the global error to increase exponentially with problem size, which is again not what we find (Section 8.6).

8.5.3 Runge-Kutta 4th order

Finally, let us consider $q = 4$, $\vec{\omega} = (\frac{1}{6}, \frac{1}{3}, \frac{1}{3}, \frac{1}{6})$, and the matrix

$$\lambda = \begin{pmatrix} 0 & 0 & 0 & 0 \\ \frac{1}{2} & 0 & 0 & 0 \\ 0 & \frac{1}{2} & 0 & 0 \\ 0 & 0 & 1 & 0 \end{pmatrix}. \qquad (8.19)$$

This choice defines the following map:

$$\mathbf{x}(t_{i+1}) = \mathbf{x}(t_i) + \frac{\Delta t}{6}[s_1 + 2s_2 + 2s_3 + s_4], \quad \textit{Runge-Kutta 4th order}, \qquad (8.20)$$

where

$$\begin{aligned} s_1 &= F(\mathbf{x}(t_i)), \\ s_2 &= F(\mathbf{x}(t_i) + \frac{\Delta t}{2} s_1), \\ s_3 &= F(\mathbf{x}(t_i) + \frac{\Delta t}{2} s_2), \\ s_4 &= F(\mathbf{x}(t_i) + \Delta t s_3). \end{aligned} \qquad (8.21)$$

These relations define the (explicit) *4th order Runge-Kutta* method.

Although cumbersome to prove, it can be shown that the *local* truncation error of this method is $O((\Delta t)^5)$, while the *global* truncation error is $O((\Delta t)^4)$. This method is then—and its name clearly gives it away—*fourth-order*.

It is considerably more accurate, for the same Δt, than both the forward Euler and the trapezoid methods. However, it requires *four* function evaluations per integration step.

8.6 Numerical comparisons

Let us then see how these explicit methods fare against each other in the solution of some combinatorial problems. For this comparison, we consider the MemComputing eqns (8.2), (8.3), and (8.4) designed for the 3-SAT. Let us choose a set of 'hard' instances for which we know the solution.[23]

In Section 9.1.1, I will show how different solvers (algorithms) compare on these instances as a function of problem size. This will also confirm that, indeed, the instances are 'hard' for traditional solvers, in the sense that, for those solvers, the time to solution grows *exponentially* as a function of problem size. The MemComputing solver instead scales with a *power law* up to the sizes tested.

In this Section, instead, I am only interested in how the solution of the 3-SAT instances is affected by the magnitude of the integration step, Δt, for the three explicit Runge-Kutta methods I discussed previously.

[23]A detailed account of how these 3-SAT instances are generated can be found in (Hartmann and Rieger, 2004) or (Bearden *et al.*, 2020).

8.6.1 A solvable-unsolvable transition

Once we settle on a given integration scheme, e.g., the forward Euler of eqns (8.11), and a set of problem instances, we expect that it would be more difficult to solve all instances with increasing time interval Δt. This is indeed the case, as it is evident from Fig. 8.4, where the solution of 3-SAT instances with 10^4 variables at clause-to-variable ratio $\alpha_r = 8$ (so there are 80,000 clauses) has been obtained by numerically integrating eqns (8.2), (8.3), and (8.4) with the forward Euler method, and different integration time intervals Δt. The plot shows 100 instances, each solved with 100 different initial conditions of the ODEs.

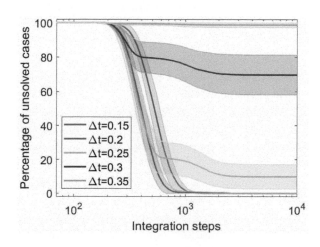

Fig. 8.4 Solution of 3-SAT instances with 10^4 variables vs. integration steps, at clause-to-variable ratio $\alpha_r = 8$, obtained by numerically integrating eqns (8.2), (8.3), and (8.4) with the forward Euler method, for different integration times Δt. The results correspond to 10^4 solution trials (100 instances and 100 different initial conditions per instance) for each Δt. The number of unsolved cases decays as integration steps increase, with the solid line representing the result for all 10^4 trials, and the shaded area representing one standard deviation over the 100 instances. Reprinted with permission from (Zhang and Di Ventra, 2021).

Figure 8.4 clearly shows that, at sufficiently small Δt, *all* instances are solved, *irrespective* of the initial conditions chosen. Increasing Δt, the percentage of unsolved instances reaches a plateau as a function of integration steps.

The value of this plateau only depends on the integration time step Δt. Similar results are found using the other two explicit methods: trapezoid (Section 8.5.2) and Runge-Kutta 4th order (Section 8.5.3).

The basin of attraction

As I have anticipated in Section 8.4, it is known that, by increasing Δt, the *basin of attraction* of the equilibrium points of the discrete map *decreases* compared to the basin of attraction of the corresponding points of the continuous-time dynamics (Stuart, 1994).[24]

[24] Assuming, of course, that the map does not change the equilibrium points of the continuous dynamics.

An estimate of the relative *size* of the basin of attraction of the DMMs's maps is then represented by the fraction of solved instances. It is clear from Fig. 8.4 that this fraction is roughly the same for all instances at a given Δt, and decreases with increasing Δt.

I show the estimate of this basin of attraction in Fig. 8.5: its relative size with respect to the ideal ($\Delta t \to 0$) case. The results are similar for all three explicit integration schemes, forward Euler, trapezoid, and Runge-Kutta 4th order. Interestingly, the results are well fitted by sigmoid-like curves, of the type $A = 1/[1 + \exp(-a(\Delta t - b))]$, with a and b fitting parameters.

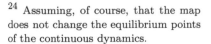
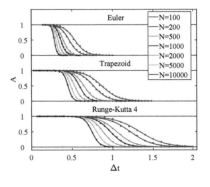

Fig. 8.5 Estimate of the relative size of the basin of attraction, A, as a function of Δt for three explicit integration schemes, and different problem size (number of variables N) and clause-to-variable ratio $\alpha_r = 8$. Each data point is calculated based on 1000 solution trials (100 instances and 10 initial conditions per instance), and the solid curves are fitted using the function *ansatz* $A = 1/[1 + \exp(-a(\Delta t - b))]$, with a and b fitting parameters. Reprinted with permission from (Zhang and Di Ventra, 2021).

[25] To understand this feature, I will provide a physical analogy with *directed percolation* in Section 12.7.5.

8.6.2 The critical integration step

In addition, the curves in Fig 8.5 become sharper as the size of the problem instance increases (namely the number of variables N increases).[25] This indicates that in the 'thermodynamic limit' of $N \to \infty$ we expect a *solvable-unsolvable* 'phase transition' to occur at a 'critical' Δt_c (cf. discussion in Section 4.3.2).

Although we do not have an analytical expression for this transition point, we can estimate from Fig. 8.5 how this 'critical' value, Δt_c, changes as a function of N. We then evaluate it from Fig. 8.5 at the point where the fraction of solved solutions (the fraction of the basin of attraction) is exactly $1/2$: $A(\Delta t_c) = 1/2$. And we do this as a function of problem size (number of variables N), at fixed clause-to-variable ratio.

This quantity is plotted in Fig 8.6 for all three explicit integration methods.

As anticipated, we would have naïvely expected Δt_c to decrease exponentially. This is because our eqns (8.2), (8.3), and (8.4) are *stiff* (Section 8.3). And as I have discussed, stiff equations typically require

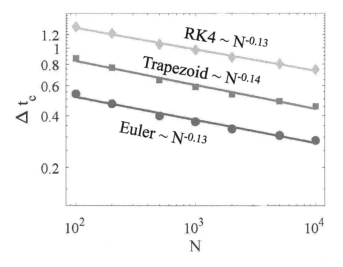

Fig. 8.6 Δt_c determined from Fig. 8.5 at the point where the relative basin of attraction $A(\Delta t_c) = 1/2$. Δt_c shows a *power-law* decay with problem size N. Adapted with permission from (Zhang and Di Ventra, 2021).

implicit methods to control the unavoidable (and large) numerical errors that accumulate in the *trajectory* $\mathbf{x}(t)$ in phase space.

Instead, Fig. 8.6 shows a *power-law* scaling with problem size N. This means that the integration time interval only needs to decrease as a power law with increasing problem size, for the (numerical) DMM to be able to solve the given instances.[26]

An important remark

Note that the power-law exponent may change not only with integration scheme, but also with problem type (see Section 9.1.1 for other examples of different clause-to-variable ratios with forward Euler and adaptive time step). Also, this power-law has been verified for *some* (although a large number of) problem instances. Therefore, it is not at all obvious it generalizes to *all* instances and *all* problems.

[26] And, as I have explained in a margin note of Section 8.5, once the Δt is comparable to the round-off error, a transition to a completely different (most likely, exponential) scalability is expected, due to a 'catastrophic' accumulations of numerical errors. The value of Δt at which this additional transition occurs should depend on the processor used, the choice of ODEs, and possibly the problem class considered.

From Fig 8.6 we also see that, as expected, the integration methods of higher order support a larger Δt_c per problem size. In other words, higher order methods are 'more forgiving' for larger Δt.

Finally, with the Δt_c empirically determined from Fig 8.6 we can now compare the *scalability* of the three different explicit methods of integration as a function of the problems size. This is shown in Fig. 8.7. For the range considered in Fig. 8.7 ($N \in [10^2, 10^5]$), Runge-Kutta 4th order requires the least *number of integration steps*. However, in terms of *function evaluations* (cf. discussion in Section 8.4.1), the forward Euler method is the most *efficient*, because it requires only one function

Fig. 8.7 Scalability curves for three different explicit integration schemes—forward Euler, trapezoid, and Runge-Kutta 4th (RK4) order—as a function of number of variables N, for the clause-to-variable ratio $\alpha_r = 8$. All three schemes show a *power-law* scaling ($\sim N^\alpha$) in both the median and 90th quantile. The values of the scaling exponents (α) are 0.34 (Euler 50%), 0.50 (Trapezoid 50%), 0.44 (RK4 50%), 0.24 (Euler 90%), 0.56 (Trapezoid 90%), 0.30 (RK4 90%). The scaling is the same, for each integration scheme, in terms of either number of steps or function evaluations; only the pre-factor changes. Reprinted with permission from (Zhang and Di Ventra, 2021).

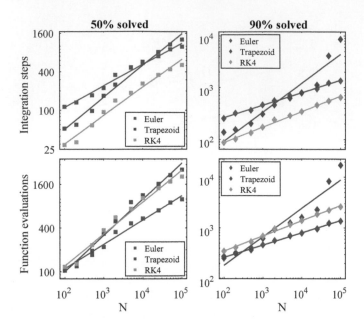

evaluation at every time step, versus two and four of the trapezoid and Runge-Kutta 4th order, respectively.

The same Fig. 8.7 also shows that the forward Euler has a slightly better scaling compared to the other two explicit methods. I will discuss in Section 12.7.5 a possible reason for this behavior. I finally conclude by noting that similar results have been obtained for other instances at different clause-to-variable ratios; see (Zhang and Di Ventra, 2021).

8.7 Chapter summary

- In this Chapter, I have first provided one possible strategy to design DMMs whose equations of motion are more conducive to numerical simulations.

- I have then discussed how to solve them with either explicit or implicit methods of integration.

- Despite *all* ODEs of DMMs are *stiff* (they have several, different time scales), implicit methods do not seem to be necessary. *Explicit* methods (even the forward Euler one) offer enough control of the numerical errors for what concerns the solution search.

- In fact, the integration time interval, Δt, needs to decay as a *power-law* as a function of problem size, in order for the DMM solver to maintain the same level of accuracy with increasing problem size.

- I will show in Chapters 11 and 12 that this robustness against numerical errors is related to the *topological* nature of the solution search.
- So long as the numerical method preserves the topological structure of the phase space, the DMM solver will be resilient against reasonable numerical errors.

Application to combinatorial optimization problems

We cannot solve our problems with the same level of thinking we used when we created them.
Albert Einstein (1879–1955)

Problems always appear to be intractable until we discover efficient algorithms for solving them.
M. R. Garey (1945–) and D. S. Johnson (1945–2016)

Takeaways from this Chapter

- Numerical simulations of digital MemComputing machines (DMMs) for a wide collection of hard combinatorial optimization problems on standard computers.
- These simulations typically outperform traditional algorithmic approaches by orders of magnitude, as the problem size increases.

In Chapter 8, I have shown that the equations of motion of DMMs can be efficiently *simulated* on our traditional computers. In fact, I have provided empirical evidence that, despite being *stiff*, namely with several time scales (Section 8.3), *explicit* methods of integration seem to be sufficient. In particular, I have shown that, for the problem instances considered, the integration time interval, Δt, does not need to decay exponentially with problem size: a *power-law* decay is enough to maintain the same level of accuracy of the DMM solver.[1]

I will provide a better understanding of this robustness against *numerical errors* (as well as *physical noise*) in Chapters 11 and 12. In this Chapter, instead, I will just take advantage of these findings, and show

---

[1] In this Chapter I will use the word 'solver' to indicate an algorithm implemented in software.

other *numerical* results of DMM solutions of 'hard' instances of several combinatorial optimization problems, as a function of the problem size.[2]

Whenever possible, I will compare with traditional algorithms on the same instances. Indeed, the 'hardness' of the instances chosen is only determined by the difficulty with which state-of-the-art (traditional) algorithms tackles them. Therefore, no claim on 'worst-case' scenarios can be made from the results I will present in this Chapter.

Nevertheless, in all cases considered—overall encompassing *hundreds of thousands* of instances for a wide variety of problems (Traversa and Di Ventra, 2017; Di Ventra and Traversa, 2018a; Traversa *et al.*, 2018; Traversa and Di Ventra, 2018; Bearden *et al.*, 2018; Manukian *et al.*, 2019; Sheldon *et al.*, 2019; Bearden *et al.*, 2019; Bearden *et al.*, 2020; Manukian *et al.*, 2020; Zhang and Di Ventra, 2021; Pei and Di Ventra, 2021)—the *simulations* of DMMs (typically done with *forward Euler*, eqn (8.11), on a *single* core/thread with no parallization) show great advantages versus traditional algorithmic approaches in finding the solution of these hard problems in terms of speed per problem instance, and scalability with problem size.

In fact, the main limitations of these simulations seem to be the random access memory (RAM) of the processor used (which determines the largest size of the problem instance one can effectively fit in the processor employed), and the actual *wall time* one is willing to wait to obtain a solution.[3]

An important remark
I stress once more that these empirical results *do not* prove the famous NP = P conjecture (Section 2.3.2)! However, they show that for a wide-range of problem instances—that traditional algorithms struggle to solve—the DMM approach offers a powerful alternative.

9.1   Search problems

In Section 8.4.1, I have shown the results of simulations of DMMs for a well-known *search problem*: the subset-sum problem (see Fig. 8.3).[4] In that example I have shown simulations of DMMs's ordinary differential equations (ODEs) using an *implicit* method of integration (Di Ventra and Traversa, 2018a). The simulations compare very favorably with a traditional algorithm, which clearly scales *exponentially* with problem size, even for relatively small instance sizes (see Fig. 8.3). The corresponding DMM's simulations, instead, fit well with a *polynomial* function, up to the sizes considered. Let's see then if such a striking difference persists for other types of problems.

[2]See Sections 2.3.4 and 7.9 for the meaning of 'solving' a problem in the optimization class.

Processor, core, and thread
By *processor* we typically intend the central processing unit (CPU) of a computer hardware. A CPU may have one (single-core CPU) or several *cores* (multi-core CPU), namely several (hardware) processing units that work in parallel (cf. discussion in Section 3.6). Each core can run one or several computational *threads*, namely many tasks at the same time while sharing the resources of the same core (Hennessy and Patterson, 2006).

[3] For instance, the solution of some instances of the subset-sum problem as implemented in (Traversa and Di Ventra, 2017) may take days on a single processor (see, e.g., Fig. 8.3). This is better than the orders of magnitude longer times it would require traditional algorithms, but still not as fast as it could be done in *hardware*!

[4]Recall from Section 2.3.4 that a search problem is a combinatorial problem in which we are searching for an assignment of Boolean variables that solves it (satisfies *all* clauses).

9.1.1   3-SAT

In this Section I will look at the solution of another search problem, the *3-SAT*, for which I have introduced the eqns (8.2), (8.3), and (8.4). I will still employ the 'hard' instances I have used for the numerical analysis in Section 8.6.[5]

These are instances for which we know there is at least one solution, since such a solution has been *planted*. However, the solutions are planted in such a way as to 'fool' known *heuristic* solvers (see Section 2.6 for their definition). We thus expect that if we choose these instances 'hard enough', these heuristic solvers would have a hard time finding the solution.

Solution planting

To *benchmark* optimization solvers, it is common to *plant solutions* in the instance formulation of the problem, in such a way that creating such instances is 'easy', but finding the solution is still 'hard', and yet known in advance (Mezard and Montanari, 2009). For example, instances with planted solutions are typically used in SAT competitions. However, this is not always possible (or done). I will show examples of these cases (with unknown optima) in Sections 9.2.4 and 9.2.5. In addition, all instances in Chapter 10 do not have planted solutions.

An important remark

Of course, once one or more solutions have been *planted* in the formulation of a problem, its complexity may not be necessarily the same as the one of the original 'non-planted case'. However, for all practical purposes, it is enough that the problem instances with planted solutions be typically 'hard' for known algorithms.

This is indeed the case, as it is evident in Fig. 9.1, where I compare the DMM simulations—performed with *forward Euler*, eqn (8.11), with an *adaptive* time step (Bearden *et al.*, 2020)—with two typical algorithms: a *stochastic local-search algorithm* (WalkSAT), and a *survey-inspired decimation procedure* (SID).

Fig. 9.1 Main panel: simulation of DMMs to solve 100 planted-solution instances of 3-SAT per pair of α_r (clause-to-variable ratio) and N (number of variables). The median (typical case) number of integration steps shows a *power-law* scalability as the number of variables, N, grows. The **insets** show *exponential* scalability for a stochastic local-search algorithm (WalkSAT) and a survey-inspired decimation procedure (SID) on the same instances. The parameters b and c for the exponential curves can be found in (Bearden *et al.*, 2020). Reprinted with permission from (Bearden *et al.*, 2020).

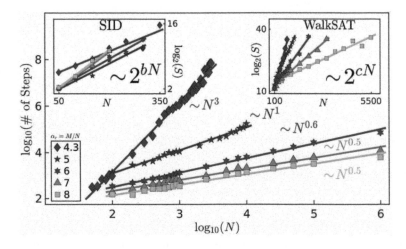

It is not the scope of this book to discuss the details of these two algorithms. I refer the reader to the literature for a full account (Arora and Boaz, 2009; Mezard and Montanari, 2009).

The important point to consider is that these algorithms are representative of the class of *heuristic* algorithms often used for benchmarks and in SAT competitions.

Note that the simulations reported in Fig. 9.1 have been done on a *single* core of a state-of-the-art server, and implemented within interpreted MATLAB (Bearden *et al.*, 2020). No attempt at parallelizing or optimizing the code has been done.

In addition, the hardest cases—corresponding to the clause-to-variable ratio of 4.3 in Fig. 9.1—took about 15 hours per instance to solve. For comparison, if we extrapolated the other two algorithms used in that figure, it would take them longer than the estimated age of the Universe ($> 10^{10}$ years) to solve the same instances.

To further support the advantage of the simulations of DMMs over these algorithms, it is instructive to show other important measures of scalability of the simulations of DMMs.

Average integration time step

As mentioned, the results presented in Fig. 9.1 have been obtained with an *adaptive* time step procedure, not a constant time step Δt, like the one employed in Chapter 8.[6]

In that Chapter, I have shown that the time step to control truncation errors needs to decay as a *power-law* with problem size N in order to have the same percentage of success in the solution of the instances (see Fig. 8.6). Does this result still hold for an adaptive time step?

In Fig. 9.2(a), I show the *average* time step, $\overline{\Delta t}$, of the adaptive procedure. As is evident from the figure, this average time step also scales as a *power law* with problem size (number of variables N, at fixed clause-to-variable ratio).[7]

In other words, as the problem size increases, the average time step decreases with a lower *polynomial* bound, rather than an exponential one.

[6] An adaptive time step helps to reduce the truncation errors only when necessary (e.g., when the voltage variables vary too quickly). This speeds up the actual CPU time to solve the instances.

[7] Of course, with different power-law exponent, compared to the results obtained in Fig. 8.6 with a constant time step.

Different measures of time to solution

In addition to integration steps as a measure of the 'time' it takes the numerical simulation to find a solution, we can also check other measures of *time to solution* (TTS) (cf. discussion in Section 2.2.1).

For instance, from the simulations we can extract the integration variable, t, and determine how it scales with problem size when the solution is found.[8] The median of this quantity is plotted in Fig. 9.2(b). Like the number of integration steps, it also shows a *power-law* (albeit with a smaller exponent) increase with problem size.

Another measure of TTS is the CPU time (measured in seconds by

[8] In some sense, this is the 'closest' quantity we can compare with the actual physical time.

Fig. 9.2 Typical-case analysis for several indicators for the solution of the instances used to generate Fig. 9.1 at $\alpha_r = 4.3$. Each data point is the median value of 100 instances. (a) The average time step, $\overline{\Delta t}$ (arbitrary units). (b) The median TTS for the integration variable, t (arb. units). (c) The CPU time (seconds). (d) The median of the maximum value of the long-term memory, $\mathbf{x}_l$ (arb. units), when the solutions were found. All indicators showcase a *power-law* behavior. Reprinted with permission from (Bearden *et al.*, 2020).

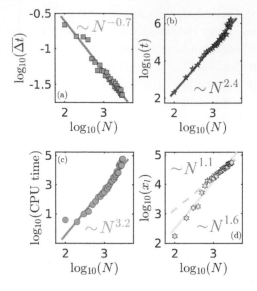

[9]The wall time showcases a similar power-law behavior (Bearden *et al.*, 2020).

MATLAB). This is plotted in Fig. 9.2(c) and shows a *power-law* trend similar to the other measures, with slightly different exponent.[9]

Scalability of the long-term memory and polynomial energy cost

Finally, let me comment on another important quantity: the 'long-term memory', $\mathbf{x}_l$, in eqns (8.4). I have briefly mentioned in Section 8.2.5 that the dynamics of this variable can be chosen so that it never reaches its upper bound (see eqn (8.4)).

In a *hardware* implementation of these machines, this requirement would be related to how the *energy* in the system varies as the problem size increases. In fact, an exponential divergence of such a quantity with problem size would mean that the *practical* realization of these machines would not be 'efficient' (cf. discussion in Section 2.2.2).

This is because memory in a physical system is related to the dynamical behavior of some of its physical properties, e.g., spin polarization, atomic displacements, etc. (Pershin and Di Ventra, 2011*a*). And the dynamics of these physical quantities would require energy consumption.

Note that I am not discussing the voltage variables or the 'short-term memory', $\mathbf{x}_s$, in eqns (8.3) on purpose. The reason is because these variables have bounds whose value is *independent* of the size of the problem. Therefore, if the solution to the problem is found in a *finite* time, with this time bounded by a *polynomial* in the size of the problem, the *energy expenditure* associated to these variables can only grow at most *polynomially* with the problem size (Traversa and Di Ventra, 2017).

Instead, this conclusion is not so straightforward in the case of $\mathbf{x}_l$ for which we chose its bound to be related to the size of the problem.[10]

In Fig. 9.2(d), I show the growth of $\mathbf{x}_l$. In particular, for each instance, we collect the maximum value of $\mathbf{x}_l$, and then find the *median* of those values. Figure 9.2(d) confirms that the typical growth of the *maximum* value of $\mathbf{x}_l$ follows a *power law*. In fact, the power-law fit appears to decrease in exponent at larger problem sizes, approaching a *linear* growth which is in agreement with the approximate linear growth of this variable as dictated by eqn (8.4).

9.1.2 Prime factorization

As an additional example of a search problem, let me show some preliminary results on *prime factorization* obtained by integrating numerically eqns (8.2), (8.3), and (8.4), appropriately modified for this particular problem. In this case, one transforms prime factorization into a SAT problem written in the conjunctive normal form (CNF), similarly to Example 7.2.

This transformation can be done in various ways. For instance, one can generate a CNF formula that contains 2-, 3-, and 4-terminal OR clauses. Equations (8.5) and (8.6) for the gradient-like and rigidity terms, respectively, would then need to be modified to account for the different numbers of literals per clause.

An example of the numerical solution of such equations for prime factorization is shown in Fig. 9.3 for several integers ranging in size from 20 to 47 bits. The integers have been generated so that they have only two primes (they are *semiprimes*), and these primes only differ in size by 1 bit.

[10]We chose the upper bound of $\mathbf{x}_l$ in eqns (8.4) to be proportional to the number of *clauses* M. However, equivalently, we could have related them to the number of variables N, while keeping the clause-to-variable ratio, $\alpha_r = M/N$, constant.

Semiprime numbers
Natural numbers that can be factored by only two primes are called *semiprime* numbers. They are typically the most difficult to factor with existing algorithms, especially if their primes have comparable bit size, e.g., differing by 1 bit (Goldreich, 2001).

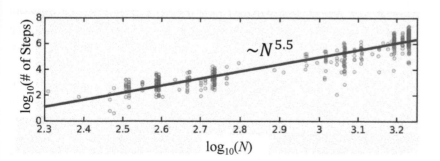

Fig. 9.3 Numerical solution of the ODEs of a DMM that solves prime factorization. Several semiprime numbers ranging in size from 20 to 47 bits (with their primes differing in size by only 1 bit) have been factored. The plot shows number of integration steps vs. the number of logical variables of the corresponding CNF formula. The median number of steps is indicated as a continuous line. Results courtesy of S. Bearden.

Figure 9.3 shows the number of steps to solution vs. the number of variables in the CNF formula. The solution is found whenever the division of the integer to factor by any one of the two numbers gives as remainder zero (hence the machine has found one of the two primes, and, as a consequence, the other can be easily obtained).

[11]The simulations have been done by integrating the DMM's ODEs using the *forward Euler* method (Section 8.5.1) within MATLAB and run on a single processor.

[12]As I will show in Section 12.3.1 using Topological Field Theory, the *hardware* implementation of such a circuit would instead scale *linearly*, or *sub-linearly*, with the number of variables.

[13]If an algorithm is *complete* (Section 2.6), we could even wait until it reports to have found the optimum. However, for most of the (difficult) problem instances of interest, that time could be astronomically large, so it is of little practical utility.

The median of the time steps to solution[11] appear to scale as a *power law* up to the sizes tested. No attempt has been made in optimizing parameters and/or the CNF formula (cf. discussion in Section 13.3.2).

However, note that the degree of such a power law is quite large, hence, computationally demanding (in terms of CPU time) if very large (e.g., RSA size) numbers are to be factored (cf. Example 7.2). Therefore, more work needs to be done to improve the *software* solution of prime factorization for such application.[12]

9.2   Optimization problems

After the discussion on the applicability of the simulations of DMMs to search problems, let us consider the *optimization* ones. Recall from Section 7.9, that for these problems we have two ways to proceed.

- We can assign a *timeout* to the machine, and maximize (or minimize) an *objective function*, subject to some constraints, within that time. Comparisons between different algorithms would then determine which one has found the best assignment of the variables (e.g., the lowest number of unsatisfied clauses) within that time.[13]

- Alternatively, one can determine the time to solution (TTS), i.e., the time (in some unit; see Section 2.2.2) it takes to find the optimum of the problem. Of course, for this task, one has to know the optimum before hand, namely one has to generate instances with *planted solutions* (cf. Section 9.1.1).

I will show results for both procedures. As in Section 9.1, I will start by providing *one* possible realization of DMMs suitable for numerical optimization of 'hard' problems for which we know the optimum in advance ('planted solution').

I will then highlight a variety of results on other optimization problems obtained using similar ideas, but implemented with a MemComputing commercial solver. This collection of results will further reinforce the usefulness of *simulating* DMMs for the solution of combinatorial optimization problems. It will also give us confidence we can employ them in the fields of Machine Learning and Quantum Mechanics, where optimization problems reign supreme (Chapter 10).

9.2.1   Spin glasses

Let us start by considering the following problem: find the *ground state(s)*—the lowest energy state(s)—of a physical system, known as the *Ising spin glass* (Goldenfeld, 1992). This is a model in which a set of *spins*, namely vectors that can acquire two values, either $+1$ or -1, are placed on a lattice and interact with a strength J_{ij}, which is *random*, and is either positive (*ferromagnetic* coupling: parallel spins are favored) or negative (*antiferromagnetic* coupling: anti-parallel spins are favored); see Fig. 9.4.[14]

The energy (cost) function, E, (or Hamiltonian H) of this model is then:[15]

$$E(\mathbf{s}) \equiv H = -\sum_{i>j} J_{ij} s_i s_j, \quad s_i \in \{-1, 1\}, \tag{9.1}$$

where the interaction may involve only *nearest neighbor* spins (the interaction would then be *short ranged*), or any type of interaction between spins (e.g., long range), and $\mathbf{s}$ is a particular spin configuration.

Since the coupling constants J_{ij} are drawn from a *probability distribution*, the energy E defines a *non-convex* landscape in the spin variables with several local minima, in addition to the global optima/ground states (cf. discussion in Section 2.3.4). The problem of finding the ground state[16] is then far from trivial.

By now, however, it should be clear how to proceed and transform this problem into a physical system representing a DMM.

Linear relaxation of the spins

As I have explained in Chapters 6, 7, and 8, the first step is to transform the *discrete* problem defined by eqn (9.1) into a *continuum* one. As I mentioned in the caption of Fig. 7.7, this step could also be called *linear relaxation* of the discrete variables (spins in this case).

This means that linearly relaxed spin variables vary in the entire range between $+1$ and -1, namely $s_i \in [-1, 1]$.[17] Note, however, that after this step, the continuous dynamics defined by the Hamiltonian (9.1) may still have a lot of local minima (cf. Fig. 7.7).

Coupling with memory variables

The next step is then to transform these local minima into saddle points, by leaving invariant the ground state(s) of the original Hamiltonian (9.1). This is accomplished by coupling the continuous spin variables with memory variables.

One way (although not unique) to do this is to first introduce the continuously relaxed spin glass Hamiltonian,

$$H = -\sum_{i>j} \left(J_{ij} s_i s_j - \frac{1}{2} x_{ij} (s_i^2 + s_j^2) \right), \quad s_i \in [-1, +1], \tag{9.2}$$

where $\mathbf{x}$ are some *memory variables* to be determined.[18]

[14]The random constants, J_{ij}, do *not* vary with time. Therefore, the Ising spin glass is an example of a system with *quenched disorder* (see also Section 12.7.5).

A spin-glass configuration

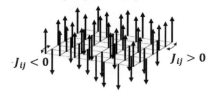

$J_{ij} < 0$ $J_{ij} > 0$

Fig. 9.4 Schematic of a 2D spin-glass configuration with random ferromagnetic ($J_{ij} > 0$) and anti-ferromagnetic interactions ($J_{ij} < 0$).

[15]The symbol $i > j$ indicates that we are not double counting the spin interactions. Alternatively, a factor of $1/2$ can be put in front of the summation.

[16]At least one, if it is degenerate.

[17]Referring to a hardware implementation using electrical circuits, we could also call the spin variables 'voltages'.

Monte Carlo methods
There are several numerical methods that employ *sampling* from a probability distribution to estimate quantities of interest that are difficult to obtain with other approaches. A prototypical example is the *Markov Chain Monte Carlo* (MCMC) method, which can be implemented in a variety of ways, such as the *Metropolis-Hastings algorithm* or the *Gibbs sampling* (see Section 10.2.1). See, e.g., (Gamerman and Lopes, 2006).

[18]Note that, written this way, the memory variables act as dynamic *Lagrange multipliers* (Goldstein, 1965) to enforce constraints on the spin magnitude.

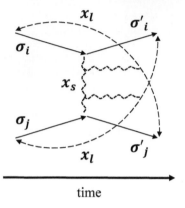

Fig. 9.5 The short-term memory, x^s, (which acts on spins, σ_i, interacting with strength J_{ij}) couples *directly* to the long-term memory, x^l. This is different from the choice of memory dynamics in Chapter 8 (cf. Fig. 8.2).

[19] Again, although important to bound all variables in the system, the upper limit L of $x^l_{i,j}$ in eqn (9.6) is *arbitrary*, and it may never be reached, if chosen appropriately. Cf. discussion in Section 8.2.5.

Metropolis-Hastings algorithm

Consider a (target) distribution $\pi(x)$ of a stochastic variable x and a (transition) distribution $Q(x|y)$ that provides a sample x, given a 'previous' sample y. Consider an arbitrary value x_i and take a sample, x', from $Q(x'|x_i)$ (a *Metropolis sample*). Draw a random number, r, between 0 and 1. The *Metropolis-Hastings algorithm* accepts such a sample with probability

$$P = \min\left(1, \frac{\pi(x')Q(x_i|x')}{\pi(x_i)Q(x'|x_i)}\right),$$

and sets $x_{i+1} = x'$, if $r \leq P$; $x_{i+1} = x_i$, otherwise.

[20] To be precise, as soon as we 'plant' a solution in the Hamiltonian (9.1), the latter is not really describing a 'glass'. Rather it should be properly called a 'benchmark' instance.

[21] cf. Section 8.2.5.

If the memory variables, **x**, were fixed in time, then from eqn (9.2) we could obtain the dynamics

$$\dot{s}_i = -\nabla_{s_i} H = \sum_j \left(J_{ij}s_j - x_{ij}s_i\right), \tag{9.3}$$

by simply differentiating the Hamiltonian with respect to the (linearly relaxed) spin variables.

However, the dynamics of these continuous spins (with fixed memory) would still suffer from local minima, and most likely would end up into one of them (see, e.g., Fig. 12.8). Therefore, in order to efficiently escape these states, we can *continuously* deform the local minima, and transform them into saddle points as I have done in Section 8.2.5, namely by letting the memory variables evolve.

Following the procedure in Section 8.2.5, I then introduce both *short-term memory*, x^s_{ij}, and *long-term memory*, x^l_{ij}, variables, which, however, are here coupled *directly* (cf. Fig. 9.5 with Fig. 8.2). The equations of motion of a DMM describing the spin memory coupled dynamics can then be written as[19]

$$\dot{s}_i = \alpha \overbrace{\sum_j J_{ij}s_j}^{\text{gradient-like}} - 2\beta \overbrace{\sum_j x^s_{ij}s_i}^{\text{rigidity}}, \tag{9.4}$$

$$\dot{x}^s_{ij} = \gamma C_{ij} - x^l_{ij}, \quad x^s_{ij} \in [0,1] \rightarrow \textit{short-term memory}, \tag{9.5}$$

$$\dot{x}^l_{ij} = \delta x^s_{ij} - \zeta, \quad x^l_{ij} \in [1,L] \rightarrow \textit{long-term memory}, \tag{9.6}$$

where $C_{ij} = \frac{1}{2}(J_{ij}s_is_j + 1) \in [0,1]$, is a *clause function* (similar to the one in eqn (8.1)), and $\alpha, \beta, \gamma, \delta, \zeta$ are time-scale parameters, fixed for all system sizes.

One can then simulate the coupled eqns (9.4), (9.5), and (9.6) until the system reaches a fixed timeout, or a planted solution. Either way, at the end of the computation, we can take $s_i = \text{sgn}(s_i)$ (*orthant* dynamics) as the state that minimizes the Ising energy in eqn (9.1).

Spin glasses with planted solutions

In Fig. 9.6, I report simulations of the previous equations, using the *forward Euler* integration scheme, eqn (8.11), for a class of three-dimensional spin glasses with planted solutions.[20]

(See (Hamze *et al.*, 2018) for how these instances are generated (and why they are particularly difficult to solve), and (Pei and Di Ventra, 2021) for more details on the numerical implementation and choice of parameters.)

The DMM simulations are compared to those done using *Monte Carlo*-based methods for *partially* frustrated[21] instances. In particular, Fig. 9.6 compares the MemComputing dynamics with *simulated annealing* (SA)

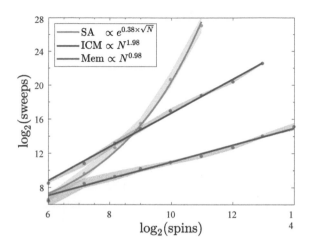

Fig. 9.6 Scalability of the median TTS (sweeps) as a function of number of spins for the DMM dynamics (Mem), simulated annealing (SA), and isoenergetic cluster moves (ICM) on *partially* frustrated 3-dimensional Ising spin glasses with planted solutions (see (Pei and Di Ventra, 2021) for how the TTS is calculated). Statistics are collected over 400 runs, with the shaded region denoting the 40-th to 60-th percentile. The confidence interval for the scaling factors are: 0.38 ± 0.01 (SA); 1.98 ± 0.05 (ICM); 0.98 ± 0.04 (Mem). Results courtesy of Yan Ru Pei. The results for the *fully* frustrated case can be found in (Pei and Di Ventra, 2021).

and the *isoenergetic cluster-move* (ICM) method. The latter one, in particular, is considered the state-of-the-art approach for simulating spin glasses.[22]

From Fig. 9.6 we see that the scaling of SA is well fitted by a sub-exponential function, the one for ICM by a quadratic function, while the simulations of the DMM's equations of motion (9.4), (9.5), and (9.6) scale *linearly* up to the maximum size tested. In (Pei and Di Ventra, 2021) the *fully* frustrated instances (frustration ratio 1/4) have been solved, showing an exponential scalability for both SA and ICM, while a *polynomial*, $O(N^{4.1})$, scalability for the simulations of the DMM's ODEs.

9.2.2 Spin glasses, QUBO, and weighted MAX-SAT problems

Let me now demonstrate that the previous Ising spin glass problem is a particular case of a more general optimization problem: the *quadratic unconstrained binary optimization* (QUBO). In turn, the latter one can be transformed into a *weighted MAX-SAT problem* (Section 2.4).

Even though this transformation is not necessary to solve the Ising spin glass—as I have shown in the previous Section 9.2.1—it will come in handy in Chapter 10, when I will consider applications to Machine Learning and Quantum Mechanics. In fact, in Section 10.4.2, I will provide one possible way of mapping the MAX-SAT that originates from the QUBO formulation of a neural network training to a DMM.

QUBO problems

If we consider s_i in eqn (9.1) as a general *binary* variable (call it v_i), and the J_{ij} some arbitrary *weights* (call them w_{ij}), the quantity E (the

[22]The partially frustrated instances have been generated by randomly mixing 80% of fully frustrated ones (frustration ratio 1/4) with 20% with lower frustration (ratio 1/6). See (Hamze *et al.*, 2018) and (Pei and Di Ventra, 2021) for more details.

Simulated annealing (SA)
A particular case of MCMC useful to find a good approximation of the global optimum is *simulated annealing*. As an example, let us start from a spin state, **s**, of the Ising spin glass, eqn (9.1). The energy of such a state is $E(\mathbf{s})$. Let us then flip one random spin of that state to obtain another state, $\mathbf{s}^*$, and energy $E(\mathbf{s}^*)$ (a 'Metropolis sample'). Draw a random number, r, between 0 and 1, and consider a parameter, T (a 'temperature'). The acceptance probability of the new state is 1 if $E(\mathbf{s}^*) < E(\mathbf{s})$, (setting the Boltzmann constant $k_B = 1$)

$$e^{-(E(\mathbf{s}^*)-E(\mathbf{s}))/T} > r,$$

otherwise. This procedure is repeated for the whole lattice (a *sweep*). The temperature T is then lowered, according to some *annealing schedule*, and the process repeated till a desired result or a timeout are reached.

Hamiltonian) can be interpreted as an *objective function* (a quadratic polynomial over binary variables) that needs to be minimized (cf. discussion in Section 2.5).

The Ising spin glass problem is then an example of a much more general (NP-hard) combinatorial optimization problem known as:

Quadratic unconstrained binary optimization (QUBO)

Given a set of *binary* variables ($v_i = 0, 1$) and the objective (cost) function:

$$E = \sum_i \sum_j w_{ij} v_i v_j, \qquad (9.7)$$

with $w_{ij} \in \mathbb{R}$ a set of weights, a QUBO problem is to find the assignment of the variables that minimizes (or maximizes) E (Arora and Boaz, 2009).

If we indicate with $W = \{w_{ij}\}$ the matrix of constants, and $\vec{v}$ the vector of binary variables, we can re-write the same problem in matrix notation as:

QUBO problem

$$\text{minimize/maximize: } E = \vec{v}^T W \vec{v}, \qquad (9.8)$$

where the symbol $\vec{v}^T$ indicates the *transpose* of the vector $\vec{v}$.

From QUBO to weighted MAX-SAT

We can further manipulate a QUBO problem so as to transform it into a *weighted* MAX-SAT (see definition in Section 2.4).[23]

To be specific, I will show the transformation for the Ising spin glass problem (9.1). First, we note that finding the ground state of an Ising spin glass is equivalent to solving a weighted MAX-SAT problem, where each interaction is expressed as an exclusive-OR (XOR), between the associated binary variables (Sheldon *et al.*, 2019):

$$-J_{ij} s_i s_j \quad \leftrightarrow \quad 2|J_{ij}|\, b_i \oplus b_j \equiv j_{ij} \qquad (9.9)$$

where $b_i = (s_i + 1)/2$; $2|J_{ij}|$ is the weight associated with the constraint; $\oplus$ is the XOR symbol. This relation can be easily verified by taking into account the XOR truth table (see Fig. 7.2).

We now want to transform the obtained weighted MAX-SAT into a traditional CNF formula (see Section 2.4 for a refresh of what this means).

This can be done by mapping each interaction into two OR clauses (as it can be checked by referring to the OR truth table in Fig. 2.7).

Parallel tempering

Replica exchange MCMC (also known as *parallel tempering*) is a variant of the MCMC method where one runs N copies (replicas) of the system, randomly initialized, at different temperatures, T_j ($j = 1, \dots N$). One step of the algorithm consists in a single sweep of Metropolis sampling over all replicas, followed by a single proposed *exchange*, where replicas at neighboring temperatures have their configurations switched according to the probability,

$$P = \min\left(1, e^{\{(\beta_i - \beta_j)(E_i - E_j)\}}\right),$$

where E_i and E_j are energies of the instances, with $\beta_i = 1/T_i$ and $\beta_j = 1/T_j$. This cycle of Metropolis sweeps and exchanges is repeated until a solution (or a timeout) is reached.

---

[23] This is an example of the (polynomial) transformation (reduction) from one NP-hard problem to another, I anticipated in Section 2.3.3.

Depending on the sign of the interaction we then have:

$$-J_{ij}s_i s_j \quad \leftrightarrow \quad \begin{cases} \begin{smallmatrix} 2|J_{ij}| \ b_i \vee \bar{b}_j \\ 2|J_{ij}| \ \bar{b}_i \vee b_j \end{smallmatrix}, & \mathrm{sgn}(J_{ij}) = 1, \\[1.5em] \begin{smallmatrix} 2|J_{ij}| \ b_i \vee b_j \\ 2|J_{ij}| \ \bar{b}_i \vee \bar{b}_j \end{smallmatrix}, & \mathrm{sgn}(J_{ij}) = -1. \end{cases} \quad (9.10)$$

In this relation each constraint carries a weight, and the symbol $\bar{b}_i$ means the negation of b_i. The transformation (9.10) can be easily verified by keeping in mind that while in the QUBO problem we need to *minimize* the energy (10.3), in the MAX-SAT we need to *maximize* the sum of the weights of *satisfied* clauses (cf. discussion in Section 10.4.2).

An important remark

For each sign of the interaction strength, the OR clauses in eqn (9.10) need to be related by *conjunctions* (Section 2.4). For instance, for $\mathrm{sgn}(J_{ij}) = 1$, the two clauses are related as $-J_{ij}s_i s_j \leftrightarrow 2|J_{ij}|(b_i \vee \bar{b}_j) \wedge 2|J_{ij}|(\bar{b}_i \vee b_j)$. Note also that the factor of 2 may be dropped since it is only a *global* scaling of the energy E in eqn (9.1).

A numerical example

This transformation from the Ising spin glass problem to a weighted MAX-SAT written in CNF has the advantage that it can be easily fed into commercial solvers that read this type of format by default.

One such comparison has been reported in (Sheldon *et al.*, 2019), where different solvers have been compared to a MemComputing *commercial solver*, provided by MemComputing, Inc.[24]

Those tests have been performed again on instances with planted solutions, but different from the ones used in the previous example (Fig. 9.6). Although the details of how these instances have been generated can be found in (Sheldon *et al.*, 2019), the idea is simple.

Consider an hypercubic lattice (e.g., in three dimensions) with periodic boundary conditions, and fill such a cube with *loops* of spins all aligned with each other (namely with interaction $J_{ij} = 1$), except for one bond ($J_{kl} = -1$). This bond is then *frustrated* (Section 8.2.5), hence the name 'frustrated loops'.

The ground state energy of each one of these *frustrated-loop instances* is then the sum of the energies of the individual loops, and can be easily determined.[25] However, if the loop length is chosen appropriately (namely larger than some finite length), any heuristic solver (as defined in Section 2.6) would have a hard time to find the ground state configuration, because the loops are highly *non-local* objects (cf. also discussion in Section 4.3.4).

Isoenergetic cluster moves (ICM) algorithm
The *ICM algorithm* combines the parallel tempering (PT) with appropriate cluster moves. First of all, as in PT, a number of replica pairs are initialized, with the pairs spaced geometrically in temperature. After one sweep in every replica, the following update is performed for every replica pair. A cluster update flips a nontrivial cluster of spins with negative overlap between two replicas to increase the mixing rate. Finally, the PT routine attempts to exchange again the temperature between two random replicas of neighboring temperatures. See, e.g., (Houdayer and Hartmann, 2004) for a more in-depth description of this method.

[24]Even though the ODEs in this commercial solver are different than the ones I presented in this book (which are published in the literature), the overall strategy is the same, as the one I have discussed in Chapters 6 and 7. The solver is available either as *software as a service* (SaaS) or for demonstration at www.memcpu.com.

[25]If $H_{FL,i}$ is the Hamiltonian of one such loop, the full lattice Hamiltonian is then $H = \sum_i H_{FL,i}$.

[26]This is a *Mixed Integer Linear Programming* solver. See Section 9.2.5.

In Fig. 9.7, I report the results of the numerical simulation of the ODEs of DMMs solving these instances, versus two standard algorithms—SA and parallel tempering (PT)—as well as a well-known commercial solver, the IBM CPlex (CPlex, 2018).[26]

Fig. 9.7 Scaling of floating-point operations necessary for different solvers to reach the ground state of 3-dimensional frustrated-loop instances as a function of the total number of spins N. All calculations were performed on a single core. The solid lines are the best fits of the 95th quantile TTS for all four solvers. The exponential fits have the following parameters: for IBM CPlex, $b = 0.12$ and $c = 0.46$, for SA, $b = 0.069$ and $c = 0.67$, and for PT $b = 0.32$ and $c = 0.46$. The commercial MemComputing solver used is called Falcon, and its scaling is best fitted with a power law (sub-quadratic scaling). Reprinted with permission from (Sheldon *et al.*, 2019).

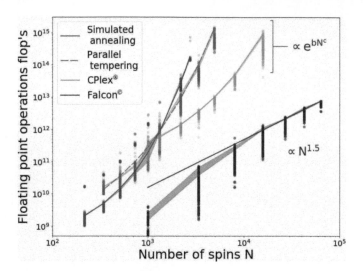

[27]Namely, an exponential fit did not converge for the MemComputing solver (Sheldon *et al.*, 2019).

To account for the different implementations of the various (sequential) solvers, Fig. 9.7 reports *floating point operations* to reach the ground state, versus the number of spins in the system.

All solvers, except the MemComputing one, appear to show an *exponential* scaling. For the MemComputing solver, only a *power-law* fit converges, up to the problem sizes considered ($N = 64,000$ spins).[27] More details of all simulations can be found in (Sheldon *et al.*, 2019).

> **Floating point operations per second (FLOPS)**
> In order to better manipulate real numbers on our modern computers, a *floating-point* representation is used. A floating-point number, $r \in \mathbb{R}$, is expressed with a mantissa (significant digits or *significand*), a base (typically 2), and an exponent:
>
> $$r \equiv \text{mantissa} \times \text{base}^{\text{exponent}}.$$
>
> *FLOPS* count how many operations on these numbers are performed by the processor in a second.

9.2.3 Decreased spread in solution time with increasing problem size

The last example also shows an interesting effect observed in other cases as well: the *spread* in the TTS—however we measure such a 'time'—for the MemComputing solver *decreases* with increasing problem size. This is clearly seen in Fig. 9.7, where the spread shrinks with increasing number of spins, N.

Similar results have been shown, for instance, in the case of the 3-SAT problem I have discussed in Section 9.1.1; see (Bearden *et al.*, 2018). It is also somewhat visible in Fig. 9.6, for another class of instances.[28]

[28]The effect is less evident from Fig. 9.6, because that figure reports only the 40-th to 60-th percentile range.

Although there is no proof that such a behavior is universal, it seems to indicate that, at least for the cases studied so far, the larger the problem size, the smaller the differences in solution time between the instances.

This could be related to how such instances are generated, in such a way that the bigger the problem size, the less noticeable their variability is. However, if that were the case, we would expect other solvers to showcase a similar behavior.

Instead, this is not the case, as it is clearly evident from, e.g., Fig. 9.7. All the other solvers show the *opposite* behavior: the spread in TTS typically *increases* with problem size.

If the instances themselves do not contribute to such an effect, then the latter must be related to how DMMs operate in the *thermodynamic limit*, namely when their size increases, while keeping, say, the clause-to-variable ratio fixed (cf. discussion in Section 4.3.2).

As I will show in Section 11.7, in that limit, the *dynamical long-range order* that DMMs develop (which is at the origin of the efficiency of these machines in solving these instances) covers the 'bulk' of the machine, and 'boundary effects' are of measure zero. In other words, the machines themselves become more 'uniform' with increasing size, in the sense that, given a particular problem, the way in which they solve its instances, is essentially unrelated to how 'different' those instances are 'microscopically' (cf. also discussion in Section 11.8).[29]

[29]In a sense, this is similar to the 'universality' that certain physical systems showcase at a continuous phase transition. See Section 4.3.3 for the concept of 'universality classes', and the discussion in Section 13.2 on how it may relate to DMMs.

Having discussed possible applications of DMMs to both *search* and *optimization* problems, for which the solution is known, in the final Sections of this Chapter I will show additional results of optimization problems for which the optimum in *not* known.

As I have discussed at the beginning of Section 9.2, in this case, we need to set a *timeout*, and check which solver has found the best assignment of variables (e.g., the lowest number of unsatisfied clauses) within that allotted time.

Equivalently, we could set a 'quality bound' (say, a certain percentage of unsatisfied clauses), and check how long it would take the solver to surpass (improve on) such a bound. This is the type of tests I will show next, again using the commercial solver provided by MemComputing, Inc.

9.2.4 Maximum satisfiability (MAX-SAT)

The first problem I want to discuss is the *maximum satisfiability* (MAX-SAT) that I have defined in Section 2.4. I remind the reader that the MAX-SAT aims at finding an assignment of the Boolean variables that *maximizes* the number of *satisfied* clauses—equivalently, *minimizes* the number of *unsatisfied* clauses—in a given logic formula.

The logic formula is again written in CNF (Section 2.4), and the self-organizing circuit is similar to the one in Fig. 8.1. However, now

[30]In fact, in the infinite time limit it would settle into a (quasi-)periodic orbit; see Fig. 7.20 and discussion in Section 7.9.

Dynamic RAM
Dynamic random access memory (DRAM) is a type of *volatile* memory that stores bits of information in memory cells consisting of a capacitor and a transistor (Hennessy and Patterson, 2006). When the capacitor is charged it represents the logical 1; when discharged the logical 0.

[31]Any MAX-E*k*SAT, with $k \geq 2$ belongs to the class of NP-hard problems (Section 2.3.3).

[32]Equivalently, any (heuristic) algorithm should require exponentially increasing time to improve on their approximation beyond this limit, unless NP = P.

[33]www.maxsat.udl.cat.

[34]A further confirmation of this can be obtained by estimating the optimum using an ensemble of small instances for which it is easier to find a fairly good approximation by brute force. For example, for the instances used in Fig. 9.8, using 300 variables and a clause-to-variable ratio of 5, the global optimum was estimated to be at about 1.3% of unsatisfied clauses, which is indeed in the expected threshold range (Traversa *et al.*, 2018).

there are clauses that are *unsatisfied*. Therefore, the dynamics do not necessarily reach an equilibrium point.[30]

Indeed, even if the dynamics reached the *global* optimum, we would not know if that is really the optimum, or some other 'sub-optimal' solution (cf. discussion in Section 2.3.4).

However, as I have anticipated, we formulate this problem slightly differently as follows. Let us consider a percentage of *unsatisfied* clauses, call it α_{th}. We can then ask the following question: how long does it take the solver to reach the threshold α_{th} as a function of problem size?

Note that, in general, this is still not an easy problem to solve, in the sense that it may take an exponential time to reach such a threshold. The reason is related to the discussion in Section 2.3.4.

There are many optimization problems for which, up to a certain 'distance' from the optimum, it is relatively easy (takes polynomial time) to find an approximation to the optimum. However, improving on such an approximation beyond such a threshold may take a time growing exponentially with input size (see Fig. 2.6). Such problems have an *inapproximability gap*.

A problem that showcases such an inapproximability gap is the MAX-E3SAT, where 'E3' means that this MAX-SAT problem has exactly 3 literals per clause.[31] It has been proved that if NP $\neq$ P, then there is no algorithm that can provide, in polynomial time, an approximation better than 7/8 of the *optimal* number of satisfied clauses (Håstad, 2001).[32]

In (Traversa *et al.*, 2018), MAX-E3SAT instances have been generated that are known to be difficult for local search solvers and message-passage algorithms, e.g., for solvers like the ones I have used for the benchmarks in Fig. 9.1. In particular, in (Traversa *et al.*, 2018) two state-of-the-art solvers, winners of the 2016 MAX-SAT competition,[33] have been used as comparison. The results are shown in Fig. 9.8.

All simulations have been done on a single thread of an Intel Xeon processor with 128 Gb DRAM shared on 24 threads. The clause-to-variable ratio was chosen to be 5, and $\alpha_{th} = 1.5\%$.

For the two state-of-the-art solvers it takes an *exponentially* increasing time to overcome the set threshold, supporting the notion that indeed the instances chosen fall within the *inapproximability gap* of the MAX-E3SAT.[34]

The MemComputing solver, instead, overcomes the same threshold in *linear* time up to 2×10^6 variables. As indicated in Fig. 9.8, for the same size, it would take the other two solvers more than the age of Universe to overcome the threshold, while the MemComputing solver could reach that threshold in about 10^4 seconds (less than 3 hours).

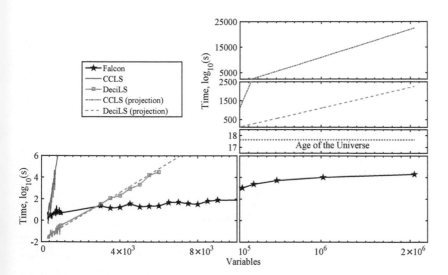

Fig. 9.8 CPU time comparison between two winners of the 2016 MAX-SAT competition (CCLS and DeciLS) against the MemComputing solver (named Falcon), for the MAX-E3SAT. It is the time it takes to overcome a threshold of 1.5% of unsatisfiable clauses with increasing number of variables. All calculations have been performed on a single thread of an Intel Xeon processor with 128 Gb DRAM shared on 24 threads. The heuristic solvers require an *exponentially* increasing time to reach that threshold. The MemComputing solver instead scales *linearly* up to 2×10^6 variables. The estimated time (dashed and dashed-dotted lines) the other two solvers would have required to run up to 2×10^6 variables is also shown. Reprinted with permission from (Traversa *et al.*, 2018).

An important remark

Although it would be tempting to conclude from the previous results that NP = P, that is not the case. The theorems on the inapproximability gap of optimization problems relate to *all* possible instances of a problem class, not just a few (Håstad, 2001).

Stress testing the MemComputing solver

To 'stress test' the simulations of the ODEs of DMMs, namely to push them to the limit of the processor used, several runs up to the largest size of the problem instance that could fit in the chosen processor DRAM have been performed (Sheldon *et al.*, 2020). The results are shown in Fig. 9.9, where the MemComputing solver is again compared to one of the winners of the 2016 MAX-SAT competition (DeciLS).

The interesting thing to notice from Fig. 9.9 is that the numerical simulations of the ODEs of DMMs scale *linearly* up to the largest case considered, which corresponds to 64×10^6 variables.[35]

For these tests, the clause-to-variable ratio is also 5. Therefore, the largest case corresponds to over 3×10^8 clauses. Since each clause has exactly 3 literals, the largest case corresponds to about 10^9 literals.

As I have discussed in Chapter 8, each literal is represented by a voltage variable or its negation. In addition, one introduces memory degrees of freedom and their associated equations of motion.

In total, then, the largest case considered in Fig. 9.9 corresponds to

[35]The state-of-the-art solver used for comparison requires an *exponentially* increasing time, and it is already not competitive when the problem instances surpass a few thousand variables.

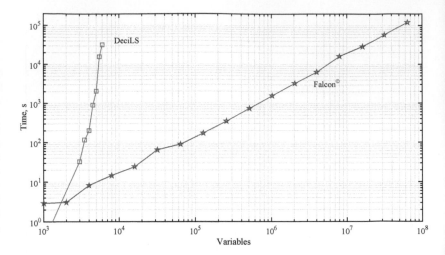

Fig. 9.9 CPU time comparison between the DeciLS solver and the Mem-Computing solver (Falcon) of Mem-Computing, Inc., to cross the 1.5% threshold of unsatisfied clauses for MAX-E3SAT instances. All tests have been done on a single thread of an Intel Xeon processor with 128 GB dynamic random access memory (DRAM) shared on 24 threads. Reprinted with permission from (Sheldon *et al.*, 2020).

a simulation of over 1 billion coupled ordinary differential equations! This is quite remarkable if one considers that these equations have been integrated using *forward Euler* (Section 8.5.1), implemented in interpreted MATLAB, and running for about 24 hours on a single thread of a standard processor.

These results further support the notion that the numerical simulations of the ODEs of DMMs are very *robust* against numerical noise. I will discuss this point in detail in Section 12.7.

RAM requirements

I want to conclude this Section with an explicit example of what I have anticipated in Section 3.3: the RAM requirement of the simulations of ODEs scales *polynomially* with problem size.[36]

This polynomial (in fact, linear) scalability is shown in Fig. 9.10 for the same instances used to generate Fig. 9.9. Since the particular processor employed in those simulations has 128 gigabytes of dynamic random access memory (DRAM), it is clear that the largest case considered fits within that limit.[37]

9.2.5   Mixed Integer Linear Programming

Let me conclude this Chapter with a brief discussion of applications of DMMs to *algebraic* problems (Traversa and Di Ventra, 2018). In this case, one would follow a similar strategy in designing self-organizing gates, with the notable exception that these gates have to satisfy an algebraic relation, not a Boolean one (see Section 7.2.2).

An important (NP-hard) optimization problem in this class is known as *Integer Linear Programming* (ILP) (Garey and Johnson, 1990). The word 'integer' implies that all variables can only take integer values. If

[36]In fact, *linearly* with the number of coupled ordinary differential equations considered.

[37]If the problem size requires more RAM than the one of the processor, the latter will 'swap' data between the RAM and the hard disk, with considerable slowing down of the execution of the program.

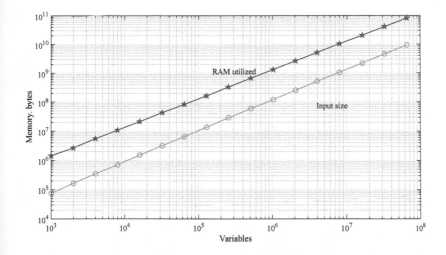

Variables

Fig. 9.10 Memory (RAM) requirements of the MemComputing solver (Falcon) as a function of variables for the MAX-E3SAT instances used to generate Fig. 9.9. Both the input size memory (open circles) and the RAM used during computation are provided. Reprinted with permission from (Sheldon *et al.*, 2020).

some of the variables are allowed to be non-integers then the corresponding NP-hard problem is known as *Mixed Integer Linear Programming*.

For simplicity, let us consider a particular version of this problem for which we have only *binary* variables. The problem is then oftentimes called $0 - 1$ *Linear Programming*.[38]

Consider then a set of n binary variables, $x = \{x_1, ..., x_n\}$ with $x_j \in \mathbb{Z}_2$ for any $j = 1, ..., n$. Consider also a set of n *weights*, $f_j \in \mathbb{R}$. We can further introduce matrices and vectors that define *algebraic constraints* between the variables x_j. Let us then introduce $A_{eq} \in \mathbb{R}^{m_{eq} \times n}$, $b_{eq} \in \mathbb{R}^{m_{eq}}$, $A_{ineq} \in \mathbb{R}^{m_{ineq} \times n}$ and $b_{ineq} \in \mathbb{R}^{m_{ineq}}$ with m_{eq} and $m_{ineq} \in \mathbb{N}$. These quantities define the following constraints:

$$A_{eq}x = b_{eq},$$
$$A_{ineq}x \leq b_{ineq}. \tag{9.11}$$

The problem then seeks an assignment, $\bar{x}$, of the variables that minimizes:

$$\min_{\{x_j\}} \sum_j f_j x_j,$$
$$x_j \in \mathbb{Z}_2, \quad \forall j, \tag{9.12}$$

subject to the constraints (9.11).

The solution $\bar{x}$ is said to be *sub-optimal* if $\sum_j f_j \bar{x}_j \geq \min_{\{x\}} \sum_j f_j x_j$ where x satisfies the constraints (9.11). We can also define the *objective*, O, of the problem as[39]

$$O = \min_{\{x\}} \sum_j f_j x_j, \quad objective. \tag{9.13}$$

There is a wide variety of algorithms, both *heuristic* and *complete* (Section 2.6) to tackle this problem. It is not the scope of this book to

[38] The 0-1 Linear Programming problem is also *NP-hard* (Section 2.3.3).

> **Feasibility**
> An instance in the class of Mixed Integer Linear Programming may not support any assignment of the variables x_j that satisfies *all* constraints (9.11). This instance is said to be *unfeasible*, otherwise it is *feasible*.

[39] Note that if we 'relax' the integrality constraint of each variable in (9.12) and replace it with $x_j \in [0, 1]$ for each j, we transform ILP into a problem that can be efficiently solved (in polynomial time). This *Linear Programming* problem (which belongs to the P class) provides a lower bound O_{LP} to the original ILP, satisfying $O_{LP} \leq O$.

discuss all these algorithms. Rather, I will only highlight some results using a MemComputing solver on particularly hard instances in the ILP class, where the traditional algorithms struggle.

Problems in the Mixed Integer Linear Programming library

To this aim, instances can be taken from the Mixed Integer Linear Programming Library (MIPLIB)[40] that contains a wide range of hard instances, and keeps track of the success various solvers have in finding a better objective O, eqn (9.13), for these instances.

In fact, for many instances in MIPLIB, the optimum is not known (they do not have 'planted solutions' as those I discussed in Section 9.1.1). In some cases, it is not even known if the instance is *feasible*, namely that there is an assignment of the variables that satisfies all constraints. These instances are classified as 'open' in MIPLIB.

A commercial MemComputing solver provided by MemComputing, Inc., was employed to solve a variety of open instances in MIPLIB (Traversa and Di Ventra, 2018). The results were then compared to a renowned commercial solver, Gurobi,[41] that employs a collection of algorithms to tackle optimization problems.

In (Traversa and Di Ventra, 2018), it was shown that the MemComputing solver outperforms Gurobi in terms of finding a better objective within a given time limit. Importantly, the MemComputing solver was the first to find the objective of two problems whose feasibility was long unknown.[42]

The 5th Airbus problem

In 2019 the Airbus company launched a *quantum-computing challenge* to solve a set of five problems relevant to the aircraft industry, varying from design to deployment of aircrafts.[43]

The challenge requires the development of a suitable algorithm for each problem that can be implemented on a *quantum computer*.[44]

The 5th problem in this challenge, formulated as an ILP problem, was solved using the MemComputing solver of MemComputing, Inc. (Traversa, 2019). It consists in optimizing the cargo placement of a subset of containers picked from an available payload formed by n different containers.

Each container is described by the triplet (k, m_k, s_k) with k the container identification number, m_k its mass, and s_k its size. In addition, there are several other (real-life) constraints that need to be satisfied.

These are the positions and size of the containers with respect to the different compartments of the cargo; the center of gravity position of the carried freight (plus the aircraft) must be within certain limits; the mass of the freight should be maximized (within a limit), while optimizing

---

[40] miplib.zib.de.

[41] www.gurobi.com.

[42] These two problems are called 'f2000' and 'pythago7824', and can be found at miplib.zib.de.

[43] www.airbus.com under the January 22, 2019 press release.

[44] As a consequence, an actual verification of the efficiency of such algorithms would require the *hardware* realization of a quantum computer on which such algorithms would need to be run (cf. discussion in Section 3.1).

the center of gravity of the carried freight (plus the aircraft) as close as possible to a target center of gravity, etc. In view of all these constraints, the problem is far from trivial.

By including *all* constraints required by the Airbus problem, with no exception and no approximation, the simulations of DMMs show a *sub-quadratic* scaling as a function of the problem size (Traversa, 2019). This is quite a remarkable result since it does not even require building these machines in hardware, let alone a quantum computer. A software solution is enough.

9.3 Chapter summary

- In this Chapter, I have provided several examples of *simulations* of DMMs applied to *search* and *optimization* problems.

- In all cases considered, the MemComputing simulations (mostly done with *forward Euler* within MATLAB) running on a single core/thread of a modern computer, typically outperform by orders of magnitude traditional algorithms on the same instances.

- In fact, the MemComputing simulations show a *polynomial* scaling in the time to solution up to the problem sizes considered, vs. the typical *exponential* scaling of the traditional algorithms tested.

- The memory (RAM) requirements of the MemComputing simulations instead grows always *linearly* with the number of ODEs that need to be integrated.

- An interesting, empirical result from these simulations is that the *spread* in the solution time, due to the variability of the instances per given size, *decreases* with the problem size.

- However, whether this trend is universal, and the spread tends to zero or to a constant value in the thermodynamic limit, is not obvious.

Application to Machine Learning and Quantum Mechanics

10

Before we work on artificial intelligence why don't we do something about natural stupidity?
Stephen Polyak (1889−1955)

All models are wrong, but some are useful.
George E.P. Box (1919−2013)

In God we trust, all others must bring data.
William Edwards Deming (1900−1993)

Takeaways from this Chapter

- How to use digital MemComputing machines (DMMs) to do *unsupervised training* of neural networks (NNs) at *optimality*.

- How to use DMMs to do *unsupervised training* of *deep* NNs as *a single unit* and at *optimality*.

- How to accelerate the *supervised* training of NNs with DMMs.

- How to do *quantum state tomography* and find the *ground state* of quantum Hamiltonians with DMMs.

[1] As I have already mentioned in Chapters 8 and 9, the simulations of DMMs provide no proof of optimality, hence the DMM solvers I have discussed in those Chapters are *incomplete* (cf. discussion in Section 2.6).

In Chapter 9, I have shown several simulations of DMMs on modern computers, and their efficient solution of a wide variety of problem instances.

Of particular note is the fact that DMMs can *efficiently* find a very good approximation of the *global optimum* (or the optimum itself) in a highly *non-convex landscape* (see also discussion in Section 2.3.4).[1]

Such a realization opens up a lot of research venues, and suggests new methods to tackle other important problems. In this Chapter, I will discuss the application of MemComputing to two classes of problems: those pertaining to *Machine Learning* (Goodfellow *et al.*, 2016) and *Quantum Mechanics* (Messiah, 1958).

Machine Learning is a vast field with applications that touch upon several aspects of science and technology. As I have anticipated in Section 2.5, it is a sub-field of *Artificial Intelligence* (see the 'four waves of AI' in the same Section) that studies algorithms that learn features from a set of data provided by the user, so that they can infer other features, without having been directly trained on them (Goodfellow *et al.*, 2016).

It is not my intention to discuss this vast field and all of its applications. The interested reader is urged to look at the books dedicated to the subject. In this book, I will only discuss how to use MemComputing as an aid in Machine Learning.

Quantum Mechanics (Messiah, 1958) is arguably the most successful description, we have so far, of various natural phenomena. Its range of applicability surpasses basic science, as it is also at the core of a wide range of technologies that we employ on a daily basis, including our modern computers.[2]

Of fundamental importance in Quantum Mechanics is finding the *ground state*, and often the *excited states* as well (namely, the entire *energy spectrum*), of *quantum Hamiltonians*.[3]

I will show in Section 10.7, that this problem as well as the reconstruction of a quantum state from data (*quantum tomography*) can be formulated as the training of a *neural network*, hence they can be tackled with similar techniques I will discuss in the case of Machine Learning. Again, I will make no attempt to exhaustively discuss all the nuances of Quantum Mechanics. These can be found in specialized books. My goal here is simply to show how MemComputing can be used in this field.

[2] For instance, a proper description of the operation of the transistor—the main basic unit in our processors—can only be accomplished by means of Quantum Mechanics (Horowitz, 2015).

[3] From the energy spectrum, in principle, *all* dynamical properties of a quantum-mechanical system can be obtained (Messiah, 1958).

10.1 Artificial neural networks

The particular aspect of Machine Learning I will be discussing in this book is the *training* of *artificial neural networks*. As I have anticipated in Section 5.1, these are computing models, inspired by the operation of the neural networks in biological brains (Haykin, 1998).

In Section 12.10, I will briefly discuss the relation between MemComputing and *biological* brains. In this Chapter, instead, all the neural networks (NNs) I will consider will be 'artificial', namely models. Therefore, we can drop the word 'artificial' in front of them, since there can be no confusion.

Generative vs. discriminative models
Consider a set of (observable) input variables **x** and labels (target variables) **y**. In statistical classification (and Machine Learning), the term *generative model* indicates any model able to learn the *joint* probability distribution, $p(\mathbf{x}, \mathbf{y})$. A *discriminative model*, instead, attempts to learn the *conditional* probability $p(\mathbf{y}|\mathbf{x})$. See also Section 10.1.2, and, e.g., (Goodfellow *et al.*, 2016).

Representational power of RBMs
Consider a restricted Boltzmann machine (RBM) and an arbitrary *discrete probability distribution* of the data to learn, with n_d visible configurations with non-zero probability. An RBM with $n_d + 1$ hidden nodes can represent such a probability distribution, *if given appropriate weights and biases* (parameters) (Le Roux and Bengio, 2008). This is an example of a *universal approximation theorem*. The $n_d + 1$ bound is sufficient but not necessary. In practice, a much smaller number of hidden nodes of the RBM is enough for a reasonable approximation (with the appropriate parameters) for many types of data sets.

Bipartite structure of RBMs
A graph is said to be *bipartite*, if its vertices can be divided into two disjoint and independent sets, such that every edge connects a vertex in one set to one in the other. An RBM is a bipartite graph (Fig. 10.1, right panel), with the visible nodes forming one set, the hidden nodes the other.

10.1.1　Why neural networks?

Neural network models are very powerful since they can *learn* features from a set of data *without* a rule-based programming (which is instead what our modern computers follow).

For this reason, in addition to being used as *discriminative models* of conditional probability distributions—namely they can discriminate between different kinds of data—they can be used as *generative models* of complex joint probability distributions (even without any labels). In other words, they can generate/predict new data, by taking advantage of only the features learned during training. Both these models are sometimes called *statistical models*.

However, as I will discuss in a moment, *training* NNs (so that they can work as either generative or discriminative models) is far from being a simple task. In fact, the training itself seems to be the limiting factor to the applicability of a given NN, rather than its structure and topology. In other words, if we find an *efficient* and *robust* method to train NNs, the choice of NN to do statistical classification may be of secondary importance. This is where MemComputing can help.

MemComputing machines (not necessarily digital ones), viewed as networks of memprocessors, *generalize* the concept of NNs (see Section 5.1 for a formal demonstration of this statement). One can then view the *training* of NNs as a special application of MemComputing.

To be specific, in this book I will discuss only the training of one type of NNs: *Boltzmann machines*. In fact, I will consider their *restricted* version: *restricted Boltzmann machines* (RBMs) (Goodfellow *et al.*, 2016). However, the same ideas can be applied to a wide range of other NNs.

The reason I will focus on these models is because they satisfy a *universal approximation theorem* for discrete probability distributions (Le Roux and Bengio, 2008). This means that if the NN has 'enough' (hidden) nodes it can represent *any* discrete probability distribution, *provided* (!) one is able to find the appropriate parameters (weights and biases), namely provided one can efficiently train them. This is such an important point that I will return to it in Section 10.1.4.

An important remark
Boltzmann machines (restricted or not) are an example of *undirected graphs*, namely their edges do not have any particular direction: their *logical topology* (see definition in Section 6.2) is *bidirectional*, so that each edge can be traversed in both directions. This class of NNs is sometimes called *recurrent*. Instead, NNs that are based on *directed acyclic graphs* (where the edges have a direc-

tion, and no cycles are allowed) are called *feedforward* NNs, and the *logical topology* of their network is *unidirectional* (from input layer to output layer). See, e.g., (Goodfellow *et al.*, 2016).

Let us now make all these points more precise by first defining these machines. Later on, I will show how they are typically *trained*, before discussing how MemComputing can improve their training.

10.1.2 (Restricted) Boltzmann machines

A *Boltzmann machine* consists of n *visible* units (*nodes* or *neurons*), $v_j, j = 1 \ldots n$, (forming a *visible layer*) each *fully connected* to a layer of m *hidden* nodes,[4] $h_i, i = 1 \ldots m$, (forming a *hidden layer*) both usually taken to be *binary* variables (0 and 1, or -1 and $+1$).

This is shown schematically in the left panel of Fig. 10.1.

[4]The visible nodes are fixed by the data, while the hidden nodes provide *expressivity* to the model.

Boltzmann Machine

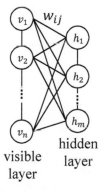

Restricted Boltzmann Machine

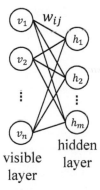

Fig. 10.1 Left panel: Schematic of a *Boltzmann machine* where all visible and hidden nodes are connected to each other. **Right panel:** Schematic of a *restricted Boltzmann machine* where there are *no intra-layer connections* in either the visible or the hidden layers. The connections between the layers represent the weights, $w_{ij} \in \mathbb{R}$ (biases not shown).

In the *restricted* model, which is called a *restricted Boltzmann machine*, *no intra-layer connections* are allowed (the RBM has a *bipartite structure*); see the right panel of Fig. 10.1. As mentioned, I will focus only on RBMs, and their extensions to *deep Boltzmann machines* (Section 10.1.3).

A (restricted) Boltzmann machine has a *joint probability distribution* between the visible and hidden nodes given by the *Gibbs distribution*,

$$p(\mathbf{v}, \mathbf{h}) = \frac{1}{\mathcal{Z}} e^{-E(\mathbf{v}, \mathbf{h})}, \tag{10.1}$$

with an appropriate energy function. The quantity, $\mathcal{Z}$, is a normalization constant, and is known in Statistical Mechanics as the *partition function* (Goldenfeld, 1992). In the present case, it is

$$\mathcal{Z} = \sum_{\{\mathbf{v}\}} \sum_{\{\mathbf{h}\}} e^{-E(\mathbf{v}, \mathbf{h})}, \tag{10.2}$$

Boltzmann (or Gibbs) distribution

The probability that a system at equilibrium with a bath of temperature T is in state E_i (over a number, M, of states E_j, $j = 1, \ldots, M$) is:

$$p(E_i) = \frac{e^{-\frac{E_i}{k_B T}}}{\mathcal{Z}},$$

where the *partition function* is

$$\mathcal{Z} = \sum_{j=1}^{M} e^{-\frac{E_j}{k_B T}},$$

and k_B is the Boltzmann constant.

[5] The attentive reader may have noticed that the first term on the right-hand side (RHS) of eqn (10.3) is formally the same (with a different connectivity though) as the one appearing in the Hamiltonian (9.1) for *spin glasses*, with the visible and hidden nodes representing spins, s_i, while the weights represent their interactions, J_{ij}. The biases of the RBM would then represent *local fields* acting on the spins.

[6] The biases are very important for the representational power of RBMs: they allow RBMs to represent also non-symmetric distributions (Goodfellow *et al.*, 2016).

Training vs. testing
Given an NN, a *training* set is a set of data on which the NN is trained. Once the weights and biases of the NN have been obtained from the training—a *model* has been generated—a *testing* set (which typically excludes the training data) is used to test the predictive power of the model.

MNIST dataset
Modified National Institute of Standards and Technology (MNIST) dataset is a database of of 28 × 28 handwritten digits in grey scale, that is typically used for training and testing neural networks. It contains 60,000 training images and 10,000 testing images.

---

where the sum runs over all possible configurations of the visible and hidden nodes.

Note that the Gibbs distribution has the same *form* for both the Boltzmann machine and its restricted version. The energy and partition functions, instead, are different.

For an RBM, the *energy function* is[5]

$$E(\mathbf{v}, \mathbf{h}) = -\sum_{i>j} w_{ij} h_i v_j - \sum_i b_i v_i - \sum_j c_j h_j, \qquad (10.3)$$

where $w_{ij}(\in \mathbb{R})$ is the *weight* between the i-th hidden neuron and the j-th visible neuron, and b_j, c_i are real numbers indicating the *biases* of the neurons.[6]

By defining the 'bias vectors' $\mathbf{b} \in \mathbb{R}^n$, $\mathbf{c} \in \mathbb{R}^m$, and 'weight matrix' $\mathbf{W} \in \mathbb{R}^{n \times m}$, we can also write eqn (10.3) as:

$$E(\mathbf{v}, \mathbf{h}) = -\mathbf{v}^T \mathbf{W} \mathbf{h} - \mathbf{b}^T \mathbf{v} - \mathbf{c}^T \mathbf{h}, \qquad (10.4)$$

where the symbol T stands for transpose.

Since there is *no* intra-layer connectivity (in both the visible and hidden layer of an RBM), given the hidden variables, each visible node is *conditionally independent* of all the others:

$$p(v_i, v_j | \mathbf{h}) = p(v_i | \mathbf{h}) p(v_j | \mathbf{h}). \qquad (10.5)$$

This is an important simplification, which does *not* hold for deep networks, even those constructed from RBMs (Section 10.1.3).

Training data

Having defined the structure of an RBM, the next step is to provide a set of data to train it on. The set of data can vary widely. In this book, I will focus only on *synthetic* data sets, which are mainly used for testing, and the MNIST data set, which is a Machine Learning standard on which different methods are trained and tested on (Goodfellow *et al.*, 2016).

The data distribution over the visible layer, $q(\mathbf{v})$, is typically specified by a data set of samples. The unique elements of this data set can be indicated by $\mathcal{D} = \{\mathbf{v}_1, \cdots, \mathbf{v}_{n_d}\} \subset \Omega$, where Ω is the space of all binary sequences of length n (the number of visible nodes).

Given the set of data, we can now define what we mean by 'training'.

KL divergence

Suppose first that you know the data distribution, $q(\mathbf{v})$. The training of an RBM then consists in matching its *marginal* probability distribution over the visible layer, $p(\mathbf{v}) = \sum_{\{\mathbf{h}\}} p(\mathbf{v}, \mathbf{h})$, to the data distribution,

$q(\mathbf{v})$. The marginal distribution can be written as (Goodfellow *et al.*, 2016),

$$p(\mathbf{v}) = \frac{1}{\mathcal{Z}} \prod_{i=1}^{n} e^{b_i v_i} \prod_{j=1}^{m} \left(1 + e^{c_j + \sum_{i'=1}^{n} w_{i'j} v_{i'}} \right). \qquad (10.6)$$

Training an RBM then amounts to a search for the appropriate weights and biases that will minimize the quantity known as the *Kullback-Leibler* (KL) *divergence* (or *relative entropy*) between the two distributions,

$$\mathrm{KL}(q||p) = \sum_{\{\mathbf{v}\}} q(\mathbf{v}) \log \frac{q(\mathbf{v})}{p(\mathbf{v})}, \qquad KL\ divergence, \qquad (10.7)$$

which quantifies how different the RBM marginal distribution is with respect to the data one. It is clear from eqn (10.7) that when the RBM marginal distribution equals *exactly* the data distribution the KL divergence is zero. If we could achieve such an ideal training we would say that we have trained the RBM *at optimality* (i.e., there is no loss of information in approximating the data distribution with the model distribution, without *overfitting*—Section 10.1.4). Therefore, we can also define:

> **Training an RBM**
> Find a set of weights and biases that *minimizes* the KL divergence (10.7), ideally as close as possible to *zero* (optimality).

Log-likelihood

Here, however, we need to make an important remark. For certain data sets—synthetic data sets, in particular—the data distribution, $q(\mathbf{v})$, is indeed known explicitly. For other data sets (e.g., the MNIST one),[7] instead, $q(\mathbf{v})$ is not known explicitly, even though it is known 'implicitly' through the data themselves.

Therefore, since the KL divergence (10.7) requires the data distribution, $q(\mathbf{v})$, if the latter is not known explicitly, another measure of 'quality of training' is needed. This can be provided by the *log-likelihood* of the probability distribution over the visible nodes:

$$\mathrm{LL}(p) = \sum_{\mathbf{v} \in \mathcal{D}} \log p(\mathbf{v}), \qquad log\text{-}likelihood, \qquad (10.8)$$

where $p(\mathbf{v})$ is explicitly written in eqn (10.6), and $\mathcal{D}$ contains the unique elements of the data set.

Of course, the larger the log-likelihood, the better the training. Therefore, we can equally say:

Joint, marginal, and conditional distributions
Consider two sets of stochastic (random) variables, like the visible and hidden nodes of an RBM, $\mathbf{v}$ and $\mathbf{h}$. Their *joint* probability distribution, $p(\mathbf{v}, \mathbf{h})$ gives the probability density for the variables $\mathbf{v}$ and $\mathbf{h}$ to acquire certain values (*realizations* of the stochastic variables).
If we care only about the probability that the variables $\mathbf{v}$ acquire certain values, *irrespective* of the values acquired by the second set $\mathbf{h}$, we obtain the *marginal* probability distribution, $p(\mathbf{v}) = \sum_{\{\mathbf{h}\}} p(\mathbf{v}, \mathbf{h})$, by *integrating out* the realizations of the variables $\mathbf{h}$. Finally, if we want to know, say, what is the probability of obtaining certain values of the variables $\mathbf{h}$, *given* that the other variables, $\mathbf{v}$, have acquired specific values, we obtain the *conditional* probability distribution, $p(\mathbf{h} \,|\, \mathbf{v})$.
The three distributions are related by *Bayes' rule*:

$$p(\mathbf{v}, \mathbf{h}) = p(\mathbf{v})p(\mathbf{h} \,|\, \mathbf{v}).$$

See, e.g., (van Kampen, 1992).

[7]And, in fact, for the majority of 'real-world' data sets.

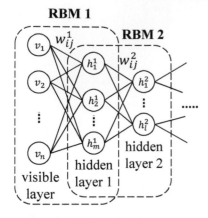

Fig. 10.2 Deep Boltzmann machine (DBM) made of two RBMs. The connections between the layers represent the weights, $w_{ij}^k \in \mathbb{R}$ (biases not shown).

[8] An example of NNs employed in *Deep Learning*.

Deep Learning
The subfield of Machine Learning that employs neural networks with three or more layers (including the visible layer) is typically called *Deep Learning* (Goodfellow *et al.*, 2016). Its success has been tied to the availability of large amounts of data ('big data') and substantial computational resources to train the networks. However, the ever-increasing computational resources (with consequent economic and environmental costs) needed to improve *accuracy* (cf. Section 10.3), even marginally, are now becoming a serious bottleneck for further advancements in this field; see, e.g. (Thompson *et al.*, 2021).

Training an RBM
Find a set of weights and biases that *maximizes* the log-likelihood (10.8).

10.1.3 Deep Boltzmann machines

All of the previous considerations can be extended to *deep Boltzmann machines* (DBMs), namely NNs with several hidden layers, each pair of them a distinct Boltzamnn machine.[8] This is schematically shown in Fig. 10.2, for the case of two RBMs.

More precisely, DBMs are *undirected* weighted graphs that differentiate between n visible nodes, $\mathbf{v} \in \{0,1\}^n$ and ℓ layers of n_ℓ 'latent', or 'hidden', nodes, $\mathbf{h}^{(\ell)} \in \{0,1\}^{n_\ell}$, not directly constrained by the data (Goodfellow *et al.*, 2016).

If we choose, like for an RBM, to have no connections *within* a layer, then each state of the machine corresponds to an energy of the form (cf. eqn (10.4))

$$E(\mathbf{v}, \mathbf{h}^{(1)}, \cdots, \mathbf{h}^{(\ell)}) = -\mathbf{b}^T\mathbf{v} - \mathbf{v}^T\mathbf{W}^{(1)}\mathbf{h}^{(1)} \tag{10.9}$$

$$- \sum_{i=1}^{\ell} \mathbf{c}^{(i)T}\mathbf{h}^{(i)} - \sum_{i=2}^{\ell} \mathbf{h}^{(i-1)T}\mathbf{W}^{(i-1)}\mathbf{h}^{(i)},$$

where the biases $\mathbf{b} \in \mathbb{R}^n$, $\mathbf{c}^{(\ell)} \in \mathbb{R}^{n_\ell}$, and weights $\mathbf{W}^\ell \in \mathbb{R}^{n_{\ell-1} \times n_\ell}$ are the learnable parameters.

The probability distribution of this DBN is again given by the Gibbs distribution, eqn (10.1), with the partition function (10.2) summed over all visible and hidden nodes of all layers.

We can then employ either the KL divergence (10.7) or the log-likelihood (10.8) to quantify the quality of the training of DBMs. In analogy to the RBM case, we then define

Training a deep Boltzmann machine
Find a set of weights and biases that *minimizes* the KL divergence (10.7), or *maximizes* the log-likelihood (10.8).

Before discussing the actual training of these NNs, it is worth stressing one important point.

10.1.4 On the representational power of NNs

As I have anticipated in Section 10.1, I chose restricted Boltzmann machines as model NNs because they enjoy a *universal approximation theorem*. In the case of RBMs this theorem says that an RBM with $n_d + 1$

hidden nodes (where n_d is the number of visible configurations with non-zero probability) can represent an arbitrary discrete probability distribution, *if the RBM has appropriate weights and biases* (Le Roux and Bengio, 2008). Similar theorems hold also for some other types of NNs (Goodfellow *et al.*, 2016).

However, notice that these theorems concern only the *existence* of a set of parameters (weights and biases) such that the NN with *such* parameters can represent a given distribution. It does *not* specify *how* those parameters are obtained.

This is *not* a minor point. In fact, even if, say an RBM, may, in principle, approximate *any* discrete probability distribution, it is not at all obvious (and, indeed, it is *not* typically the case) that it does so when trained using the most common method, Gibbs sampling (Section 10.2.1). The quality and efficiency of the training is then as important, if not *more* important, than the model NN we choose to learn data.

With these preliminaries, I am now ready to discuss the actual training of an NN. There are essentially two ways to train NNs: an *unsupervised* one and a *supervised* one. I will discuss them separately.

> **An important remark**
> As we will see in a moment, *supervised* learning is considerably 'easier' to accomplish than the *unsupervised* one (Goodfellow *et al.*, 2016). In fact, due to its difficulty (with available methods), unsupervised learning has been all but abandoned in favor of the supervised one, which is at the core of the *second* wave of AI (as defined in Section 2.5). However, in order to realize the *third* wave of AI, one really needs efficient methods to do *unsupervised* learning. This is because, for a machine to adapt to a changing and complex environment 'on the fly', it needs to be trained without the help of 'labels' (as it is done in the supervised setting). This is where the MemComputing approach is particularly helpful.

10.2   Unsupervised learning

The unsupervised learning of an RBM or a DBM[9] is framed as a *gradient ascent* over the *log-likelihood* of the observed data. Formally, this is done by taking the gradient of (10.8), with respect to the weights and biases,[10] which gives eqns (10.10), (10.11), and (10.12).

Overfitting
If the NN model has more parameters than the data, then *overfitting* (sometimes called 'over training' in the context of Machine Learning) may occur, which leads to a significant deterioration of the model ability to generalize. In other words, if overfitting occurs, the NN has 'memorized' the training data, and has then a hard time recognizing the testing data.

Capacity of an NN
The *capacity* of an NN refers to its ability to fit a wide range of functions, and is related to the amount of (syntactic) information (Section 1.3) that can be stored in the network. If the capacity is too small, the network may not be able to learn the training data (*underfitting*). If the capacity is too large, it may simply store the training data set, hence it would be unable to generalize (*overfitting*). See, e.g., (Goodfellow *et al.*, 2016) for all these notions.

[9] In the context of supervised learning this is also called *pre-training* (Section 10.3).

[10] Namely, one computes $\nabla_{\mathbf{w}}\mathrm{LL}(p)$, $\nabla_{\mathbf{b}}\mathrm{LL}(p)$, and $\nabla_{\mathbf{c}}\mathrm{LL}(p)$ (Goodfellow *et al.*, 2016).

Hebbian learning
The rule that prescribes an *increase* of weight between two neurons (nodes) that activate simultaneously is called *Hebbian learning*. It tends to *decrease* the energy in eqn (10.4). The data term in eqn (10.13) realizes such a learning rule.

Hyper-parameters
Apart from the weights and biases that have to be 'learned' during training, an NN model is characterized by a few additional parameters, called *hyper-parameters*. Examples include the number of nodes or layers of the network (determining its architecture), or the 'learning rate' and 'momentum' in the gradient optimization procedure (eqn (10.10)), etc. Tuning of these parameters may affect the training substantially, in terms of its speed and quality. This tuning is typically performed on a 'trial and error' basis (Goodfellow *et al.*, 2016).

(Mini-)batches
It is common practice to calculate the updates of parameters (weights and biases) after *all* training data in the training set have been fed to the NN. This is called a *batch*. One can further split the training data set into smaller *mini-batches*, and compute an update of the parameters after all mini-batches have been evaluated. The advantage of this method is that it typically reduces the fluctuations in the evaluation of the averages in eqns (10.10), (10.11), and (10.12) (Goodfellow *et al.*, 2016).

Starting from, e.g., a *random initialization* of the weights and biases in eqn (10.3), one *updates* these quantities by performing what is known as a *stochastic gradient optimization*. If the weights at the n-th iteration changed by Δw_{ij}^n, the weights at the $(n+1)$-th iteration (an *update*) would then change according to (see (Goodfellow *et al.*, 2016) for a derivation of this form):

$$\Delta w_{ij}^{n+1} = \alpha \Delta w_{ij}^n + \epsilon[\langle v_i h_j \rangle_{\text{DATA}} - \langle v_i h_j \rangle_{\text{MODEL}}], \qquad (10.10)$$

where $\alpha \geq 0$ is called the *momentum* (because it 'accelerates' the training), and $\epsilon > 0$ is the *learning rate* (because it determines how fast the learning occurs; cf. with the parameter ζ in eqn (8.2)).

A similar update procedure is applied to the biases b_i and c_j:

$$\Delta b_i^{n+1} = \alpha \Delta b_i^n + \epsilon[\langle v_i \rangle_{\text{DATA}} - \langle v_i \rangle_{\text{MODEL}}], \qquad (10.11)$$

$$\Delta c_j^{n+1} = \alpha \Delta c_j^n + \epsilon[\langle h_j \rangle_{\text{DATA}} - \langle h_j \rangle_{\text{MODEL}}]. \qquad (10.12)$$

Since in these expressions we have a parameter, α, called 'momentum', this form of the weight updates is often called a 'stochastic gradient optimization with momentum'.

Apart from the hyper-parameters α and ϵ, the most important terms in eqns (10.10), (10.11), and (10.12) are the *expectation values* (averages); these averages are indicated with the symbols $\langle \cdot \rangle$.

The first one is called 'data' term, the second is the 'model' term. Let's look at them separately.

Data term
The first expectation value on the RHS of eqns (10.10), (10.11), and (10.12) is taken with respect to the *conditional probability distribution* with the data *fixed* at the visible layer. For instance:[11]

$$\langle v_i h_j \rangle_{\text{DATA}} \equiv \langle v_i h_j \rangle_{q(\mathbf{v})p(\mathbf{h}\,|\,\mathbf{v})}, \qquad (10.13)$$

and similar expressions for the biases.

The average in eqn (10.13) is then relatively easy to compute because the bipartite structure of the RBM makes the hidden nodes *conditionally independent*, given the visible nodes (and vice versa). This structure produces a closed form for the conditional distribution given by (Goodfellow *et al.*, 2016):

$$p(h_j = 1\,|\,\mathbf{v}) = \sigma(\sum_i w_{ij} v_i + c_j), \qquad (10.14)$$

where $\sigma(x) = (1 + e^{-x})^{-1}$, namely it is a *sigmoid* (or *logistic*) function. Given the data on the visible nodes, one can then sample the hidden layer configurations, that need to enter eqn (10.13), by means of the sigmoid function (10.14), and this is a trivial procedure.

---

[11]The symbol $\langle v_i h_j \rangle_{q(\mathbf{v})p(\mathbf{h}\,|\,\mathbf{v})}$ means $\sum_{\{\mathbf{v}\},\{\mathbf{h}\}} v_i h_j q(\mathbf{v}) p(\mathbf{h}\,|\,\mathbf{v})$.

An important remark

It is important to stress once more that the closed form (10.14) for the conditional probability distribution is *no longer valid* for *deep* NNs, even those made of RBMs (see Section 10.5).

Model term

On the other hand, the second expectation value on the right-hand side (RHS) of eqns (10.10), (10.11), and (10.12) is very difficult to evaluate.

It requires obtaining independent samples from a high-dimensional model distribution:[12]

$$\langle v_i h_j \rangle_{\text{MODEL}} \equiv \langle v_i h_j \rangle_{p(\mathbf{v}, \mathbf{h})}, \tag{10.15}$$

and similarly for the biases. Here, $p(\mathbf{v}, \mathbf{h})$ is the *joint* probability distribution over the visible and hidden nodes.[13],[14]

This problem easily becomes (*exponentially*) prohibitive with increasing size of the NN (number of its nodes).

An important remark

Note that the *joint* probability distribution of the visible and hidden nodes, $p(\mathbf{v}, \mathbf{h})$, appears in the model term (10.15). Instead, for the KL divergence (10.7) or log-likelihood (10.8), the *marginal* distribution over the visible nodes, $p(\mathbf{v})$, needs to be evaluated.

This is where statistical methods are required, that provide a good approximation of this model distribution, and are computationally advantageous. Several methods can be devised, but they are all essentially a variation of *Gibbs sampling*, which is a special case of the *Markov Chain Monte Carlo* (MCMC) method I have anticipated in Section 9.2.

Since this is the most popular approach employed by the Machine Learning community, I will briefly review it here for completeness and to distinguish it from the MemComputing one I will discuss in Sections 10.4, 10.5, and 10.6.

The reader who already knows this method (or is not particularly interested in learning it), may skip the following Section.

Anti-Hebbian learning
The rule that prescribes a *decrease* of weight between two neurons (nodes) that activate simultaneously is called *anti-Hebbian learning*. It tends to *increase* the energy in eqn (10.4). The model term in eqn (10.15) realizes such a learning rule.

[12]Here, the term $\langle v_i h_j \rangle_{p(\mathbf{v}, \mathbf{h})} = \sum_{\{\mathbf{v}\}, \{\mathbf{h}\}} v_i h_j p(\mathbf{v}, \mathbf{h})$.

[13]Essentially, this term originates from the partition function (10.2), whose evaluation requires resources that increase exponentially with increasing NN size (total number of nodes).

[14]Taken together, eqns (10.13) and (10.15), then realize a competition between a *Hebbian* and an *anti-Hebbian* learning rule in eqn (10.10). The training ends when these two learning rules balance.

Sampling from a distribution
Given a probability distribution, $p(\mathbf{x})$, over a set of stochastic variables, $\mathbf{x}$, we call *sampling* from such a distribution the act of drawing values of $\mathbf{x}$ with a likelihood given by such a distribution.

Stochastic process

Given a set of *stochastic variables*, $\mathbf{x} = \{x_j\}$, with $j = 1, \ldots, m$, and a *deterministic* variable, t, called 'time' (but not necessarily corresponding to the physical, continuous time), *any* function $f(\mathbf{x}, t)$ defines a *stochastic process*, $Y_\mathbf{x}(t) \equiv f(\mathbf{x}, t)$.

Markov process

Consider an arbitrary set of n consecutive times, $t_1 < t_2 < \cdots < t_n$, and a *stochastic process* $Y_\mathbf{x}(t)$. If the *conditional* probability density that the stochastic process acquires a value y_n at time t_n depends *only* on the value y_{n-1} it acquired at the immediate previous time t_{n-1}, and *not* on the values at the other times, i.e.,

$$p(y_n, t_n | y_1, t_1; \ldots; y_{n-1}, t_{n-1}) = p(y_n, t_n | y_{n-1}, t_{n-1}),$$

then the process is called *Markovian*, and $p(y_n, t_n | y_{n-1}, t_{n-1})$ is its *transition probability*. The process has *no* memory of its past dynamics.

Markov chain

A *Markov chain* is a particular class of Markov processes such that *i)* the *range* of values of $Y_\mathbf{x}(t)$ is a *discrete* set of states; *ii)* the time variable t is *discrete*, and takes only integer values ($t \in \mathbb{Z}$); *iii)* the transition probability depends only on time differences (it is a *homogeneous* process). See (van Kampen, 1992) for a more in-depth discussion of stochastic processes.

10.2.1 Gibbs sampling and contrastive divergence

Let me start with a clarification on the nomenclature, which I hope will help the reader when navigating the specialized literature. *Gibbs sampling* and *constrastive divergence* (CD) are fundamentally the *same* algorithmic procedure.

The only difference between the two is that CD—which is the term often used in Machine Learning in lieu of Gibbs sampling—employs as a starting point, data from a data set.

Gibbs sampling

Gibbs sampling is a special case of the Metropolis-Hastings algorithm I have briefly introduced in Section 9.2. It can be summarized as follows (Gamerman and Lopes, 2006).

Suppose we want to sample from a *joint* probability distribution $p(x_1, x_2, \ldots, x_l)$, which depends on l stochastic variables. However, it is somewhat easier to sample from appropriate *conditional* distributions (which is the case for RBMs; see the following example). We can then proceed as follows.

Let us denote with $\mathbf{x} = \{x_1, x_2, \ldots, x_l\}$ the vector of all variables, and initialize $\mathbf{x}$ with a given initial value for all variables: $\mathbf{x}^{(i)}$.

In the next step, $\mathbf{x}^{(i+1)}$, we sample from the *conditional* distribution of a given variable, say x_j, conditioned to all the other ones as follows: $p(x_j^{(i+1)} | \{x_{\alpha<j}^{(i+1)}\}, \{x_{\alpha>j}^{(i)}\})$, where $\alpha = 1, \cdots, l$.

In other words, we sample from a distribution over x_j conditioned, up to j variables, to the values of those variables at the $(i+1)$-th iteration (updated values), and conditioned to variables, from $j+1$ to l, evaluated at the i-th iteration. And we do this for all variables.

We then continue the previous procedure k times, by always using the most recent updated values. This process creates a *Markov chain* of length k, and the samples so obtained approximate the *joint* distribution of all variables.

In fact, in the limit of $k \to \infty$, the Markov chain converges to the joint stationary distribution, independently of the initial state of the chain (provided the chain is 'irreducible' and 'aperiodic'; see (Gamerman and Lopes, 2006) for a formal demonstration of this statement and the meaning of these terms).

However, this convergence can be painfully slow in high dimensional spaces (when l is large). This is because this type of sampling is essentially a 'random walk' in the variables' space.

And if the distribution $p(\mathbf{v})$ has modes that are quite distinct from the modes of the data distribution, $q(\mathbf{v})$, (called *spurious modes*) it would be very difficult for Gibbs sampling to find the probability of these modes and, as a consequence, the full probability distribution.

This is, indeed, the problem faced by this procedure applied to RBMs.

Contrastive divergence

In the case of RBMs, Gibbs sampling is called *contrastive divergence* (CD) (Goodfellow *et al.*, 2016), because the Markov chain is initialized from a point in the *data set*, $\mathbf{v}$, and the hidden and visible nodes are sequentially re-sampled a k number of times.

In practice, the sequential sampling of each layer is given by the sigmoidal conditional probabilities as in eqn (10.14). For instance, starting from initializing the visible nodes with the data, one samples the hidden ones from

$$p(h_i = 1|\mathbf{v}) = \sigma \left(\sum_j w_{ij} v_j + c_i \right). \tag{10.16}$$

One can then sample the visible nodes, from the values of the hidden nodes, as

$$p(v_j = 1|\mathbf{h}) = \sigma \left(\sum_i w_{ij} h_i + b_j \right). \tag{10.17}$$

This procedure is repeated k times, and is known as CD-k. The visible and hidden nodes configurations one obtains are then used to determine the average (10.15).

However, as anticipated, the data distribution may have spurious modes, so that the convergence is slow. In practice, usually only one iteration, CD-1, is used (Goodfellow *et al.*, 2016).

This means that although choosing some finite k introduces a 'bias' (in the sense that the approximate distribution is skewed compared to the exact one), it is empirically found that using $k = 1$ gives a sufficient signal for *supervised* learning. However, I will show in Section 10.4, and 10.5, that with this procedure the *unsupervised* training may diverge for both RBMs and deep NNs (and increasing k does not seem to help).

Spurious modes
Modes of the NN probability distribution $p(\mathbf{v})$ that are 'far away' from the modes of the data distribution, $q(\mathbf{v})$, are sometimes called *spurious modes*. Since the CD procedure initializes the Markov chain at the data distribution, quite often, these modes will not be sampled effectively. This may lead to the divergence of the NN training with Gibbs sampling. See, e.g., Fig. 10.6 for an explicit example of this issue, and (Goodfellow *et al.*, 2016).

Backpropagation
Backward propagation of errors, or simply *backpropagation*, is a relatively fast procedure (an algorithm) to compute the parameters (weights and biases) of a *feedforward* NN, by following the gradient of an appropriate *cost function* (Goodfellow *et al.*, 2016).

[15]For instance, the labels in the case of the MNIST data set are the digits from 0 to 9.

Feedforward Network

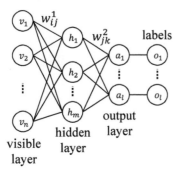

Fig. 10.3 Schematic of a feedforward network (where the edges have a direction, and no cycles are allowed) with an *output* layer for *supervised learning*, with respect to some *labels*, through *backpropagation* of errors.

[16]It is analogous to having a deep NN (with two 'hidden' layers), but with the last hidden-layer configuration 'forced' to match the appropriate labels.

10.3 Supervised learning

Since the previous unsupervised learning procedure is really difficult to accomplish (in general, not just for RBMs), it is sometimes used only to initialize the weights and biases (to *pre-train* the network), and then run a *backpropagation* over *labels* to fine tune the parameters of the network.[15]

The backpropagation procedure accomplishes what is known as *supervised learning*, and is oftentimes done even without pre-training (from a randomly initialized set of weights and biases). However, as I will show in Section 10.6, pre-training with MemComputing improves considerably the supervised learning procedure by providing a very good starting point for the downstream supervised task.

In addition, since backpropagation is a relatively fast algorithm, it has become the workhorse of training NNs, practically replacing unsupervised learning almost completely. However, as I have mentioned in Section 10.1.4, transitioning from the second to the third wave of AI (cf. definitions in Section 2.4.1), one really needs efficient ways of doing *unsupervised* learning, namely without relying on labels.

Since the latter ones are needed in supervised learning, the network is then augmented by an *output* layer that contains the labels with respect to which we want to perform backpropagation.[16] This is shown schematically in Fig. 10.3 for a *feedforward network* for which the backpropagation procedure—I will describe later in this Section—was first introduced. I will then discuss how to modify it for RBMs (which are *undirected graphs*).

The main idea behind backpropagation is to compute the errors made in the training of an NN, as evaluated on the output layer with respect to some labels, and propagate them 'back' into the network so as to appropriately change the weights and biases, till a desired low error (or high *accuracy*), is achieved (by avoiding, however, *overfitting*).

I will now provide a minimal outline of how this is done in practice, and refer the reader to more specialized textbooks for a complete discussion, including several nuances of how to apply it in certain contexts (Goodfellow *et al.*, 2016). The reader not particularly interested in this procedure can instead directly jump to Section 10.4.

Cost functions and backpropagation

For the sake of simplicity, let us first focus on the feedforward NN schematically shown in Fig. 10.3, with only weights (no biases), and one hidden layer. The extension to the biases and more than one hidden layer is straightforward.

Now, given a *training sample* (from a data set) that we feed into the visible layer, we obtain an output value in each node of the output layer. This output will, most likely, be different than the desired one, as given by the labels. This difference quantifies an error (or 'cost').

Given this cost, we want to modify the weights (and biases) of the NN in proportion to how much this error varies with respect to them. Therefore, we need to choose a *cost function*, C, to quantify how much we need to change the weights, w_{ij}, and biases, b_j, in order to match the output labels. This means that we need to compute the *gradients* of C: $\partial C/\partial w_{ij}$ and $\partial C/\partial b_j$.

The choice of cost function is not unique, and several forms can be used, some providing better performance in specific cases (Goodfellow *et al.*, 2016). As an example, we can choose a common one (for one training sample; one can then average over many training samples):

$$C = \frac{1}{2}||o - a||^2 \equiv \frac{1}{2}\sum_k (o_k - a_k)^2, \quad \textit{cost function}, \quad (10.18)$$

where o_k is the desired output (label) of the output node k (with k running over all output nodes), and a_k is an *activation function* of the k-th output node.

For easy analytical manipulations, this activation function can be chosen as (if the biases, d_k, of the output layer are also present):

$$a_k = \sigma(\sum_j w_{jk}^2 h_j + d_k) \equiv \sigma(z), \quad (10.19)$$

where the sum runs over the hidden layer adjacent to the output layer (Fig. 10.3), h_j are the values of these hidden nodes, and $\sigma(z)$ is the sigmoid function of its argument, z.

Whatever the choices of cost and activation functions, one can then update the weights, w_{ij}, (and biases) using a similar *gradient* approach I have defined in Section 10.2, eqn (10.10). For each iteration of the backpropagation procedure, we can then compute

Accuracy
In supervised learning, the quality of the training can be determined by the (classification) *accuracy*, namely the fraction of predictions that the model correctly makes out of all the testing samples:

$$\text{Accuracy} = \frac{\#\text{ correct predictions}}{\text{total }\#\text{ of test samples}}$$

This is, however, not the only measure of training quality for supervised learning. See, e.g., Section 10.6 and (Goodfellow *et al.*, 2016).

Training samples
A pair of *input* vector and *output* ('supervisory') value is called a *training sample* (or example). In supervised learning, several training samples are typically employed, and the cost function is averaged over them. This means that, if we have n training samples, and a cost function C_i for each one of them ($i = 1, \ldots n$), then the total cost function is $C = \sum_i C_i/n$. In this context, a function for a single training sample is sometimes called the *loss function*, while the term 'cost function' is reserved for the whole training set, namely for the *average* loss over the entire training set.

the change in weights as (if we do not have any 'momentum' term, see, e.g., eqn (10.10)):

$$\Delta w_{ij} = -\eta \frac{\partial C}{\partial w_{ij}}, \tag{10.20}$$

where $\eta > 0$ is a *learning rate* (typically distinct from the one appearing in eqn (10.10)).

From eqn (10.20) we then see that whatever is the sign of $\partial C / \partial w_{ij}$, the cost function C always *decreases* with every weight update.

Now, with the choice of cost function (10.19), we can easily find the gradient variation with respect to the weights, w_{ij}^1, between the visible and hidden layers, by applying the *chain rule*:

$$\frac{\partial C}{\partial w_{ij}^1} = \sum_{lk} \frac{\partial C}{\partial a_k} \frac{\partial a_k}{\partial h_l} \frac{\partial h_l}{\partial w_{ij}^1}, \tag{10.21}$$

where I have indicated with h_j the activations of the hidden layers (e.g., the sigmoid function (10.14)). This expression is very easy to compute:

$$\frac{\partial C}{\partial w_{ij}^1} = \sum_{lk} (a_k - o_k) \sigma'(h_l) w_{lk}^2 \sigma'(v_i) w_{ij}^1, \tag{10.22}$$

where $\sigma'(h_l)$ is the derivative of the activation function (10.19) with respect to h_l, and $\sigma'(v_i)$ is the derivative of (10.14) with respect to v_i. (Recall that the derivative of a sigmoid function, $\sigma(z)$, with respect to its argument, z, is $\sigma'(z) = \sigma(z)(1 - \sigma(z))$.)

Similar considerations hold for the other weights and biases, and, by applying the chain rule as before, for more than one hidden layer.

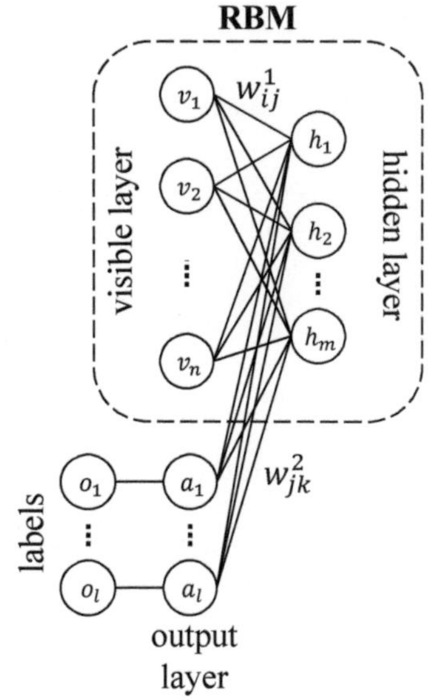

Fig. 10.4 *Semi-supervised training* of an RBM with an *output* layer for *labels*.

Semi-supervised training of RBMs and DBMs

As I already mentioned, the backpropagation procedure has all but supplanted the unsupervised training of NNs, and it is now applied to a wide variety of NN models (Goodfellow *et al.*, 2016). However, such a procedure is strictly valid for *feedforward networks*. An RBM, instead, is an *undirected graph*, not a feedforward network.

By treating them as directed graphs during the supervised phase of the training does not take full advantage of their topology, in particular some correlations between the nodes.

Therefore, this procedure would, ideally, need to be somewhat modified for RBMs or DBMs. Here, I will briefly show how this could be done.

Let us start by re-interpreting the feedforward network of Fig. 10.3

as the one of Fig. 10.4. Although Fig. 10.4 may seem like a trivial re-drawing of Fig. 10.3, it is actually quite different.

For the network in Fig. 10.3 we seek the *conditional* probability distribution of the NN, given the visible layer, and improve on it by backpropagation (treating the RBM as a feedforward network). Instead, for the network in Fig. 10.4 we look for the *joint* probability distribution of *both* the visible layer *and* the output layer (constrained by the labels).

One could then do a 'semi-supervised' training of the RBM, by employing the gradient updates as in eqns (10.10), (10.11), and (10.12), where the *data* term in those equations is computed with respect to *both* the data on the visible layer *and* the labels on the output layer, while the *model* term is estimated with some sampling procedure (e.g., Gibbs sampling or the MemComputing one I will discuss soon).

Although I will not show any results obtained with this approach, it is worth noting that MemComputing, as discussed in Sections 10.4, 10.5, and 10.6 could potentially take advantage of this way of interpreting the NN, and lead to better 'semi-supervised' training compared to backpropagation. I will leave this for future work.

After a discussion on the difference between unsupervised and supervised methods to train NNs, I can now discuss how MemComputing can be used in both contexts. I will first start by showing how to employ it in the *unsupervised training* of RBMs with a KL divergence (10.7) that can be made as small as possible, namely training them *at optimality*.

In Chapter 10.5, I will show that this method extends to the case of *deep networks* as well, and allows these NNs to be trained *jointly* (as a single unit), with much less number of parameters than the single RBM. In Chapter 10.6, I will show how to use MemComputing as a better starting point (*pre-training*) for a *supervised* learning phase, thus accelerating considerably the training.

10.4 Unsupervised training of RBMs at optimality with MemComputing

Let us first focus on a single RBM, and see how MemComputing can be employed advantageously in this case. Let us recall from Section 10.2 that the stochastic gradient optimization—to either minimize the KL divergence (10.7), or maximize the log-likelihood (10.8)—contains two terms: a *data* term (10.13), and a *model* term (10.15).

As I already mentioned in Section 10.2, the data term is easy to estimate for an RBM, while the model term is the challenging one. In fact,

Mode, mean, and median of a distribution
Consider a continuous probability distribution, $p(x)$, over a stochastic variable, x. The *mode* of such a distribution is the most probable outcome (and may occur at different values of x). The *mean* is the *average* (or expected value): $\mu = \int dx\, x\, p(x)$. The *median* is that value of x (call it m) that separates the higher half from the lower half of the distribution: $\int_{-\infty}^{m} dx\, p(x) = 1/2$. For a Gaussian distribution, all these quantities are the same, but they can be substantially different for arbitrary distributions. All these notions can be extended to multivariate distributions, whether discrete or continuous.

[17]In turn, the mode carries *global* information on the distribution the NN is attempting to learn. Therefore, the main issue with the currently used training methods for Deep Learning is their over-reliance on *local* information—provided by the gradient (10.10)—and lack of *global* (*topological*) information. This is the same problem that plagues *all* difficult combinatorial optimization problems, as I have discussed in Section 2.6.

[18]The (topological) reason for this efficiency will be discussed in Chapters 11 and 12.

it requires computing the *partition function* of the RBM, eqn (10.2), which is hard even to approximate. We are then left to *sample* from the high-dimensional joint probability distribution, $p(\mathbf{v}, \mathbf{h})$, of visible and hidden nodes (the *model distribution*).

However, as I have anticipated in Section 10.2.1, Gibbs sampling, i.e., contrastive divergence (CD), is very slow at this task. This issue is further exacerbated if the model distribution represented by the RBM contains *modes* where the data distribution has negligible probability, namely the two distributions have quite different modes (*spurious modes*). CD then has a hard time to find and correct the probability of these modes.

Therefore, the *most difficult* part of a sampling procedure is to sample the mode (or modes) of a probability distribution.[17] But sampling the mode of an arbitrary distribution is like finding a good approximation to the optimum (or the optimum itself) of a *non-convex landscape*, a problem which is NP-hard (cf. Section 2.3.3).

However, I have shown in Chapter 9 several examples where Mem-Computing is able to find the optimum, or a very good approximation of the optimum, of a non-convex landscape very *efficiently* (namely with a cost that scales *polynomially* with problem size).[18]

In addition, in Section 9.2.2, I have shown how to transform a problem from QUBO to weighted MAX-SAT. This transformation comes really handy now, in view of the fact that finding the *ground state* of the RBM energy (10.3) can be formulated as a QUBO problem, which, in turn, can be formulated as a weighted MAX-SAT problem.

10.4.1   From RBM to QUBO to MAX-SAT

To see this transformation more explicitly, let us re-write the *energy function* (10.4) as follows (Manukian *et al.*, 2019). First, let us consider the visible and hidden nodes as elements of a common vector $\{x_k\} = \mathbf{x} \equiv (\mathbf{v}, \mathbf{h})$, with each element $x_k = \{0, 1\}$. With this notation, we can then re-write eqn (10.4) as

$$E(\mathbf{x}) = -\mathbf{x}^T \mathbf{Q} \mathbf{x}, \qquad (10.23)$$

where $\mathbf{Q}$ is the matrix

$$\mathbf{Q} = \begin{bmatrix} \mathbf{B} & \mathbf{W} \\ \mathbf{0} & \mathbf{C} \end{bmatrix}, \qquad (10.24)$$

with $\mathbf{B}$ and $\mathbf{C}$ diagonal matrices that contain the biases b_j for the visible layer, and c_i, for the hidden layer, respectively; $\mathbf{0}$ is the null matrix (all entries are zero), and $\mathbf{W}$ is the weight matrix.

Given the matrices $\mathbf{B}$, $\mathbf{C}$, and $\mathbf{W}$, the problem of finding the configuration $\mathbf{x}$ that *minimizes* (10.23) is precisely the *quadratic unconstrained optimization* (QUBO) problem I have formulated in Section 9.2.2.

In fact, in that Section, I have shown how to transform it into a *weighted* MAX-SAT problem which consists in finding an assignment of **x** that *maximizes* the total weight of satisfied clauses (cf. discussion in Section 2.4).[19]

In Section 10.4.2 I will show *one* possible way to map this particular weighted MAX-2-SAT (with, at most, 2 literals per clause) to a DMM. As usual, the reader should keep in mind that such a mapping is not unique, and different mappings may be more efficient, for either the *software* or the *hardware* implementation of these machines.

10.4.2   Sampling the mode with MemComputing

The equations I will present here are similar to the ones I introduced in Section 8.2.5 for the 3-SAT. They are adapted to the particular weighted MAX-2-SAT that is generated from the QUBO problem of an RBM as follows.

Following the discussion of Section 9.2.2, the transformation of the QUBO problem defined by eqn (10.23) into a MAX-2-SAT will generate *both* 2-literal clauses *and* clauses with one literal. The reason for 1-literal clauses is related to the presence of the biases.[20]

2-literal clauses

Let us indicate with $W_{2,m} \equiv |w_{ij}|$ the absolute value of the weights of the RBM, where m stands for the clause m of the MAX-2-SAT CNF formula[21] that relates, say, the visible node v_j to the hidden node h_i (see, e.g., Fig. 10.1).

The first term on the RHS of eqn (10.3)—the one containing the weights—can then be transformed into a MAX-2-SAT as I have done in Section 9.2.2 (see eqns (9.9) and (9.10)). However, in Section 9.2.2, the spin values were chosen to be either $+1$ or -1.

In the present case, instead, the node variables are either 0 or 1. The transformation of the visible and hidden nodes interaction term is then:

$$-w_{ij}h_iv_j \quad \leftrightarrow \quad \begin{cases} \begin{array}{l} |w_{ij}|\ h_i \vee \bar{v}_j \\ |w_{ij}|\ \bar{h}_i \vee v_j, \\ |w_{ij}|\ h_i \vee v_j \end{array} & \mathrm{sgn}(w_{ij}) = 1, \\[6pt] |w_{ij}|\ \bar{h}_i \vee \bar{v}_j, & \mathrm{sgn}(w_{ij}) = -1, \end{cases} \quad (10.25)$$

whose validity, as in Section 9.2.2, can be easily checked by keeping in mind that while in the QUBO problem we need to *minimize* the energy (10.3), in the MAX-2-SAT we need to *maximize* the sum of the weights of *satisfied* clauses.[22]

For instance, for $\mathrm{sgn}(w_{ij}) = -1$, the energy is minimized when $h_i = v_j = 0$ or $h_i = 1$ and $v_j = 0$ (or vice versa: $h_i = 0$ and $v_j = 1$). In any of these three cases, the clause $(\bar{h}_i \vee \bar{v}_j)$ is satisfied, because at least one

[19] Note that there is actually no need to go from a QUBO problem to a weighted MAX-SAT. We could have equally formulated DMMs directly for the QUBO problem.

[20] 1-literal clauses could also be generated by the weights, if a different transformation from QUBO to MAX-2-SAT is chosen (Manukian *et al.*, 2020).

[21] See Section 2.4 for a refresh of how to construct a CNF formula.

[22] Note that for $\mathrm{sgn}(w_{ij}) = 1$, eqn (10.25) means $-w_{ij}h_iv_j \leftrightarrow |w_{ij}|(h_i \vee \bar{v}_j) \wedge |w_{ij}|(\bar{h}_i \vee v_j) \wedge |w_{ij}|(h_i \vee v_j)$.

of the literals is 1 (see the OR truth table in Fig. 2.7).

If this were the whole story, I could simply use eqns (8.2), (8.3), and (8.4) adapted for 2-SAT clauses, rather than 3-SAT clauses, with the weights reinforcing specific terms in those equations.

A 2-SAT clause consists of two literals, $\{l_{i,m}, l_{j,m}\}$, where a literal in the m-th clause, $l_{i,m}$, is either a negated, $\bar{x}_i$, or un-negated, x_i, variable.[23]

In analogy with eqn (8.1), I can then define a 2-SAT *clause function* for continuous *voltages* v_i as follows

$$C_m(v_i, v_j) = \frac{1}{2}\min[(1 - q_{i,m}v_i), (1 - q_{j,m}v_j)], \qquad (10.26)$$

where the terms $q_{i,m}$ contain the information about the relation between the literal in the m-th clause, $l_{m,i}$, and its associated variable, x_i. They are $+1$ if $l_{m,i} = v_i$, and -1 if $l_{m,i} = \bar{v}_i$ (negation of v_i).[24]

As in the case of the 3-SAT, the present cost function is also bounded, $C_m \in [0,1]$. Furthermore, a clause is satisfied when $C_m(v_i, v_j) < 1/2$ (cf. discussion in Section 8.2.4). However, this is not enough.

1-literal clauses

As already mentioned, in addition to two 2-literal clauses, the presence of biases in RBMs introduces 1-literal clauses (with just one literal) in the corresponding MAX-2-SAT. These can be handled in a variety of ways.

For instance, we could define another cost function for these clauses, and let them evolve according to their own equations of motion. This route, however, would necessarily increase the ODEs of the problem, with concomitant increase in numerical (as well as hardware) complexity. Instead, we can proceed as follows.

Since the biases of an NN essentially tend to 'force' a given node toward a 0 or a 1 configuration, we can choose to include them simply as *currents* in the voltage dynamics to guide the voltages (in arbitrary units) either toward -1 (logical 0) or $+1$ (logical 1).[25]

Let's indicate with $W_{1,i}$ the weight of a 1-SAT clause for the i-th variable, x_i (with x_i being either a visible or a hidden node), and with Q_{ii} the diagonal elements of the QUBO matrix (10.24), which contains the biases: $Q_{ii} = \{b_i\}$ or $Q_{ii} = \{c_i\}$.

If we choose the transformation (10.25) for the weights of the RBM, then the transformation for the biases is:

$$-Q_{ii}x_i \quad \leftrightarrow \quad \begin{cases} |Q_{ii}|\,(x_i), & \mathrm{sgn}(Q_{ii}) = 1, \\[2mm] |Q_{ii}|\,(\bar{x}_i), & \mathrm{sgn}(Q_{ii}) = -1, \end{cases} \qquad (10.27)$$

where (x_i) is the 1-SAT clause for the variable x_i, and $(\bar{x}_i)$ is the 1-SAT clause for the negation of that variable.[26]

[23] And, in the present context, the variables x_i are either the visible or the hidden nodes.

[24] Do not confuse the voltage symbols v_i in (10.26) with the visible nodes symbols. The voltages represent *both* the visible *and* the hidden variables.

[25] The biases could then be interpreted as additional voltage-controlled voltage generators (VCVGs) at the appropriate terminals in an actual hardware implementation of these machines. See Example 7.3.

[26] The 1-SAT clause (x_i) is satisfied (true) when $x_i = 1$, and $(\bar{x}_i)$ is satisfied when $x_i = 0$.

With this choice of transformation, then the extra current for a given variable i could then simply be $W_{1,i} = Q_{ii}$, so that, when $Q_{ii} > 0$, it would drive the corresponding voltage toward $+1$ (logical 1); when $Q_{ii} < 0$, it would drive it toward -1 (logical 0).[27]

The DMM equations for weighted MAX-2-SAT

With these preliminaries we can now suggest a DMM to tackle this type of weighted MAX-2-SAT. Let's then consider a MAX-2-SAT with N variables, M_1 1-SAT clauses, and M_2 2-SAT clauses.

Following the same strategy I have outlined in Section 8.2.5, an example of equations of motion of a DMM that solve such a problem can then be chosen as[28]

$$\dot{v}_i = \overbrace{W_{1,i}}^{\text{local current}} + \overbrace{\sum_m \left\{ W_{2,m} x_{s,m} x_{l,m} G_{i,m} \right\}}^{\text{gradient-like}} + \overbrace{\sum_m \left\{ \rho(1 - x_{s,m}) R_{i,m} \right\}}^{\text{rigidity}},$$

(10.28)

$$\dot{x}_{s,m} = \beta(x_{s,m} + \epsilon)(C_m - \gamma), \quad x_{s,m} \in [0,1] \to \textit{short-term memory},$$

(10.29)

$$\dot{x}_{l,m} = \alpha(1 + W_{2,m})(C_m - \delta), \quad x_{l,m} \in [1, 10M_2] \to \textit{long-term memory},$$

(10.30)

with $i \in [[1, N]]$ and $m \in [[1, M_2]]$.[29]

The voltages, $v_i \in [-1, 1]$, are again *continuous* representations of the N Boolean variables of the problem, which in the case of the RBM are both the visible and the hidden nodes, x_i, (elements of the vector $\mathbf{x} = (\mathbf{v}, \mathbf{h})$). We represent a false assignment as $v_i < 0$, and a true assignment as $v_i > 0$, while $v_i = 0$ would be ambiguous, were it not for the thresholding of the clause function (10.26).

Similar to eqns (8.2), (8.3), and (8.4), each clause has a 'short-term', $x_{s,m}$, and a 'long-term', $x_{l,m}$, memory variable that keep track of the *history* of the state of the cost function $C_m(v_i, v_j)$.

The short-term memory contains information on the *recent* history of C_m. The long-term memory, instead, contains information on the *entire* history of C_m. Both memory variables are bounded, and the small offset, $0 < \epsilon \ll 1$, in eqn (10.29) again removes the spurious solution $x_{s,m} = 0$.

The gradient-like term in eqn (10.28) is $G_{i,m} = 0$ if the variable x_i is not associated with any literal in clause m. Otherwise, it is

$$G_{i,m} = q_{i,m} \frac{1}{2}(1 - q_{j,m} v_j),$$

(10.31)

where v_j is the value of the voltage corresponding to the other literal in the clause. The 'rigidity' term in eqn (10.28) is

$$R_{i,m} = \begin{cases} q_{i,m} \frac{1}{2}(1 - q_{i,m} v_i), & C_m(v_i, v_j) = \frac{1}{2}(1 - q_{i,m} v_i), \\ 0, & C_m(v_i, v_j) = \frac{1}{2}(1 - q_{j,m} v_j). \end{cases}$$

(10.32)

[27] A different transformation from QUBO to MAX-2-SAT was chosen in (Manukian *et al.*, 2020), leading to a different form for the 1-SAT clause weight $W_{1,i}$. The results, of course, would be the same.

[28] Note that the extra parameter η multiplying the long-term memory in eqn (8.2) has been replaced by a constant ρ in eqn (10.28), as it does not seem to matter for sampling.

[29] The symbol $[[a, b]]$ (with a and b integers) means the set of *all* integers between a and b, included the end values.

This term only influences the voltage that is closest to the satisfying assignment in the clause, and it is weighted by a constant strength ρ in eqn (10.28).

The parameters β and α determine the rate of variation for the short-term and long-term memory variables, respectively. Each memory variable may also have a threshold parameter (γ for the short-term memory, and δ for the long-term memory).[30]

[30]In (Manukian *et al.*, 2020), γ was chosen to be $1/4$ and $\delta = 0$. However, other values of the parameters may also be appropriate, provided they satisfy $\delta < \gamma < 1/2$ (cf. discussion in Section (8.2.5)).

The weight of each 2-SAT clause, $W_{2,m}$, is incorporated in the dynamics of the long-term memory variable and the dynamics of the voltages. They reinforce both the gradient-like term (essentially, by 'injecting a larger current', the larger the weight), and the long-term memory (by increasing the weight of 'unsatisfaction' of that particular clause). The biases, instead, enter in eqn (10.28) through the 'extra currents' $W_{1,i}$, as I have discussed previously.

The DMM described by eqns (10.28), (10.29), and (10.30), would then attempt to *maximize* the total weight of the satisfied clauses, as its dynamics progress, namely it will attempt to converge toward the optimum (or optima, if more than one are present).

An important remark

Unlike a SAT problem (Section 2.4.1), the eqns (10.28), (10.29), and (10.30), are fed with an *optimization* problem, via the CNF formula they receive as input, namely the CNF formula that determines the interaction between the different variables (the topology of the network of variables; cf. discussion in Section 2.4.1). In such a case, some clauses would be *unsatisfiable* by construction. Therefore, the physical system they describe does *not* have equilibrium points, only saddle points. This means that even if the DMM reached the optimum (or one of the optima) it would not stay at that point forever. Instead, the dynamics would drive the system *away* from the optimum. However, since the DMM described by (10.28), (10.29), and (10.30), will *always* attempt to *maximize* the sum of the weights of the satisfied clauses, it would quickly try to reach the optimum again, from which it will escape, and so forth. In the long-time limit the system will then settle into a (quasi-)periodic orbit.

Given the weights and biases of the RBM, we can then employ eqns (10.28), (10.29), and (10.30) to *sample the mode* of the RBM probability distribution, by running them for a fixed time interval $\bar{T}$. The configuration that maximizes the sum of the weights of satisfied clauses within that time $\bar{T}$ can then be taken as the sample of the optimum or mode of the RBM joint probability distribution $p(\mathbf{v}, \mathbf{h})$, required to evaluate the model term (10.15).

An important remark

Due to the fact that the QUBO matrix (10.24) is *sparse* (most of its elements are zero), the numerical integration of eqns (10.28), (10.29), and (10.30) to sample the optimum (mode) can be performed quite efficiently (Manukian *et al.*, 2020).

For instance, consider an RBM that has N visible nodes and N hidden nodes. When we map it into the weighted MAX-2-SAT as we have done, we would then have $2N$ voltage variables, and $O(N^2)$ clauses for a given MAX-2-SAT instance.

Suppose we use the *forward Euler* scheme (see Section 8.5.1) to integrate eqns (10.28), (10.29), and (10.30). If the interval of time $\bar{T}$ for sampling is fixed, then, by means of *sparse matrix-vector multiplication* techniques, the time complexity would be $O(N^2)$ floating point operations per second. If we allow the time $\bar{T}$ to scale *linearly* with N, then the time complexity would be $O(N^3)$.

The *memory* (storage) complexity instead scales as $O(N^2)$, since the algorithm requires the storage of $O(N^2)$ non-zero elements of the weight matrix.

Sparse matrix
A matrix A is *sparse* if many of its elements are zero. Otherwise, it is *dense* or *full*.

Sparse matrix-vector multiplication
Consider an input vector x and an output vector y related by $y = Ax$ with A a *sparse* matrix. Due to the sparcity of A, efficient numerical methods can be devised to treat such a multiplication, with a cost that scales roughly as the number of non-zero elements of A (see, e.g., (Sauer, 2018)).

10.4.3 Performance comparison between MemComputing and contrastive divergence

I have anticipated at the beginning of Section 10.4, that the most difficult part of estimating the model term (10.15) is the sampling of the *mode* of the joint probability distribution, $p(\mathbf{v}, \mathbf{h})$. This is equivalent to sampling the lowest energy state (*ground state*) of the RBM energy (10.3).

The question is how better does a DMM sample such an energy compared to the traditional contrastive divergence (CD) (which I discussed in Section 10.2.1). Here, I am asking not just in terms of *quality* of the energy (i.e., how *far* the sampled state is from the ground state), but also in terms of how *efficiently* MemComputing can do such a sampling compared to CD.

In this Section, I show explicitly that the RBM energy sampled by the MemComputing approach is consistently better than the one found by CD. Importantly, this result is *independent* of the size of the RBM.

Also, the comparison I will show here refers to the *simulation* of the equations of motion of the DMMs I have introduced in Section 10.4.2. We expect their *hardware* implementation to perform significantly better, in view of the fact that the physical machine does not need to discretize time, hence will be immune from numerical errors, which are more detrimental than physical noise (cf. discussion in Section 12.7).

The 'exchange rate' between MemComputing and CD

First of all, to have a fair comparison, let us estimate the 'exchange rate' between one iteration of the numerical integration of the DMM eqns (10.28), (10.29), and (10.30), and k steps of CD (cf. discussion in Section 10.2.1).

In other words, given an integration method of such equations (e.g., forward Euler; Section 8.5.1), how many steps of CD (k steps of the Markov chain) are needed to match the computational cost (measured, e.g., by the wall time on the same processor) of one step of the numerical integration of those DMMs?[31]

It was found empirically in (Manukian *et al.*, 2020), across a large range of RBM sizes, that this exchange rate is about 30, namely 30 steps of CD (CD-30) can be performed during one iteration time of the dynamics of the DMMs described by eqns (10.28), (10.29), and (10.30), and integrated using the forward Euler method.

After we have determined this 'exchange rate' we can meaningfully compare the simulations of DMMs with CD. For this comparison we can choose a set of randomly initialized $N \times N$ RBMs, with all weights sampled from a normal distribution with average $\mu = 0$ and variance $\sigma^2 = 0.01$.

The results are shown in Fig. 10.5 for system sizes ranging from $N = 100$ to 1000. In particular, Fig. 10.5 shows the relative energy differences, in percentage, $\Delta\epsilon\% = 100 \times (E_{Mem} - E_{CD})/E_{CD}$, between the lowest energy E_{Mem} obtained with the simulations of the DMMs described previously, and the lowest energy E_{CD} obtained using CD-k.

The MemComputing simulations were run for $N_{tot} = 2N$ integration steps, scaled with the system size N. Contrastive divergence was instead run using the empirical exchange rate, $k = 30 \times N_{tot}$ ($= 60 \times N$ in the present case, since $N_{tot} = 2N$), resulting in the same computational cost (Manukian *et al.*, 2020).

It is obvious from Fig. 10.5 that the energies found by the MemComputing dynamics are substantially lower than those found by CD-k within the same amount of time. The relative energy differences are consistently above 400%, often showing an improvement of more than 1000%.

In terms of time complexity, the best fit to the asymptotic behavior of both algorithms was found to be almost cubic (Manukian *et al.*, 2020), consistent with the complexity analysis I have discussed in Section 10.4.2.

[31]Equivalently, how many steps of CD can we do (on the same processor) during the (wall or CPU) time we perform one integration step of the equations of motion of DMMs?

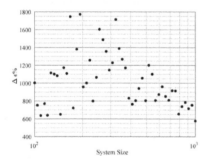

Fig. 10.5 The median relative energy differences, in percentage, $\Delta\epsilon\% = 100 \times (E_{Mem} - E_{CD})/E_{CD}$, between the energies obtained by the MemComputing simulations and CD-k across 20 randomly initialized $N \times N$ RBMs, with system size, N, ranging from 100 to 1000, with $k = 60 \times N$. E_{Mem} and E_{CD} are the lowest energies achieved during a run of MemComputing dynamics and CD-k, respectively. Both calculations have been done on a single core of the same processor. Reprinted with permission from (Manukian *et al.*, 2020).

10.4.4 MemComputing-assisted unsupervised learning

We can now take advantage of the previous results and see how the *unsupervised* training of RBMs (and other types of NNs as well) can be improved using MemComputing.

As I mentioned in Section 10.1.2, training an RBM requires minimizing the KL divergence (10.7) or maximizing the log-likelihood (10.8). However, those quantities contain the *marginal* probability distribution over the visible nodes, $p(\mathbf{v})$, *not*, the *joint* probability distribution, $p(\mathbf{v}, \mathbf{h})$, whose mode can be efficiently sampled by the DMM dynamics, as I have shown previously.

However, it was found in (Manukian *et al.*, 2020) that the *mode* of the marginal and the joint probability distributions are *equivalent* (with high probability) during the training of an RBM. In addition, it was shown in (Pei *et al.*, 2020) that the *frustration*[32] of an RBM increases, as the training progresses.

Similar to what we have seen in Section 8.2.5, the more *frustrated* a problem instance is, the more difficult it is to find its solution (or optimum) with traditional approaches. The MemComputing dynamics, instead, do not seem to suffer from this issue.

The reader interested in an explicit analytical and numerical demonstration of these facts is urged to look at the referenced papers (Manukian *et al.*, 2020; Pei *et al.*, 2020). Here, I will just exploit these findings and proceed as follows.

> **Mode correspondence of marginal and joint distributions**
> Although not necessarily a universal feature, it was found in (Manukian *et al.*, 2020), that the mode of the *marginal* probability distribution over the visible nodes, $p(\mathbf{v})$, and the mode of the *joint* probability distribution, $p(\mathbf{v}, \mathbf{h})$, are equivalent with high probability under scenarios typical for different stages of the unsupervised training of RBMs and DBMs. This equivalence, however, holds with 'high probability', meaning that there could very well be particular situations (i.e., data sets) for which the equivalence may not hold.

[32]See Section 8.2.5, for a quick introduction to this concept.

Improving Gibbs sampling with MemComputing

Since sampling the mode of a distribution is the most difficult part of the training, we can utilize MemComputing only when such estimate is necessary. In other words, we may still want to perform CD (Gibbs sampling) to sample the 'tails' of the distribution, and let MemComputing do the hard work of sampling the mode.

The part of the training accomplished by CD then remains as in eqns (10.10), (10.11), and (10.12). For convenience, let me re-write eqn (10.10) with no momentum term (similar expressions hold for the biases) as

$$\Delta w_{ij}^{\mathrm{CD}} = \epsilon^{\mathrm{CD}} \left[\langle v_i h_j \rangle_{q(\mathbf{v}) p(\mathbf{h} \,|\, \mathbf{v})} - \langle v_i h_j \rangle_{p^k(\mathbf{v}, \mathbf{h})} \right], \qquad (10.33)$$

where ϵ^{CD} is the learning rate for any update that involves CD.

The expectation value in the second term on the RHS of eqn (10.33) is again taken over the reconstructed distribution over a Markov chain initialized from the dataset after k Gibbs samples (see Section 10.2.1). This is the 'data-driven CD term'.

However, according to some schedule, see eqn (10.35), we can replace this update with one that samples the mode of the joint probability distribution, $p(\mathbf{v}, \mathbf{h})$. Namely, we can update the weights according to:

$$\Delta w_{ij}^{\text{mode}} = \epsilon^{\text{mode}} \left[\langle v_i h_j \rangle_{q(\mathbf{v})p(\mathbf{h}\,|\,\mathbf{v})} - [v_i h_j]_{p(\mathbf{v},\mathbf{h})} \right], \qquad (10.34)$$

where $[\cdot]_p$ means the *mode* of $p(\mathbf{v}, \mathbf{h})$, which we can sample using the DMM of Section 10.4.2. This is the 'mode-driven term'.

This *mode update* may also have its own learning rate, ϵ^{mode}, which may differ from the CD learning rate, ϵ^{CD}. Similar updates hold for the biases.

Replacing the model average $\langle v_i h_j \rangle_{p(\mathbf{v},\mathbf{h})}$ with the mode $[v_i h_j]_{p(\mathbf{v},\mathbf{h})}$ is a type of *saddle-point approximation*. In fact, the largest contribution to the average value $\langle v_i h_j \rangle_{p(\mathbf{v},\mathbf{h})}$ comes precisely from the most likely state of the Boltzmann distribution of the RBM (10.1), which is the *ground state* of its energy (10.3).

By making the saddle-point approximation we essentially replace the problem of sampling from the full probability distribution with an (NP-hard) optimization problem. This trade-off would not be advantageous were it not for an efficient method to tackle optimization problems, as the one offered by MemComputing.

Finally, let me note that the saddle-point approximation would be exact if the probability distribution were a delta function centered at the ground state configuration, $\mathbf{x}_0$: $e^{-E(\mathbf{x})}/\mathcal{Z} \sim \delta(\mathbf{x} - \mathbf{x}_0)$.

We expect it to become less accurate if the underlying distribution becomes *multi-modal*, and even worse if it is *uniform*. However, empirical results have shown that such an approximation is still advantageous over a wide range of distributions (see Example 10.1 and (Manukian *et al.*, 2020)).

> **Saddle-point approximation**
>
> Consider the expectation value of a function, $f(x)$, of a (continuous) stochastic variable, x, over a distribution of the type $p(x) = e^{-g(x)}/\mathcal{Z}$:
>
> $$\langle f(x) \rangle_{p(x)} = \frac{\int f(x) e^{-g(x)} dx}{\int e^{-g(x)} dx}.$$
>
> If the function $g(x)$ has a minimum at x_0, then one can approximate it as:
>
> $$g(x) \approx g(x_0) + \frac{1}{2}|g''(x)|(x - x_0)^2,$$
>
> around the minimum. With this approximation, the expectation value is:
>
> $$\langle f(x) \rangle_{p(x)} \approx f(x_0)$$
>
> This is a type of *saddle-point approximation* (or *Laplace's method*), and holds also for several (and discrete) variables.

> **Uniform data probability distribution**
>
> When the data distribution, $q(\mathbf{v})$, is *uniform*, then it is simply $1/n_d$ within a given range, where n_d is the number of visible configurations with non-zero probability. In this case, *any* value of the distribution, within its range, is a mode.

> **An important remark**
>
> Note that the updates in eqn (10.34) are in an *off-gradient* direction, namely they *do not* follow the gradient of the KL divergence (10.7). This is because they are not taken over the full probability distribution, $p(\mathbf{v}, \mathbf{h})$, only its *mode*. This prevents spurious modes (Section 10.2.1) from appearing in the model distribution. This is the reason such updates provide increased *stability* of the training, and guarantee its *convergence* to arbitrarily small KL divergences (Manukian *et al.*, 2020).

Optimal mode-assisted training schedule

The only thing that we need to determine is how often the mode updates should be done. From the previous discussion, we have learned that the RBM becomes more frustrated the more we proceed in the training.

This means that the sampling of $p(\mathbf{v}, \mathbf{h})$ should become more challenging with increasing number of iterations, in view of the fact that sampling the mode becomes a harder problem. This suggests that we should sample the mode of the joint probability distribution, $p(\mathbf{v}, \mathbf{h})$, more 'frequently' the more the training proceeds.

We can then suggest a *schedule* of when to perform mode updates by calculating the probability of replacing the data-driven CD term with a mode-driven term, at the iteration step n, as:

$$P_{\text{mode}}(n) = P_{\text{max}}\sigma(\alpha n + \beta), \qquad (10.35)$$

where σ is a sigmoid function, as in eqn (10.14), and $0 < P_{\text{max}} \leq 1$ is the maximum probability of a mode update. At the beginning of the training, P_{mode} is then small, but it increases gradually with the number of updates (see right panel of Fig 10.6).[33]

The quantities α and β are parameters that control how the mode updates are introduced. They are chosen in such a way that the frequency of mode updates dominates only when both large weights and frustration occur during training. I refer the reader to (Manukian *et al.*, 2020) for an in-depth discussion of this point.

[33]The choice of the sigmoid in eqn (10.35) is arbitrary. Other types of functions can be chosen that accomplish the same goal.

An important remark

Apart from the hyper-parameters, P_{max}, α and β in eqn (10.35), the learning rate ϵ^{mode} can be related to ϵ^{CD}. It was found analytically in (Manukian *et al.*, 2020) that the choice $\epsilon^{\text{mode}} = \gamma \epsilon^{\text{CD}}$ is optimal when $\gamma = -E_0/[(n+1)(m+1)]$, with $E_0(< 0)$ being the ground state of the corresponding RBM (with n visible nodes and m hidden nodes) with nodal values $\{-1, 1\}^{n+m}$. This particular choice of γ is an *upper bound* to the learning rate which *minimizes* the RBM *energy variance* over all states.

Example 10.1

MemComputing-assisted training vs. CD

Here, I will provide an explicit example of how the MemComputing-assisted training, I have just outlined, can improve significantly over CD. This is shown in Fig. 10.6 for randomly generated RBMs.

I refer the reader to (Manukian *et al.*, 2020) for more examples of this mode-assisted method applied to various data sets, including the MNIST one, showing similar advantages.

Fig. 10.6 Performance comparison between the MemComputing-assisted training and Giggs sampling, across 25 randomly generated 6×6 RBMs with a random uniform data set of size $n_d = 10$. **(a)** The KL divergence as a function of training iterations of CD-1. **(b)** The KL divergence as a function of training iterations of the mode-assisted training. The solid curves are the median KL divergences. The shaded region is defined by the maximum/minimum KL divergence at that point in training. The mode sampling probability, P_{mode}, eqn (10.35), is shown as the dotted line in (b). In both cases, the learning rate was a constant $\epsilon^{CD} = \epsilon^{\mathrm{mode}} = 0.05$. Reprinted with permission from (Manukian *et al.*, 2020).

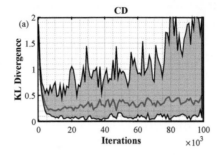

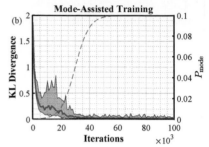

Of course, for small RBMs, both the KL divergence and the mode of $p(\mathbf{v}, \mathbf{h})$ can be computed exactly. However, as previously discussed, the MemComputing approach is of help when the size of the problem precludes an exact calculation of the mode (and when its sampling is difficult with traditional methods; see again Fig. 10.5). In addition, if the data distribution, $q(\mathbf{v})$, is not known explicitly, the evaluation of the log-likelihood (10.8) needs to replace the KL divergence (10.7), as a measure of training quality.

Irrespective, there are several noteworthy features that are evident in Fig. 10.6, and which have been found in the application of this method to other cases as well (Manukian *et al.*, 2020).

First of all, while the training with only CD may *diverge* (become worse) with increasing number of iterations,[34] the MemComputing assisted one does not.

Second, the MemComputing-assisted training shows a much more *stable* learning than CD alone, as indicated by the fact that the maximum and minimum KL divergences tend toward a common value (the variance decreases with number of iterations).

Finally, in the MemComputing-assisted training the KL divergence converges toward zero, namely the training leads to solutions with arbitrary high quality: *optimal solutions*.

This is indeed the most desirable feature we were after, and cannot be achieved with Gibbs sampling only.

---

[34]I mentioned this issue in Section 10.2.1 in connection with the presence of *spurious modes*.

> **An important remark**
> Note that training at optimality is *not* overfitting (see Section 10.1.4). The latter occurs when the NN model has more parameters than the data. The former, instead, happens if the model is able to match the data distribution, *without* overfitting.

It is also worth noting that the MemComputing-assisted approach is not limited to RBMs. It can be equally applied to other types of NNs such as convolutional, fully connected Boltzmann machines, and *deep* NNs (Goodfellow *et al.*, 2016).

In fact, unsupervised learning of deep networks is considered a more difficult task than the training of single-layer NNs. In Section 10.5, I will then show the advantages of the MemComputing-assisted approach for the unsupervised learning of deep NNs.

10.5 Joint unsupervised training of DBMs with MemComputing

In Section 10.1.3, I discussed DBMs, which are composed of two or more concatenated Boltzmann machines (restricted or not) joined to form a deep network.[35] These machines are fully *undirected graphs*.

[35]Refer to Fig. 10.2 for an example with two concatenated RBMs.

However, in view of the difficulty to train them *jointly* as a single unit, it was suggested to *pre-train* them instead *one layer at a time* (Goodfellow *et al.*, 2016). In other words, one trains the Boltzmann machine with the visible nodes first. Then, the hidden nodes of this trained machine are used as the 'visible' nodes of the next one, and so forth, for as many layers as there are in the network. In fact, without this 'greedy' layer-wise pre-training, it was found that DBM models could not successfully learn even the MNIST digits (Salakhutdinov and Hinton, 2012).

DBMs vs. DBNs
Deep Boltzmann machines (DBMs) are *undirected graphs* consisting of concatenated Boltzmann machines (see, e.g., Fig. 10.2, for the case of RBMs). If they are trained *layer-wise* (namely one layer at a time), they are called *deep belief networks* (DBNs), and are then (partly) treated as *directed graphs*. DBNs do *not* take full advantage of the topology of DBMs. Hence, we expect them to have less *capacity* (Section 10.1.4) than DBMs.

From undirected to directed graph

It is clear that this type of training changes *considerably* the topology of the original network. In fact, it treats the network as a partly *directed graph*. Therefore, to distinguish it from the DBMs, when the latter ones are trained in this so-called *greedy layer-wise* manner, this model NN is often called a *deep belief network* (DBN) (Goodfellow *et al.*, 2016).

We then expect that many *correlations* among nodes—which are present in the original *undirected* DBM—are *lost* in the training of the corresponding *directed* DBN. Therefore, ideally, we would like to avoid training DBNs, and rather keep the full undirected structure of the DBM from the outset.

However, dealing with DBMs brings with it further complications as far as Gibbs sampling (CD) is concerned. I will provide in the following shaded box a minimal outline of these extra technical details.

The reader already familiar with these procedures (or not particularly interested in them) can directly skip to Section 10.5.1. Further details can, of course, be found in specialized textbooks (Goodfellow *et al.*, 2016).

Approximations relevant to the joint training of DBMs with Gibbs sampling (CD)

There are three major approximations that need to be done when jointly training DBMs with Gibbs sampling.

i) The log-likelihood (10.8) is not directly maximized as in the case of RBMs. Instead, a *variational lower bound* to the log-likelihood is optimized.

ii) The most common variational form chosen leads to *mean-field* updates for the *data* term of the gradient (see Section 12.5.3 for the meaning of 'mean-field approximation').

iii) In order to evaluate the log-likelihood lower bound, the partition function is approximated.

I will now briefly describe these three approximations separately.

Log-likelihood lower bound

Since the data expectation in eqn (10.10) can no longer be evaluated in closed form for the DBM, data-dependent statistics must be approximated with a sampling technique over the conditional distribution, $p(\mathbf{h}|\mathbf{v})$, where $\mathbf{h} = \{\mathbf{h}^{(1)}, \cdots, \mathbf{h}^{(\ell)}\}$. In practice, a variational lower bound to the log-likelihood is maximized instead, which is tractable and is found to work well (as in the model term) (Salakhutdinov and Hinton, 2009).

The variational approximation replaces the original conditional distribution, $p(\mathbf{h}|\mathbf{v})$ with an approximate distribution, $r(\mathbf{h}|\mathbf{v})$, where the parameters of r follow the gradient of the resulting *lower bound* on the original log-likelihood:

$$\log p(\mathbf{v}) \geq \sum_{\mathbf{h}} r(\mathbf{h}|\mathbf{v}) \log p(\mathbf{v}, \mathbf{h}) + H_r(\mathbf{v}) \tag{10.36}$$

$$= \log p(\mathbf{v}) - \mathrm{KL}(r(\mathbf{h}|\mathbf{v})\|p(\mathbf{h}|\mathbf{v})),$$

where $H_r(\mathbf{v}) = -\sum_{\mathbf{h}} r(\mathbf{h}|\mathbf{v}) \log r(\mathbf{h}|\mathbf{v})$ is the *Shannon entropy* of r. The lower bound simultaneously attempts to maximize the log-likelihood of the data set ($\log p(\mathbf{v})$) and minimize the KL divergence (cf. eqn (10.7)) between the true conditional distribution, $p(\mathbf{h}|\mathbf{v})$ and its approximation, $r(\mathbf{h}|\mathbf{v})$.

Mean-field approximation of the data term

A fully factorial mean field ansatz is often used in the variational approach, $r(\mathbf{h}|\mathbf{v}) = \prod_i r(h_i|\mathbf{v})$ with $r(h_i = 1|\mathbf{v}) = \mu_i$, with $\mu_i \in [0, 1]$ randomly initialized from a uniform distribution. Maximizing the lower bound in eqn (10.36) results in updates of the form,

$$\mu_i \leftarrow \sigma\left(\sum_j W_{ij}^{(1)} v_j + \sum_{i \neq j} J_{ij}\mu_i + b_i\right). \tag{10.37}$$

Here J is a block matrix containing the weights of the hidden nodes, and b_i is the bias of the i-th hidden node. Convergence is typically fast (typically less than 30 iterations are enough). These states are then used to calculate the data expectation in eqn (10.10) during the Gibbs phase of the training.

Partition function approximation

Computing the partition function in eqn (10.2) exactly is numerically infeasible for large DBMs. However, its value is only required to evaluate the performance of the networks and appears in the log-likelihood (10.8) as a normalizing constant.

The procedure that is often used to approximate the partition function of large RBMs and DBMs is called *Annealed Importance Sampling* (AIS) (Goodfellow *et al.*, 2016). AIS estimates the ratio of partition functions, Z_N/Z_0, using a sequence of probability distributions between a chosen initial distribution (at step 0) and the desired one (at step N).

The initial distribution, p_0, is chosen to have an exactly known partition function, and to be easy to sample from (e.g., uniform distribution, Section 10.4), and p_N is the desired distribution whose partition function one wants to compute.

The sequence in the case of DBMs is parameterized by β_k (inverse temperatures), giving $p_k(\mathbf{x}) = e^{-\beta_k E(\mathbf{x})}/Z_k$. Markov chains (Section 10.2.1) are initialized uniformly according to p_0 and $0 \leq \beta_k \leq 1$ is slowly annealed to unity according to a desired schedule, all the while allowing the chains to run.

The ratio is then approximated by the product of ratios of the unnormalized intermediate distributions $(p_j^*(\mathbf{x}))$:

$$\frac{Z_N}{Z_0} = \frac{p_1^*(\mathbf{x})}{p_0^*(\mathbf{x})} \cdots \frac{p_k^*(\mathbf{x})}{p_{k-1}^*(\mathbf{x})}. \tag{10.38}$$

> Note that AIS tends to *under*-approximate the partition function, leading to an *over*-estimate of the log-likelihood, and this effect is exaggerated in larger models.

10.5.1 Estimating the data and model terms of DBMs with MemComputing

I will now show how MemComputing can help in the *joint unsupervised training* of DBMs. First of all, as I have anticipated in Section 10.2, in the case of deep NNs, the data term does *not* have the closed form of eqn (10.14).

In other words, like the model term (10.15), also the data term needs to be estimated over some unknown probability distribution. This estimate is then as difficult as the one in the model term.

As we have done before, we can then employ DMMs to evaluate *both* the data *and* the model terms in eqns (10.10), (10.11), and (10.12) over the mode of the joint probability distribution. This is similar to the procedure as I have used in eqn (10.34), for just the model term of an RBM.

This mode update does not need to be done all the time. Rather, it will follow a *schedule* as in eqn (10.35).[36]

In other words, at certain intervals of the training—that follow the schedule (10.35)—the Gibbs sampling (over the *whole network* distribution) is replaced by the mode estimate of the same.[37]

Let us first define $\mathbf{x} \equiv \{\mathbf{v}, \mathbf{h}^{(1)}, \cdots, \mathbf{h}^{(\ell)}\}$, for a DBM with ℓ hidden nodes. For the weight updates we would then make the following substitution:

$$\Delta w_{ij}^{\mathrm{CD}} \propto \langle x_i x_j \rangle_{q(\mathbf{v}) p^k(\mathbf{h}^{(1)}, \cdots, \mathbf{h}^{(\ell)} \mid \mathbf{v})} - \langle x_i x_j \rangle_{p^k(\mathbf{x})}, \quad (10.39)$$

$$\Downarrow \quad\quad\quad\quad\quad\quad\quad\quad (10.40)$$

$$\Delta w_{ij}^{\mathrm{mode}} \propto [x_i x_j]_{p_{\mathrm{DATA}}} - [x_i x_j]_{p_{\mathrm{MODEL}}}, \quad (10.41)$$

where $p^k(\mathbf{h}^{(1)}, \cdots, \mathbf{h}^{(\ell)} \mid \mathbf{v})$ is the probability distribution of the deep network *conditioned* to the visible nodes, and $p^k(\mathbf{x})$ is the *joint* probability distribution of the deep network. Both distributions are (Gibbs) sampled with a Markov chain of length k (see Section 10.2.1).

Instead

$$p_{\mathrm{DATA}} \equiv q(\mathbf{v}) p(\mathbf{h}^{(1)}, \cdots, \mathbf{h}^{(\ell)} \mid \mathbf{v}), \quad (10.42)$$

while

$$p_{\mathrm{MODEL}} \equiv p(\mathbf{v}, \mathbf{h}^{(1)}, \cdots, \mathbf{h}^{(\ell)}), \quad (10.43)$$

and $[\cdot]$ indicates the *mode* of the corresponding probability distribution. Similar equations hold for the bias terms as well.

[36]The same schedule used for the training of a single RBM works well also for a DBM because the latter is just a collection of RBMs (Manukian and Di Ventra, 2021).

[37]Note that now the Gibbs sampling is done over the *full* network, not on the single layers. Namely, we are *not* training the network in a greedy layer-wise manner as for a DBN.

As in the case of the RBM, the mode of the distributions in eqns (10.42) and (10.43) can be efficiently sampled with MemComputing. I will now provide a few examples of this approach, taken from (Manukian and Di Ventra, 2021).

Before doing that, though, let me make another important observation.

The 'fan-in' phenomenon

In deep networks, like DBMs, we have the freedom to choose the (physical) *topology* of the network, namely we have the extra freedom of choosing how to distribute the neurons across the hidden nodes.[38]

It turns out that how these nodes are distributed makes quite a difference in the quality of the training (Manukian and Di Ventra, 2021).

To see this, let us refer to the DBM of Fig. 10.2 with just two hidden layers, and perform the MemComputing-assisted training I have explained previously, by keeping the *total* number of hidden nodes, n_h, constant, and vary the ratio between the number of nodes in the second layer, $n_{h(2)}$, with respect to the first, $n_{h(1)}$. Call this ratio $\alpha = n_{h(2)}/n_{h(1)}$.

In Fig. 10.7 I plot the *average* log-likelihood[39] for DBMs trained with and without the mode-assisted approach as a function of α. The figure refers to the training with a synthetic data set.[40]

For this type of small networks we can also calculate the (average) log-likelihood *exactly*.

First of all, we do see that the mode-assisted training performs better than its unassisted counterpart (more on this in the Example 10.2 below).

However, it is particularly noticeable that the log-likelihood *increases* as $(\alpha \to 0)$, namely when the network has a *fan-in* topology. In fact, when $\alpha \sim 0.15$, namely the second hidden layer has only about 15% of nodes than the first hidden layer, the training seems optimal.

This 'fan-in phenomenon' can be understood by noting that too many hidden nodes in the second layer act as a 'noise source' for the hidden nodes in the first layer, which is the one closer to the visible nodes. This noise, then, may slow down learning during training, if it is significant, namely if $\alpha \to 1$.

On the other hand, too few hidden nodes in the second layer, namely in the limit $\alpha \to 0$, one recovers the single-layer RBM, hence the capacity (Section 10.1.4) of the network is considerably reduced.

For the network sizes in Fig. 10.7, these two effects balance at the 'sweet spot' $\alpha \sim 0.15$. At this stage, however, it is not clear if this value is specific to DBMs or it is universally valid for all types of deep networks.

Fan-in vs. fan-out network
A deep neural network in which the number of nodes of the deeper layers (those farther from the visible layer) decreases, showcases a *fan-in* network (physical) topology. A deep network in which the number of nodes of the deeper layers increases, has a *fan-out* topology.

[38] The number of visible nodes are fixed by the data set.

[39] From eqn (10.8) it is $\langle LL \rangle(p) = \sum_{\mathbf{v} \in \mathcal{D}} \log p(\mathbf{v})/\mathrm{cardinality}[\mathcal{D}]$.

[40] Called 'shifting bar' data set (Goodfellow *et al.*, 2016).

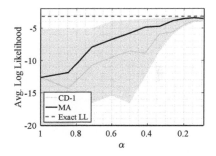

Fig. 10.7 Average log-likelihood achieved on an $n_v = 24$ dimensional synthetic data set as a function of DBM shape, α. The total number of hidden nodes are kept fixed at $n_{h(1)} + n_{h(2)} = 22$. An ensemble of 50 DBMs were trained with CD-1 and mode assistance (MA). The solid line shows the median average log-likelihood achieved after 10^5 gradient iterations with a linearly decaying learning rate $\epsilon = 1 \to 0.01$. The shaded regions denote the 95th and 5th percentiles. Reprinted with permission from (Manukian and Di Ventra, 2021).

Centering a neural network
The procedure of substracting the mean $\langle \mathbf{x} \rangle_{\text{DATA}}$ from all binary variables during the training of a neural network is called *centering*. It has been shown to improve the training of RBMs and DBMs considerably. See Fig. 10.8 and (Melchior *et al.*, 2016).

Fig. 10.8 Average log-likelihood on the MNIST data set achieved after 100 epochs with the MemComputing-assisted (MA) training compared to the recent best achieved results on DBMs using CD—with centering (C-DBMs), and without (DBMs)—as well as pre-training. DBMs trained with MA were all with fixed $\alpha = 0.15$ and an increasing number of hidden nodes, n_h. The learning rate was chosen to follow a linear decay, $\epsilon = 0.05 \rightarrow 0.0005$ and no pre-training was employed. The best log-likelihood was reported out of 10 randomly initialized runs. Reprinted with permission from (Manukian and Di Ventra, 2021).

Epoch
When the entire data set is fed through the neural network once, we say that we have completed one *epoch* (Goodfellow *et al.*, 2016).

[41]Note that, for a fair comparison, the total training iterations were set to 100 epochs in each case, and the log-likelihood results with the MemComputing approach are shown as a function of total number of hidden nodes at a fixed ratio, $\alpha = 0.15$. For the Gibbs-sampling part of the training, the partition function was approximated using *Annealed Importance Sampling*, as for the other referenced results (Manukian and Di Ventra, 2021).

Example 10.2

MemComputing-assisted training of DBMs

Having found the best topology for a DBM, we can now show more clearly the advantages of the MemComputing-assisted training compared to state-of-the-art training procedures. This is plotted in Fig. 10.8 for the MNIST data set, where the MemComputing-assisted approach is compared to state-of-the-art results of a two-hidden layer DBN pre-trained in a greedy layer-wise manner (Salakhutdinov and Hinton, 2009), and DBNs without such a pre-training but with and without *centering* (Melchior *et al.*, 2016).

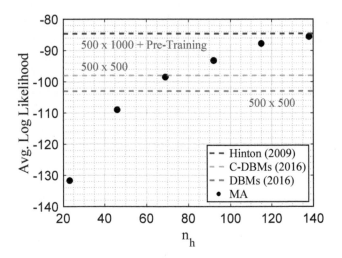

It is clear from Fig. 10.8 that, as the number of hidden nodes increases, very small MemComputing-assisted DBMs surpass the performance of significantly larger standard DBMs, centered DBMs, and eclipse the performance of much larger pre-trained DBMs. In fact, with a DBM of only 138 hidden nodes (and *no* pre-training), the MemComputing-assisted approach reaches the same log-likelihood of a DBM with 1500 hidden nodes trained with Gibbs sampling *and* pre-trained. This means a parameter (weights and biases) savings of two orders of magnitude.[41]

10.6 Accelerating supervised training of RBMs with MemComputing

After the discussion of how to do *unsupervised* training with MemComputing, let me discuss its application to *supervised* learning. In this case, one can take advantage of some form of *pre-training* (unsupervised learning) in order to provide a better initial condition for the downstream supervised phase.

Here, I will show the simplest way to accomplish this. It is similar to the MemComputing-assisted approach I discussed in Sections 10.4 and 10.5, but without ever employing Gibbs sampling during pre-training.

The idea is as follows. Instead of sampling the mode of the joint probability distribution, $p(\mathbf{v}, \mathbf{h})$, and perform an update according to eqn (10.34), with the schedule (10.35), one can do pre-training by *always* employing eqn (10.34), namely, by setting the probability $P_{\text{mode}}(n)$ in eqn (10.35) to one: $P_{\text{mode}}(n) = 1, \forall n$.

Of course, we do not expect this procedure to provide the best KL diverge or log-likelihood, as the method employed in Sections 10.4 and 10.5. However, it is enough to *accelerate* substantially the supervised phase of the training, which is what we are interested in here. It is also advantageous in terms of (classification) *accuracy* (Section 10.3), with respect to state-of-the-art supervised training methods (Manukian *et al.*, 2019).

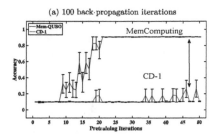

(a) 100 back-propagation iterations

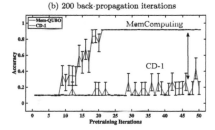

(b) 200 back-propagation iterations

An important remark

We expect that if we did pre-training as I have discussed in Sections 10.4 and 10.5—namely, a pre-training aimed at minimizing the KL divergence or maximize the log-likelihood—the downstream supervised phase should also improve. I leave this study for future work.

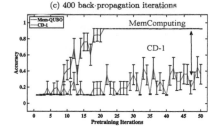

(c) 400 back-propagation iterations

These advantages again stem from the fact that the mode of the joint probability distribution, $p(\mathbf{v}, \mathbf{h})$, contains a lot of information on the distribution itself, and pre-training based on it sets the NN model in a more advantageous initial point, than starting from samples of $p(\mathbf{v}, \mathbf{h})$ away from the mode.

Fig. 10.9 Accuracy on the test set of a reduced MNIST problem with MemComputing-assisted pre-training vs. pre-training with CD, for $n = 100(a), 200(b), 400(c)$ iterations of backpropagation with mini-batches of 100. The plots show average accuracy with $\pm \sigma / \sqrt{N}$ error bars calculated across 10 DBNs trained on $N = 10$ different partitions of the training set. Adapted with permission from (Manukian *et al.*, 2019).

Supervised training acceleration

Let us first show the acceleration of supervised learning with MemComputing. In Fig. 10.9, I report the pre-training with the MemComputing-assisted procedure I just outlined, vs. the pre-training using Gibbs sampling (CD-1) only, on a reduced MNIST data set.

The MemComputing sampling of the mode has been done with a commercial solver provided by MemComputing, Inc., which is based on the general ideas I have discussed in this Chapter. (See (Manukian *et al.*, 2020) for the numerical details of the MemComputing simulations.)

It is evident from Fig. 10.9 that one needs far less backpropagation iterations (Section 10.3) to achieve the same accuracy. One also notices an improved performance (indicated by a black arrow) of the MemComputing-assisted method, which CD-1 seems to have difficulty to overcome, even with increasing number of backpropagation iterations.

Error rate

The *error rate* (ER) is the fraction of incorrect predictions the model makes, out of all the possible predictions on the testing set:

$$ER = \frac{\text{\# incorrect predictions}}{\text{total \# of predictions}}.$$

It is related to the accuracy as:

$$Accuracy = 1 - ER.$$

Batch normalization

To reduce the effect of initial random conditions and speedup convergence in training deep networks one can employ *batch normalization* of each layer, which normalizes the input of the layer by subtracting the batch mean and dividing by the batch standard deviation.

Rectified linear units

In place of the sigmoid activation functions (10.14), one can employ the *rectified linear units* (ReLU) which return the value 0 if the input is negative, and the value itself, if it is positive: $\text{ReLU}(x) = \max(0, x)$. They render the energy landscape more convex. See (Goodfellow *et al.*, 2016) for more details.

Quality advantage over state-of-the-art approaches

Having discussed how MemComputing can accelerate the supervised training, it is also worth looking at the *quality* of such a training compared with traditional methods.

To measure the quality, we can calculate the *accuracy* (Section 10.3), namely the fraction of predictions that the model *correctly* makes, with respect to the total number of testing samples. Alternatively, one can also report the *error rate*, which is defined as the total number of *incorrect* predictions made by the model on the test set, divided by all the predictions.

Of course, the accuracy and the error rate are related as: *accuracy =* $(1 - error\ rate)$.

In order to compare with the most favorable techniques, we can start from a randomly initialized network, and employ methods that both 'smooth out' the role of initial conditions (like the *batch-normalization* procedure) and render the energy landscape defined by eqn (10.3) more 'convex' (like *rectified linear units*).

It is beyond the scope of this book to discuss in detail these methods, which have found great success in supervised learning of a variety of NNs. The interested reader should look into (Goodfellow *et al.*, 2016) for an in-depth account.

The important point is that these are now common techniques used in supervised learning to improve the quality of the training and speedup ist convergence. In fact, these techniques together seem to provide an advantage compared to the network trained with CD using just sigmoidal functions.

The comparison between the accuracy obtained with these techniques and the MemComputing-assisted pre-training (with sigmoidal functions and no batch normalization) is shown in Fig. 10.10 for an RBM trained on a reduced MNIST data set.

It is clear from Fig. 10.10 that the network pre-trained with Mem-Computing has an accuracy advantage of more than 1%, and a 20% reduction in error rate on the test set out to more than a thousand backpropagation iterations. Of particular note is the fact that the RBM pre-trained with the MemComputing-assisted approach has sigmoidal activations, while the other network has rectifiers.

Therefore, the network pre-training with MemComputing should gen-

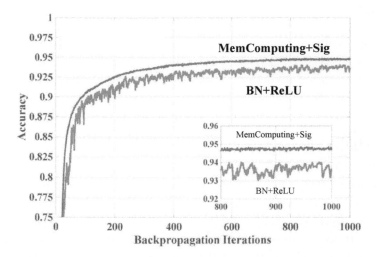

Fig. 10.10 Main panel: accuracy on a reduced MNIST test set obtained on an RBM pre-trained with the MemComputing-assisted approach and sigmoidal activation functions (Sig) versus the same size network with no pre-training but with batch normalization (BN) and rectified linear units (ReLU). The **inset** shows an accuracy advantage of MemComputing approach greater than 1% and an error rate reduction of 20% throughout the training. Adapted with permission from (Manukian *et al.*, 2019).

erate a more difficult optimization problem for stochastic gradient descent. Instead, we found an accuracy advantage for the MemComputing-assisted pre-training that persists throughout the whole supervised learning phase.

This is consistent with the advantages we have observed in the unsupervised learning of NNs with the mode-assisted training, as discussed in Section 10.4 and 10.5. It further reinforces the notion that initializing the weights and biases of the network from samples of the mode of $p(\mathbf{v}, \mathbf{h})$ is more advantageous than initializing them from arbitrary samples of the same distribution.

10.7 Application to Quantum Mechanics

Having shown how to take advantage of MemComputing to do supervised and unsupervised learning of NNs, I will conclude this Chapter with a brief introduction to a possible application of this paradigm to problems in Quantum Mechanics. In fact, among all the applications I have discussed so far, this is the least explored, hence the one with the largest potential for further developments.

I will focus on only two aspects: how to efficiently reconstruct a quantum state from a set of (experimental) data (this is known as *quantum tomography*), and how to find the ground state of quantum Hamiltonians. In both cases, I will use a representation of quantum states using NNs.

If the reader is not familiar with Quantum Mechanics and/or is not interested in this particular application, they can safely jump directly to Part IV of the book.

Quantum state tomography
The procedure of *reconstructing* a quantum state (whether pure or mixed) of a system using an ensemble of measurements on such a state is called *quantum state tomography* (Nielsen and Chuang, 2010).

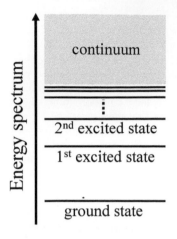

Fig. 10.11 Schematic of the energy spectrum of a quantum Hamiltonian, typically comprised of a set of *discrete* states, and a set of states in a *continuum*.

[42]The choice or guess of the *form* of this function is dictated by symmetries of the problem, boundary conditions, and, of course, the ingenuity and intuition of the researcher.

[43]A similar method can be used to obtain approximations of *excited* states by working in a subset of the Hilbert space that is *orthogonal* to the ground state. For example, for the first excited state $|\Psi_{ex}\rangle$, this can be accomplished by 'shifting' the trial state as $|\Psi_{ex}\rangle = |\Psi_\lambda\rangle - \langle\Psi_{GS}|\Psi_\lambda\rangle|\Psi_{GS}\rangle$, so that $\langle\Psi_{GS}|\Psi_{ex}\rangle = 0$. However, the approximations typically get worse with increasing energy of the excited states. Also, note that an equivalent variational theorem does *not* generally hold for the *dynamics* of quantum systems; see, e.g., (Yang, Tackett and Di Ventra, 2002).

[44]Constraints and symmetries are difficult to recover from an arbitrary trial wavefunction. An infinite (or extremely large) number of parameters, $\{\lambda\}$ could still do the job, but, of course, that would make the whole approach *practically* useless.

10.7.1   Variational representation of quantum states

Before proceeding with the actual application of MemComputing to quantum systems, I need to remind the reader that, in general, finding the ground state (and the full *spectrum* as well) of quantum Hamiltonians (Fig. 10.11) cannot be accomplished analytically. Therefore, one has to resort to approximations in the form of *variational* (also called *trial*) *wavefunctions* (Messiah, 1958).

The word 'variational' (or 'trial') means that one chooses an approximation $|\Psi_\lambda\rangle$ of the exact state (call it $|\Psi_0\rangle$) which depends on a set of parameters, $\{\lambda\}$, and then varies such parameters so as to reproduce the exact $|\Psi_0\rangle$ as close as possible.[42] The rationale behind this method rests on the *variational theorem*: if $|\Psi_\lambda\rangle$ is an arbitrary state in the Hilbert space, $\mathcal{H}$, of a physical system whose dynamics are described by an Hamiltonian operator $\hat{H}$, then the (real) values of the *energy functional*

$$E[\Psi_\lambda] = \frac{\langle\Psi_\lambda|\hat{H}|\Psi_\lambda\rangle}{\langle\Psi_\lambda|\Psi_\lambda\rangle}, \tag{10.44}$$

are always larger than or equal to the energy, E_{GS}, of the ground state (or vacuum) $|\Psi_{GS}\rangle$:

$$E[\Psi_\lambda] \geq E_{GS}, \qquad \forall\,|\Psi_\lambda\rangle \in \mathcal{H}, \tag{10.45}$$

with the equal sign valid if and only if $|\Psi_\lambda\rangle = |\Psi_{GS}\rangle$, modulo a phase.

By varying the set of parameters, $\lambda = \{\lambda_j\}$, one can check the quality of the variational wavefunction: if the new set of parameters provides a lower energy than the previous one, then it is a better approximation of the ground state.[43]

Apart from symmetries of the Hamiltonian and other constraints—which should be appropriately imposed on $|\Psi_\lambda\rangle$ *before* the application of the variational method[44]—we have quite a large freedom in the choice of variational wavefunctions. Recently, one such choice has been to use NNs as approximators of quantum states (Carleo and Troyer, 2017).

In this case, one trains the NN to find the weights and biases that best approximate the exact ground (or excited) state. In fact, since many NNs satisfy *universal approximation theorems* for discrete probability distributions (Le Roux and Bengio, 2008), they should provide good representations of quantum states as well.

However, as I have explained in Section 10.1.4, and I have shown in the rest of this Chapter, the most common method to train these NNs, Gibbs sampling (Section 10.2.1), is not always the best one to find the appropriate parameters. Instead, I have demonstrated that a MemComputing-assisted approach to sample the mode of the NN probability distribution improves substantially the quality of the training. In turn, this should reflect in an improvement in the quality of representation of quantum states, while employing similar (or less) resources. Let us then see how to apply such a method to the present problem.

Neural network trial states

Consider a given basis, $|\mathbf{x}\rangle$, in the Hilbert space, $\mathcal{H}$.[45] This could be, e.g., the position, momenta, or spin states of particles. Given a state vector, $|\Psi\rangle$, in this space, its wavefunction (probability *amplitude*) in the given basis is simply $\Psi(\mathbf{x}) \equiv \langle \mathbf{x} | \Psi \rangle$.

This wavefunction provides a *probability density*, $\rho(\mathbf{x}) = |\Psi(\mathbf{x})|^2$, and has a *phase*, $\phi(\mathbf{x})$. It can then be expressed as:

$$\Psi(\mathbf{x}) = \sqrt{\frac{\rho(\mathbf{x})}{Z}} e^{i\phi(\mathbf{x})}. \tag{10.46}$$

In eqn (10.46), I have explicitly written the wavefunction normalization factor, $Z \equiv \langle \Psi | \Psi \rangle = \int d\mathbf{x}\, \Psi^*(\mathbf{x}) \Psi(\mathbf{x})$.

A variational choice for such a wavefunction that employs NNs would then consider an NN to represent the probability density, and an NN to represent the phase. Whatever type of NNs we choose, e.g., the RBMs I have introduced in Section 10.1.2 or the DBMs in Section 10.1.3, the variational parameters we need to vary are the weights and biases of such NNs.

If $\{\lambda\}$ is the set of parameters of the NN representing the probability density, and $\{\mu\}$ the set of parameters of the NN representing the phase, an NN trial wavefunction would be:[46]

$$\Psi_{\lambda,\mu}(\mathbf{x}) = \sqrt{\frac{\rho_\lambda(\mathbf{x})}{Z_\lambda}} e^{i\phi_\mu(\mathbf{x})}, \qquad \textit{trial wavefunction}. \tag{10.47}$$

In the following, I will use RBMs trained with MemComputing, as explained in Section 10.4, namely the MemComputing approach will be used to sample the mode of the RBM probability distribution.

10.7.2 Quantum state tomography

As a benchmark example, I present here the reconstruction of the following entangled N-qubit state, known as W state (cf. Section 4.3.1)

$$|\Psi_W\rangle = \frac{1}{\sqrt{N}} \overbrace{(|100\cdots\rangle + \cdots |\cdots 001\rangle)}^{N \text{ states}}. \tag{10.48}$$

As in Section 4.3.1, the N-qubit notation $|10\cdots\rangle \equiv |1\rangle \otimes |0\rangle \otimes \cdots$, namely the tensor product, $\otimes$, of N qubits.

We perform quantum state tomography on this W state with m projective measurements (cf. Section 4.3.1), and record the reconstruction *fidelity*, namely how close the reconstructed state is to the target one.

[45] This basis satisfies the *resolution of the identity*: $\sum_{\mathbf{x}} |\mathbf{x}\rangle\langle\mathbf{x}| = \mathbf{1}$.

[46] Note that the phase does not appear in the normalization constant Z, hence the latter depends only on the parameters $\{\lambda\}$.

Fidelity of quantum states

Given two states of a quantum system, the *fidelity* measures how different these states are (Nielsen and Chuang, 2010). Mathematically, if the states are described by the statistical operators (density matrices) $\hat\rho$ and $\hat\sigma$, respectively, then their quantum fidelity is

$$F(\hat\rho, \hat\sigma) = \left(\mathrm{Tr} \left\{ \sqrt{\sqrt{\hat\rho}\hat\sigma\sqrt{\hat\rho}} \right\} \right)^2.$$

If the two states are pure ($\hat\rho = |\Psi_\rho\rangle\langle\Psi_\rho|$ and $\hat\sigma = |\Psi_\sigma\rangle\langle\Psi_\sigma|$), then the fidelity reduces to

$$F(\hat\rho, \hat\sigma) = |\langle\Psi_\rho|\Psi_\sigma\rangle|^2.$$

Fig. 10.12 Number of measurements needed to reach (**left panel**) 0.9 fidelity, and (**right panel**) 0.98 fidelity for the W state with three approaches: maximum likelihood, RBM trial wavefunctions, and mode-assisted RBM trial wavefunctions. N^2 samples for each problem size (N qubits) have been used to train the RBMs. Results courtesy of Y. Zhang.

Due to its particular form, we can reconstruct the W state by performing all measurements with the Pauli $\hat{\sigma}^z$ operator.

This means that we choose to write the wavefunction in eqn (10.46) with $|\mathbf{x}\rangle \rightarrow |\boldsymbol{\sigma}^z\rangle \equiv |\sigma_1^z, \sigma_2^z, \dots, \sigma_N^z\rangle$, the basis of the tensor product of N individual Pauli $\hat{\sigma}^z$ matrices: $\hat{\sigma}^z \otimes \cdots \otimes \hat{\sigma}^z$. Since we have the prior knowledge that this W-state wavefunction is real and has no phase factor, we can use a single RBM to represent the amplitude in eqn (10.47).

As a proof of concept, in the following, the m 'measurements' are performed numerically by randomly drawing (from a uniform probability distribution) from the N-qubit W state, eqn (10.48), in the $\boldsymbol{\sigma}^z$ basis. These measurements are then used to train a single RBM to find the appropriate weights and biases. Once the training has been achieved we compute the fidelity $F = |\langle \Psi_W|\Psi_\lambda\rangle|^2$.

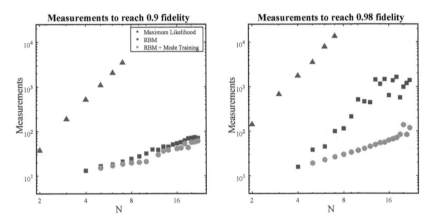

Fig. 10.12 shows how many measurements are required to reach some fixed fidelity.

The plot compares three approaches: the *maximum likelihood method* applied to quantum tomography (Hradil, 1997), the wavefunction extracted from the Gibbs sampling of an RBM, and the one extracted using the mode-assisted approach I discussed in Section 10.4.4.

As expected, at any fixed fidelity, the number of measurements required grows with the number of qubits for all three methods, although the maximum likelihood method grows much faster than the other two.

We also see that at a fidelity threshold of $F = 0.9$ there is not an appreciable advantage in using the mode-assisted training. Instead, when the fidelity is raised to $F = 0.98$, the RBM with mode training requires substantially less measurements to reach that fidelity, with that difference growing with the number N of qubits in the W state.

In addition, it is evident that the mode-assisted training shows a much smaller variance, namely not only does the mode-assisted training substantially improve over Gibbs sampling, it also stabilizes the training.

This is in agreement with the results I presented previously (see, e.g.,

Fig. 10.6).

An important remark

The maximum likelihood method for quantum state tomography was originally proposed in (Hradil, 1997). It is very accurate given enough measurements. However, it requires the full density matrix which scales exponentially with the number of qubits. For instance, in the previous example it requires measurements in different bases to estimate the off-diagonal elements of the density matrix. For the W state, these measurements need to be performed on all 3^N possible Pauli bases.

10.7.3 Ground state of quantum Hamiltonians

A similar mode-assisted approach used to reconstruct a quantum state from experimental data, can be employed to find the ground state (and, in principle, the full spectrum; see margin note [39] in Section 10.7.1) of quantum Hamiltonians.

If we again use the NN trial wavefunction (10.47), we then first need a procedure to train the NN according to the values acquired by the variational energy functional (10.44). This corresponds to the optimization problem of finding the optimum in the non-convex energy landscape defined by the energy functional $E[\Psi_\lambda]$, as we vary the parameters λ_j's, the weights and biases of the NN.

If we follow a (gradient-descent) training strategy, as the one discussed in this Chapter, then we need to determine the variations $\partial_{\lambda_j} E[\Psi_\lambda]$. With these variations in hand, we can then update the NN parameters at iteration $n + 1$, given their value at the previous iteration, n, as we did in eqn (10.10), namely (Carleo and Troyer, 2017)

$$\Delta\lambda_j^{n+1} = \alpha\Delta\lambda_j^n - \epsilon\,\partial_{\lambda_j} E[\Psi_\lambda], \tag{10.49}$$

where α and ϵ are again the momentum and the learning rate, respectively, of the gradient update (cf. Section 10.2).

The full procedure to obtain these derivatives is given in the margin note. Here, I just report the final result:

$$\partial_{\lambda_j} E[\Psi_\lambda] = \langle\hat{H}_{\mathrm{loc}}(\mathbf{x})D_{\lambda_j}^*(\mathbf{x})\rangle_{\Pi(\mathbf{x})} - \langle\hat{H}_{\mathrm{loc}}(\mathbf{x})\rangle_{\Pi(\mathbf{x})}\langle D_{\lambda_j}^*(\mathbf{x})\rangle_{\Pi(\mathbf{x})} + \text{c.c.}, \tag{10.50}$$

where 'c.c.' means 'complex conjugate' of the preceding two terms, and the ensemble averages are all taken with respect to the probability distribution $\Pi(\mathbf{x}) \equiv |\Psi_\lambda(\mathbf{x})|^2$.

Variational energy gradient
Consider a Hamiltonian $\hat{H}$ and the variational energy functional (10.44). Define the following matrix elements $\hat{H}_{\mathbf{xx}'} = \langle\mathbf{x}|\hat{H}|\mathbf{x}'\rangle$ over a basis set $|\mathbf{x}\rangle$. Using the resolution of the identity, $\sum_{\mathbf{x}}|\mathbf{x}\rangle\langle\mathbf{x}| = \mathbb{1}$, the energy functional (10.44) can be written as

$$E[\Psi_\lambda] = \frac{\sum_{\mathbf{x},\mathbf{x}'}\Psi_\lambda^*(\mathbf{x})\hat{H}_{\mathbf{xx}'}\Psi_\lambda(\mathbf{x}')}{\sum_{\mathbf{x}}|\Psi_\lambda(\mathbf{x})|^2}.$$

Let us further define a diagonal ('local') operator:

$$\hat{H}_{\mathrm{loc}}(\mathbf{x}) = \sum_{\mathbf{x}'}\hat{H}_{\mathbf{xx}'}\frac{\Psi_\lambda(\mathbf{x}')}{\Psi_\lambda(\mathbf{x})}.$$

Since $|\Psi_\lambda(\mathbf{x})|^2$ can be interpreted as a probability density, then $E[\Psi_\lambda]$ is nothing other than the *ensemble average* of this diagonal operator over the wavefunction probability distribution $\Pi(\mathbf{x}) \equiv |\Psi_\lambda(\mathbf{x})|^2$:

$$E[\Psi_\lambda] = \langle\hat{H}_{\mathrm{loc}}(\mathbf{x})\rangle_{\Pi(\mathbf{x})}.$$

Let us define

$$D_{\lambda_j}(\mathbf{x}) = \partial_{\lambda_j}\Psi_\lambda(\mathbf{x})/\Psi_\lambda(\mathbf{x}).$$

With this definition, a straightforward calculation of $\partial_{\lambda_j} E[\Psi_\lambda]$ leads to eqn (10.50).

[47] Note that if we sampled the mode in *both* expressions on the RHS of eqn (10.50) we would get zero.

Density-matrix renormalization group (DMRG)
DMRG is a numerical technique that employs variational (trial) states in the form of matrix product states (cf. Section 4.3.1) to find (typically) the ground-state of one-dimensional quantum many-body systems (White, 1992). It is a very efficient and accurate algorithm. However, its extension to dimensions larger than one is not particularly useful.

Fig. 10.13 Mode-assisted training of an RBM trial wavefunction to find the ground state of the one-dimensional transverse-field Ising Hamiltonian (10.51) with 50 spins (and $h = 1$) compared to the DMRG approach. The **main panel** shows the energy per spin (in units of $J_{ij} = J = 1$). The **top-left inset** shows the relative energy error as a function of training iterations. The **top-right inset** reports the scheduling P_{mode}, eqn (10.35). The **bottom-right inset** shows the median of the number of iterations required to reach a relative error of 10^{-3} as a function of the total number of spins. Results courtesy of Y. Zhang.

Mode-assisted training

If we neglect for a moment the momentum term (set $\alpha = 0$ in eqn (10.49)), one notes that the first term on the RHS of eqn (10.50) plays a similar role as the 'model term' on the RHS of eqn (10.10).

Following the strategy I have discussed in Section 10.5.1, I can then assist Gibbs sampling of this 'model term' by sampling the mode of the probability distribution $\Pi(\mathbf{x})$, according to some scheduling. The training then proceeds as I have explained in Section 10.5.1, where Gibbs sampling is replaced at appropriate intervals—following, e.g., the schedule (10.35)—by the sampling of the mode of $\Pi(\mathbf{x}) \equiv |\Psi_\lambda(\mathbf{x})|^2$.[47]

An example: the transverse-field Ising Hamiltonian

As an example of this approach I show in Fig. 10.13 the calculation of the ground state energy of the transverse-field Ising (TFI) Hamiltonian on a one-dimensional lattice:

$$\hat{H}_{\text{TFI}} = -\sum_{\langle ij \rangle} J_{ij} \hat{\sigma}_i^z \hat{\sigma}_j^z - h \sum_i \hat{\sigma}_i^x, \tag{10.51}$$

where J_{ij} are the nearest-neighbor spin interaction strengths, h represents the magnitude of a transverse magnetic field, and $\hat{\sigma}^z$ and $\hat{\sigma}^x$ are Pauli matrices (Section 10.7.2).

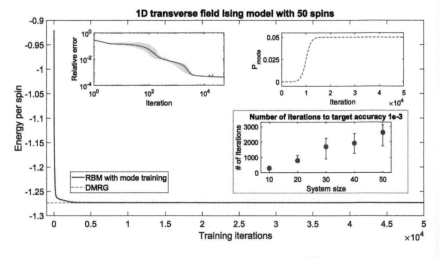

The figure compares the ground state found using the mode-assisted training versus the one found using the *density-matrix renormalization group* (DMRG) approach which is ideal to find the ground state of one-dimensional strongly correlated quantum systems with very high accuracy (White, 1992).

Of course, this is a particularly simple (yet very important) example for which the machinery I have outlined in this Chapter is not really that necessary. However, for strongly correlated quantum systems in two or three dimensions, methods like DMRG, or even the NN trial wavefunctions using just Gibbs sampling may not be enough, and a MemComputing-assisted approach may provide substantial advantages.

In fact, one can bypass the training with gradient descent altogether by mapping the NNs into DMMs directly. The training would then correspond to simply finding the optimum (or a good approximation to the optimum) of the corresponding Boolean problem 'on the fly', which I have shown in this Chapter, and in Chapter 9, can be efficiently done by DMMs. Research in this direction would then be very beneficial.

10.8   Chapter summary

- In this Chapter, I have shown how MemComputing can aid traditional Machine Learning approaches to train NNs.

- In particular, I have shown how MemComputing can improve *unsupervised training* of RBMs by allowing learning at *optimality*, namely with a KL divergence that can be made as small as desired.

- The key to this success is that DMMs can efficiently sample the *mode* of a probability distribution. Since the mode of a distribution is the most difficult part to recover, using the mode sample from MemComputing can both improve and stabilize the training of these NNs, compared to Gibbs sampling.

- I have also shown how this approach can be extended to deep NNs, in particular, deep Boltzmann machines. In this case, the MemComputing-assisted approach requires a much smaller number of parameters to reach optimal training, as it takes full advantage of the correlations between hidden layers.

- A network *pre-trained* with such a MemComputing-assisted approach also provides a better starting point for a downstream supervised learning task, thus *accelerating* considerably supervised learning, namely requiring far less number of backpropagation iterations.

- Finally, by representing wavefunctions with RBMs, I have shown one possible (and not unique) way to employ MemComputing both in quantum state tomography and in finding the ground state of quantum Hamiltonians.

- The MemComputing approach presented here can be extended to other types of NNs, to excited states of quantum Hamiltonians, as well as to quantum dynamics.

- A particularly intriguing research direction would be to map NNs directly into DMMs, so that the gradient-descent training can be eliminated altogether, and replaced by the finding of the optimum (or a good approximation to the optimum) of the NN energy landscape itself 'on the fly'.

Part IV

MemComputing and Topology

Topology is precisely the mathematical discipline that allows the passage from local to global.
René Thom (1923—2002)

In this part of the book, I will show that digital MemComputing machines (DMMs) employ objects of a *topological* nature, such as *critical points* and *instantons* (a *CW complex*) to compute. As a consequence, they are robust against noise and perturbations. It is indeed thanks to such features that the *numerical* simulations of the corresponding equations of motion are robust against the unavoidable (and large) numerical errors (noise), as I have explicitly shown in Chapter 8. I will also briefly mention the (supersymmetric) *Topological Field Theory* of dynamical systems that allows us to show that certain matrix elements on instantons are *topological invariants* (*topological quantum numbers*), hence they cannot change without changing the topology of phase space.

... therefore, since men philosophized in order to escape from ignorance, evidently they were pursuing Science in order to know, and not for any utilitarian end. ...

I will also show that DMMs *self-tune* into a state with *long-range order*, and go through a *critical branching process*, which persists throughout the dynamics, until the machines find a solution of the given problem. Long-range order and critical branching are *emerging* properties of DMMs (criterion IV of MemComputing I discussed in Section 1.6) that allow them to tackle hard combinatorial optimization problems efficiently. Next, I will discuss that the dynamics of DMMs in the presence of noise are akin to the *directed percolation* of the state trajectory in the phase space. I will finally conclude with a bird's eye view of the *brain-like* features of MemComputing machines.

- In Chapter 11, I will introduce the topological features of DMMs.
- The emerging property of critical branching and directed percolation of the state dynamics of DMMs in the presence of noise are discussed in Chapter 12. Brain-like features of MemComputing machines will also be reported in this Chapter.

11

Topological features of DMMs

Topology is the property of something that doesn't change when you bend it or stretch it as long as you don't break anything.
Edward Witten (1951−)

[1]Recall that I now employ the term *DMM* and *self-organizing circuit* interchangeably to indicate both the *ideal* machine and its *practical*, circuit realization.

[2]Implemented in interpreted MAT-LAB.

Takeaways from this Chapter

- Digital MemComputing machines (DMMs) employ objects of a topological nature to compute.

- These are *critical points* in the phase space and trajectories in between them (*instantons*).

- Instantons express the *dynamical long-range order* of DMMs in both space and time, and are *non-perturbative* objects.

- This allows DMMs to navigate efficiently their (vast) phase space.

In Chapters 8, 9, and 10, I have shown *simulations* of dynamical systems representing self-organizing circuits, which are the practical realization of DMMs.[1] All these simulations have been done on a single processor of a standard computer.

The vast majority of them has been done by integrating the ordinary differential equations (ODEs) of these machines using *explicit methods* (see Section 8.4.1).[2] In fact, due to the fact that these ODEs have different time scales, they are properly called *stiff* differential equations (Section 8.3).

As I discussed in Chapter 8, 'stiff' ODEs are notorious for requiring a higher level of precision—like the one afforded by *implicit methods* of integration—to faithfully represent the state trajectory. In other words, explicit methods should not even work properly for such a class of equations (Sauer, 2018)!

And yet, they do. Indeed, the results in Chapters 8–10 clearly demonstrate the great benefit of solving hard combinatorial optimization problems using a Physics-based approach, like MemComputing, rather than the traditional algorithmic ones. This means that, even though the numerical integration of the ODEs representing a DMM brings its own overhead, and may introduce substantial numerical errors during time evolution, it still outperforms state-of-the-art algorithms for the same problems, especially as the size of the problem increases.

An important remark

Of course, the ultimate step of integrating an ODE in time *is*, in itself, an algorithm. However, this type of algorithm is *radically* different from those typically employed in the solution of combinatorial optimization problems, where clever strategies are used to 'guess' the next step in the variables' assignment of the problem (cf. discussion in Section 2.6, and see, e.g., (Garey and Johnson, 1990)).

Since it is inconceivable that integration of ODEs may be immune from numerical errors (they are not!), and yet those representing DMMs seem to be quite efficient in solving hard problems,[3] there *must* be some properties inherent to those specific equations (dynamical systems) that carry over even when they are integrated numerically (hence discretized in time). This Chapter and Chapter 12 discuss these properties.

In fact, the same properties that render the simulations of DMMs robust against numerical errors will make them robust against the unavoidable *noise* and *perturbations* of the physical systems used to build them in *hardware* (see Section 12.7). Let's then try to identify these properties.

11.1   The phase space of DMMs

From the discussion in Chapters 6 and 7, we know that DMMs are dynamical systems whose dynamics 'live' in a phase space, X.

Suppose now we design a DMM to solve a combinatorial optimization problem, like the ones I have discussed in Chapters 8, 9, and 10. For the sake of this discussion let's consider a search problem (Section 2.3.4), like the satisfiability (SAT) problem I discussed in Section 9.1.1. We then attempt to solve this problem by letting the DMM self-organize to the solution.[4]

From the point of view of dynamical systems then, the DMM starts from an *initial state*, $\mathbf{x}(t=0)$. This is a point in the (vast) phase space,

[3] Even when integrated with the worst numerical method possible: forward Euler (Section 8.5.1).

[4] Assuming a solution exists.

[5] If we consider an optimization problem, then we either find an approximation to the optimum, or the optimum itself, within a set time (cf. discussion in Section 7.9).

Fig. 11.1 Sketch of a phase space landscape with valleys, hills, and so on.

Random walk
It is the stochastic process (see Section 10.2.1) by which a path in a multi-dimensional space is formed by a succession of random steps. Arguably the most famous physical example of random walk is the Brownian motion of a particle in a liquid (van Kampen, 1992). DMMs do *not* perform a random walk in phase space.

X, and unless we were extremely lucky, it is most likely *not* a critical point of the dynamics, namely one in which the flow vector field, $F(\mathbf{x})$, of eqn (3.1) is zero (see Chapter 3). From then on, the system somehow travels along a trajectory in this space, till it finds the solution.[5] This solution is again a point in the phase space. In fact, for a search problem, it is an *equilibrium point*, $\mathbf{x}_{eq}$, which is also a *critical point*.

As usual, this (high-dimensional) phase space can be pictured as an extremely corrugated landscape, made of valleys, hills, and saddles of varying heights (see the sketch in Fig. 11.1).

An important remark

An analysis done in (Bearden *et al.*, 2018) has shown that for the equations of motion of DMMs as reported in (Traversa and Di Ventra, 2017), the critical points are such that the *magnitude* of the real part of the eigenvalues, λ_i, of their unstable directions (see again Section 3.7 for the definition and meaning of these eigenvalues), is *much smaller* than the corresponding quantity of their stable directions. This feature seems to be shared by generic high-dimensional non-convex landscapes that have mainly saddle points surrounded by quasi-flat unstable directions (Dauphin *et al.*, 2014). In the case of DMMs, it essentially originates from the very different time scales of the voltages (fast degrees of freedom) and the memory variables (slow degrees of freedom); cf. discussion in Section 7.4.

The trajectory that leads the system from an arbitrary initial condition to the final (solution) point must explore this corrugated landscape, and, in order to perform an efficient search, it must do so without wandering aimlessly by visiting all possible valleys, saddles, and so on.

Without further information on the dynamics, we would expect that such 'aimless' search would also occur, in principle, even if we succeeded in designing the DMM as I have discussed in Section 6.3.2, namely with no additional equilibrium points, other than those corresponding to the solution of the given problem. In fact, if the DMM had to do just that, namely a *random walk* across this vast landscape, it would take an inordinate amount of time to reach the equilibrium point, the solution of the problem. Instead, the numerical results of Chapters 8–10 strongly suggest that the system does go from the initial state to the final state by following *specific paths* (trajectories) in this vast phase space, with the path 'length' ultimately growing only polynomially with system size.

How does a DMM do that? What is the Physics behind this efficient search in a vast phase space? And, does it tell us anything about their stability against noise and perturbations?

11.2 State dynamics

To answer these questions let's first look at how the state of a DMM (the collection of all the voltages/currents and internal memory variables) evolves in time. Let's first focus on a typical trajectory for the voltages at the different terminals of a given self-organizing circuit representing a DMM.[6]

I am focusing on the voltages because, ultimately, these are the ones that encode the solutions of the given problem we want to solve. As I have discussed in Chapter 7, the internal (memory) variables are meant only to allow the terminal voltages to *adapt* to satisfy the correct logical (or algebraic) propositions, by transforming any local minima of the reduced voltage space into saddle points of the full phase space X.

An example of such voltage trajectories is shown in Fig. 11.2 for the factorization of a 20-bit number using the self-organizing logic gates (SOLGs) I discussed in Section 7.2 and a Boolean circuit similar to the one shown in Fig. 7.4.[7] Figure 11.2 shows that starting from an arbitrary initial condition, the DMM evolves toward the equilibrium point, solution of the problem.

[6]The currents of the dynamic correction modules (as defined in Section 7.3.1) serve the purpose of changing the terminal voltages. Therefore, it is enough to study the latter ones as a function of time.

[7]Of course, the circuit is for 20 bits not 6, as in that figure; cf. also Fig. 7.18 and the results in (Traversa and Di Ventra, 2017).

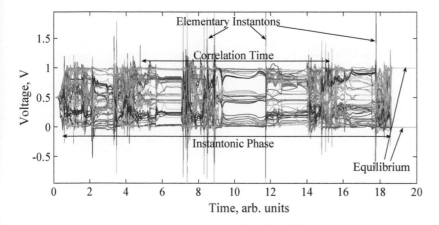

Fig. 11.2 Voltages vs. time at a randomly chosen subset of 70 literals (nodes) of a self-organizing circuit (with Boolean topology as the one in Fig. 7.4) for the factorization of a 20-bit number. The factored number is $497503 = 499 \times 997$. The instantonic phase consists of *elementary instantons*. The correlation time is defined as the time τ for which the time correlation, $C(\tau)$, defined in the text (eqn (11.26)), drops by one order of magnitude and is indicated by the double head arrow. Note that in this figure the logical 0 is represented as 0V. Reproduced with permission from (Di Ventra *et al.*, 2017).

During the *transient phase* (after the initial condition has been set, and before the equilibrium point is reached) the voltages may acquire values that differ from the logical 1 and the logical 0. In fact, those voltages can be even higher or lower than the corresponding logical values, according to how the SOLGs are designed.

For instance, in most of the examples in Chapters 8–10 the voltages are bounded between +1V and −1V, but that is not necessary, as shown in Fig. 11.2. Irrespective, due to the presence of capacitances in the corresponding SOLGs, the terminal voltages are always *bounded*.

Transient phase
The dynamics from the initial time until an equilibrium is reached is called a *transient phase* or *transient dynamics*. If no equilibrium exists then a DMM will trace a (quasi-)periodic orbit after the transient (cf. Section 7.9).

[8] The spike's (instantonic) width is difficult to resolve from Fig. 11.2. However, it can be easily seen from Fig. 7.13, where the RC time was chosen quite large, and the instantonic peak was then quite broad.

Avalanche phenomena

Many natural phenomena, such as earthquakes, solar flares, quenches, etc., display *transient* dynamics whereby energy is released in 'bursts'—events of spatiotemporal activity or *avalanches*—with the size and duration of these avalanches typically following a power-law distribution, namely they display *long-range correlations* (see Section 4.3.2).

Self-organized criticality

The power-law distribution of avalanche phenomena has been sometimes understood as originating from the fact that the system 'self-tunes' to a *critical state*, hence the name of *self-organized criticality* (Pruessner, 2012). DMMs do indeed self-tune to a long-range ordered state (Chapter 12), but I prefer to understand this phenomenon in terms of the more rigorously grounded (supersymmetric) Topological Field Theory of dynamical systems (Ovchinnikov, 2016), which also reveals the topological aspects of the dynamics (see Section 11.3).

[9] Therefore, showcasing an *intermittent* dynamics.

That said, one observes from Fig. 11.2 distinct 'spikes' in voltages, which have a time width on the order of the typical RC time of the circuit,[8] where R is the typical resistance of the resistive elements, and C is the capacitance introduced to simulate parasitic capacitive effects (see Section 7.3.1).

These voltage spikes involve some of the terminal voltages, not all of them at once, and are then followed by time intervals of practically no (or very little) voltage variations. This type of voltage dynamics is typical of *all* the dynamical systems representing DMMs, as those I have discussed in Chapters 8–10. The internal memory variables, instead, have, by construction, a much slower type of dynamics (see, e.g., Fig. 7.13 and Chapter 9). This means that during the sudden variation of the voltages, the memory variables vary relatively smoothly.

This is easy to understand. As I have mentioned before (Section 7.4), the memory variables' only role is to allow the voltages to self-organize into the correct logical or algebraic proposition. If their dynamics were too fast, on a time scale comparable to that of the voltages, the latter ones would experience too much variation at all times, even during the time intervals in between sudden jumps. This can ultimately lead to chaotic behavior and the impossibility to reach an equilibrium point. Therefore, the whole transient dynamics can be summarized as follows:

Transient dynamics of DMMs

(i) The DMM starts from an initial condition that is typically *not* a critical point of the dynamics.

(ii) After some time, a few voltages (a *cluster* of voltages) experience sudden variations, on a time scale of order RC, where R is the typical resistance of the circuit and C its capacitance.

(iii) This is followed by a period of little voltage variations, after which a new cluster of voltages varies rapidly, followed by another 'quiet' period, and so on.

(iv) This process repeats itself until an equilibrium point is reached, if such an equilibrium exists, otherwise the DMM will end up into a (quasi-)periodic orbit; Section 7.9. In the meantime, the internal memory variables vary smoothly during the whole dynamics.

The transient phase of DMMs then reminds us of an *avalanche phenomenon* in which energy (or some other physical property) is released in bursts, intercalated by quiescent periods of inactivity,[9] till the system reaches the equilibrium state of lowest energy. In our case, the *size* of these avalanches is the number of voltages belonging to it (see also

Chapter 12).

It turns out that these avalanches have a well-defined (supersymmetric) *Topological Field Theory* (TFT) description that goes under the name of *instantons* (Rajaraman, 1982).[10]

It is not my intention to discuss in this book the TFT of non-quantum dynamical systems in its full depth. In the following Section 11.3—which is adapted from (Di Ventra and Ovchinnikov, 2019)—I will only provide the main ingredients of such a theory that will help us understand the topological features of DMMs.

Arguably, this Section 'sweeps a lot of information under the rug',[11] and requires quite a bit of familiarity with techniques of Quantum Field Theory (see, e.g., (Peskin and Schroeder, 1995) for a thorough discussion of these techniques), (Algebraic) Topology (Fomenko, 1994), and even some notions of Differential Geometry (Göckeler and Schücker, 1989).

Therefore, it can be skipped at first read, without much detriment to the understanding of the subsequent Sections. However, the reader familiar with these theoretical techniques, and who wants to delve deeper into this topic, is urged to complement Section 11.3 with the review (Ovchinnikov, 2016) and references therein, where a more detailed exposition of this theory is reported.[12]

11.3   A (very) brief primer of supersymmetric TFT

In Physics (and Mathematics) TFT typically refers to a *Quantum Field Theory* where the correlation functions (observables) of interest are *topological invariants* (Witten, 1988; Schwarz, 1993).

However, this is a somewhat restrictive viewpoint. In fact, it was recently recognized that all *non-quantum* dynamical systems, as those following the eqns (3.1) (with added noise as well, as we will see), can be described by a *supersymmetric* TFT formalism (Ovchinnikov, 2016).

The starting point is the realization—which goes back to the early days of Quantum Mechanics (Koopman, 1931; von Neumann, 1932)—that, like quantum-mechanical systems, also the dynamics of non-quantum systems can be represented *algebraically*.

In other words, instead of working directly with the trajectories $\mathbf{x}(t)$ in the (non-linear) phase space X, one can work with *state vectors* $|\psi\rangle$ in a *linear* Hilbert space (see Section 3.1) and operators acting on such a space.

The price one pays for this algebraic representation of non-quantum

[10]Note that instantons emerge also in Field Theories that are *not* topological and/or supersymmetric (cf. Section 11.4).

[11]A full derivation of all the equations in Section 11.3, and, most importantly, a thorough exposition of the concepts it contains, would easily (more than) double the length of the present book.

[12]A general and fairly complete review of Topological (Quantum) Field Theory is also in (Birmingham *et al.*, 1991).

Homeomorphism
A continuous function between *topological spaces* (Section 3.2) that has an inverse, which is also continuous, is called a *homeomorphism* (Fomenko, 1994). Two spaces that are *homeomorphic* to each other have the *same* topological structure.

Topological invariant
Any property of a topological space that is *invariant* under homeomorphisms is a *topological invariant/quantum number/charge*. Namely, it depends only on the *global* features of the space, not on its geometry.

Homotopy
The equivalence relation among all maps (from one topological space to another) that can be continuously deformed into each other is called a *homotopy*. A homeomorphism is a special case of a homotopy equivalence.

Exterior (or Grassmann) algebra
It is the algebraic structure whose product (indicated with ∧) is the *exterior* or *wedge product* of vectors in a vector space. The exterior product generalizes the standard 'cross product' of two 3-dimensional vectors to multidimensional spaces (Göckeler and Schücker, 1989).

Tangent space and bundle

Given any point $\mathbf{x}$ of a differentiable manifold, X, (see Section 3.2), its *tangent space* $TX_{\mathbf{x}}$, is the space of all vectors that tangentially pass through $\mathbf{x}$ (Göckeler and Schücker, 1989). For instance, for a point on a sphere, it is the plane tangent to the sphere at that point. The *disjoint union* ($\sqcup$) of all tangent spaces of X is called the *tangent bundle* of X: $TX = \sqcup_{\mathbf{x} \in X} TX_{\mathbf{x}}$. It is equipped with the projection mapping, $\pi : TX \to X$ ($TX_{\mathbf{x}}$ is then a *fiber* of the bundle).

Gaussian white noise

It is a stationary *stochastic process* (Section 10.2.1), $\xi(t)$, with the following first two *moments* of its probability distribution:

$$\langle \xi(t) \rangle = 0,$$
$$\langle \xi(t)\xi(t') \rangle = \Gamma \delta(t - t'),$$

where $\Gamma > 0$ indicates the strength of the noise, and the symbol $\langle \cdot \rangle$ indicates *ensemble average*, namely an average over an ensemble of *realizations* of the stochastic process. All higher *odd* moments are zero, while the *even* moments can be expressed as sums of products of δ-functions (van Kampen, 1992).

Fermionic and bosonic variables

Fermionic (or Grassmann) variables, χ_i, are *anti-commuting* generators of an algebra over the complex (or real) numbers:

$$\chi_i \chi_j = -\chi_j \chi_i, \quad \forall i, j.$$

This implies that $(\chi_i)^2 = 0$. We then call *bosonic variables* those generators that *commute* with each other: $x_i x_j = x_j x_i, \forall i, j$. The state $\mathbf{x}$ is composed of bosonic variables. Finally, Grassmann variables *commute* with bosonic ones: $\chi_i x_j = x_j \chi_i, \forall i, j$.

dynamics is that, unlike Quantum Mechanics, the evolution operator—that determines the dynamics of the system—is *not* Hermitian. It is instead *pseudo*-Hermitian, namely all its entries are real, and its spectrum is composed of only real eigenvalues and pairs of complex conjugate eigenvalues (Mostafazadeh, 2002).

The Hilbert space of this algebraic representation of dynamics is the (complex-valued) *exterior algebra* (or Grassmann algebra) of X: $\Omega(X, \mathbb{C})$ (Göckeler and Schücker, 1989).

Adding physical noise

Since I will discuss in Section 12.7 the robustness of DMMs to noise and perturbations, it is convenient to generalize the theory to *stochastic* dynamical systems, namely those subject to noise. In fact, *any* physical system is always subject to some noise, so this is a very important generalization. If needed, one can always take the limit of zero noise.

We then add to the flow vector field, $F(\mathbf{x})$, of eqns (3.1) a noise term:

$$F \to \tilde{F} = F + \sum_{a=1}^{k} e_a(x)\xi^a(t), \tag{11.1}$$

where $\{e_a(x) \in TX_{\mathbf{x}}, a = 1...k\}$ are vector fields that couple the noise to the system, and $TX_{\mathbf{x}}$ is the *tangent space* of X at point $\mathbf{x}$ (F is a *section* of the *tangent bundle*). Without loss of generality, we can choose as noise variables, $\xi^a(t) \in \mathbb{R}$, *Gaussian white noise*.

Let's now consider the following object,

$$w(\xi) = \int_{p.b.c.} D\mathbf{x} \prod_{\tau} \delta(\dot{\mathbf{x}}(\tau) - \tilde{F}(\tau))J, \tag{11.2}$$

where ξ is a single realization of the noise, and the *functional integration*, $\int_{p.b.c.} D\mathbf{x}$, is performed over closed paths/periodic boundary conditions (p.b.c.), $\mathbf{x}(t) = \mathbf{x}(t')$. The Jacobian determinant, J, is (the symbol $\delta/\delta\mathbf{x}(\tau')$ means *functional differentiation*)

$$J = \text{Det} \frac{\delta(\dot{\mathbf{x}}(\tau) - \tilde{F}(\mathbf{x}(\tau)))}{\delta\mathbf{x}(\tau')}. \tag{11.3}$$

By performing the functional integration explicitly (recall that for a δ-function: $\int g(\mathbf{x})\delta(f(\mathbf{x}))d\mathbf{x} = \sum_{\mathbf{x}_\alpha, f(\mathbf{x}_\alpha)=0} g(x_\alpha)/|f'(\mathbf{x}_\alpha)|$) one obtains a summation of the signs of the Jacobian determinant over the closed solutions of the stochastic ODE, namely,

$$w(\xi) = \sum_{\text{closed paths}} \text{sign } J. \tag{11.4}$$

It turns out that $w(\xi)$ does *not* depend on the noise configuration: it is a *topological invariant* (it is just an integer). This means that also its stochastic average

$$W = \langle w \rangle,\tag{11.5}$$

is a *topological invariant*.

Bosonic and fermionic variables

At this stage, we introduce *fermionic* (or Grassmann) variables, and call the 'original' state variables $\mathbf{x}$ *bosonic*, because they follow a *commuting algebra* (the standard multiplication of two real numbers is commutative).

We then introduce what is known as a pair of (Grassmann variables) Faddeev-Popov 'ghosts', χ^i and 'anti-ghosts' $\bar\chi_i = -i\partial/\partial\chi^i$, for each element of the state $\mathbf{x}$, and the *conjugate momenta* (or *Lagrange multipliers*), $B_i = -i\partial/\partial x^i$, (which are instead bosonic variables) of the original state variables x^i (with x^i an element of the state $\mathbf{x}$).

The reason why these fermionic variables are called 'ghosts', is because they are only a *mathematical* tool (to implement the so-called *gauge fixing* procedure and a change of coordinates (Peskin and Schroeder, 1995)), and *not* physically observable particles.

Path-integral formulation

If we now substitute eqn (11.2) into eqn (11.5) and use standard *path-integral techniques* (Peskin and Schroeder, 1995), one obtains the following expression (which is still a *topological invariant*)

$$W = \left\langle \int_{p.b.c.} D\Phi e^{-\{Q,i\int_{t'}^{t} d\tau \bar\chi(\tau)(\dot{\mathbf{x}}(\tau)-\bar F(\mathbf{x}(\tau)))\}}\right\rangle,\tag{11.6}$$

where the notation Φ means the collection of the original bosonic fields, x^i, additional bosonic momenta, B_i, and the pairs of Faddeev-Popov ghosts, χ^i and $\bar\chi_i$. The quantity in (11.6) is the stochastic generalization of the *Witten index*.

The operator $\{Q,\cdot\}$ is defined via its action on an arbitrary functional A of the fields Φ as

$$\{Q,A\} = \int_{t'}^{t} d\tau \sum_i \left(\chi^i(\tau)\frac{\delta}{\delta x^i(\tau)} + B_i(\tau)\frac{\delta}{\delta\bar\chi_i(\tau)}\right) A.\tag{11.7}$$

By performing the functional integration over the Gaussian white

Field theory and path integrals

The state $\mathbf{x}(t)$ in eqns (3.1) can be interpreted as a *vector field* that associates at each time t a field amplitude $\mathbf{x}: \mathbb{R} \to X$. The dynamics described by eqns (3.1) can then be interpreted as a *field theory*. This is typically called a '(0+1)-dimensional field theory' because the 'space-time' of this theory has 0 spatial dimensions and 1 temporal dimension. *Any* field theory (indeed, *any* operator/algebraic theory) can be expressed in the language of *path integrals* (Peskin and Schroeder, 1995). Path integrals are particularly useful to compute quantities related to non-perturbative phenomena, such as the instanton matrix elements (11.24).

BRST algebra

The fields x^i, B_i, χ^i and $\bar\chi_i$ obey the (Becchi-Rouet-Stora-Tyutin) *BRST algebra*. If $\{Q,\cdot\}$ is the operator in eqn (11.7), then the algebra is: $\{Q,x^i\} = \chi^i$; $\{Q,\bar\chi_i\} = B_i$; $\{Q,\chi^i\} = 0$; $\{Q,B_i\} = 0$. The fields $(x^i, B_i, \chi^i, \bar\chi_i)$ have fermion/ghost content (0, 0, 1, -1); (Birmingham *et al.*, 1991).

Differential forms

Consider any point $\mathbf{x}$ in a manifold, X. A *differential form* of degree k (a k-form) is a mapping between every point $\mathbf{x}$ and an *alternating* k-form $\phi_{\mathbf{x}}$ on the tangent space $TX_{\mathbf{x}}$, where $\phi_{\mathbf{x}}: (TX_{\mathbf{x}})^k \to \mathbb{R}$ (or $\mathbb{C}$), and 'alternating' means that for any pairs of vectors v_i and v_j: $\phi_{\mathbf{x}}(\ldots,v_i,\ldots,v_j,\ldots) = -\phi_{\mathbf{x}}(\ldots,v_j,\ldots,v_i,\ldots)$. See, e.g., (Göckeler and Schücker, 1989).

Exterior derivative (or de Rham operator)
The *exterior derivative* $\hat{d}$ is the (unique) linear mapping between k-forms and $(k+1)$-forms, such that:
(i) $\hat{d}f$ is the usual *differential* of a smooth function f (a 0-form),
(ii) $\hat{d}$ is *nilpotent*, namely $\hat{d}(\hat{d}|\psi^{(k)}\rangle) = 0$, for arbitrary k-forms $|\psi^{(k)}\rangle$, and
(iii) $\hat{d}(|\psi^{(p)}\rangle \wedge |\psi^{(k)}\rangle) = \hat{d}|\psi^{(p)}\rangle \wedge |\psi^{(k)}\rangle + (-1)^p |\psi^{(p)}\rangle \wedge \hat{d}|\psi^{(k)}\rangle$.

Interior multiplication
It is the *contraction* of a differential form, $|\psi^{(k)}\rangle$ with a vector field, F, so that it maps k-forms into $(k-1)$-forms. For example, if $|\psi^{(1)}\rangle$, is a 1-form on $\mathbb{R}$ (or $\mathbb{C}$), $\hat{\imath}_F|\psi^{(1)}\rangle = \psi^{(1)}(F) \in \mathbb{R}$ (or $\mathbb{C}$). See, e.g., (Göckeler and Schücker, 1989) for an in-depth discussion of all these notions.

Lie derivative
The *Lie derivative* L_F expresses the change of a (tensor) field it acts upon, over the flow defined by the field F.

Spectrum of the evolution operator
Since the evolution operator in eqn (11.13) is $\hat{d}$-exact (namely $\hat{H} = [\hat{d}, \cdot]$), its *eigenstates* are *supersymmetric singlets* (with zero eigenvalue) and *non-supersymmetric doublets* (with eigenvalues that are real or complex conjugate pairs). The latter ones are related to each other via the exterior derivative $\hat{d}$. For instance, if $|\phi\rangle$ is one such non-supersymmetric state, then $\hat{d}|\phi\rangle$ is the other.

noise exactly we obtain:

$$W = \int_{p.b.c.} D\Phi\, e^{-\{Q,\Psi\}}, \tag{11.8}$$

where

$$\Psi(\Phi) = \int_{t'}^{t} d\tau \left(i \sum_j \bar{\chi}_j(\tau)\dot{x}^j(\tau) - \bar{d}(\Phi(\tau)) \right), \tag{11.9}$$

is called the *gauge fermion* and

$$\bar{d}(\Phi) = i \sum_j \bar{\chi}_j \left(F^j - \Theta e_a^j \{Q, i \sum_k \bar{\chi}_k e_a^k\} \right), \tag{11.10}$$

can be identified as a 'probability current' (Ovchinnikov, 2016).

Time-evolution operator
The evolution operator of the theory can be obtained by simply removing the periodic boundary conditions in eqn (11.8):

$$\hat{M}_{tt'}(\mathbf{x}\chi, \mathbf{x}'\chi') = \iint_{\substack{\mathbf{x}(t)=\mathbf{x},\mathbf{x}(t')=\mathbf{x}' \\ \chi(t)=\chi,\chi(t')=\chi'}} D\Phi\, e^{-\{Q,\Psi\}}, \tag{11.11}$$

where the path integration in eqn (11.11) is now over trajectories connecting the arguments of the evolution operator.

The operator representation of eqn (11.11) is

$$\hat{M}_{tt'}(\mathbf{x}\chi, \mathbf{x}'\chi') = e^{-(t-t')\hat{H}} \delta(\mathbf{x}-\mathbf{x}')\delta(\chi-\chi'), \tag{11.12}$$

where (without noise) $\hat{H} = \hat{L}_F$ and the operator L_F is the *Lie derivative* over the flow vector field F.

This *pseudo*-Hermitian operator can also be written as

$$\hat{H} = [\hat{d}, \hat{\bar{d}}], \tag{11.13}$$

where $\hat{d}$ is the *exterior derivative* (or *de Rham operator*), and the current (without noise) $\hat{\bar{d}} = \hat{\imath}_F$, with $\hat{\imath}_F$ the *interior multiplication*.

The square bracket $[\cdot, \cdot]$ denotes the *bi-graded commutator* of any two arbitrary operators: $[\hat{A}, \hat{B}] = \hat{A}\hat{B} - (-1)^{deg(\hat{A})deg(\hat{B})}\hat{B}\hat{A}$, where $deg(\hat{A})$ is the difference between the number of exterior and interior multiplication operators in $\hat{A}$.

The exterior derivative $\hat{d}$ is the operator version of the functional Q in the path integral (11.8). It defines what is known as a *de Rham cohomology* on the exterior algebra of the phase space (Fomenko, 1994).

The exterior algebra of the phase space X, $\Omega(X, \mathbb{C})$, is then the vector space of differential k-forms of all degrees

$$|\psi^{(k)}(\mathbf{x})\rangle = \frac{1}{k!} \sum_{i_1 \ldots i_k} \psi^{(k)}_{i_1 \ldots i_k}(\mathbf{x}) dx^{i_1} \wedge \cdots \wedge dx^{i_k} \in \Omega^{(k)}(X, \mathbb{C}),$$

(11.14)

where $\psi^{(k)}_{i_1 \ldots i_k}(\mathbf{x})$ is a smooth *antisymmetric* tensor, and $\wedge$ is the *wedge product* (or multiplication) of differentials dx^i, e.g., $dx^1 \wedge dx^2 = dx^1 \otimes dx^2 - dx^2 \otimes dx^1$. The space $\Omega^k(X, \mathbb{C})$ is that of all differential forms of degree k, and $\Omega(X, \mathbb{C}) = \bigoplus_{k=0}^{D} \Omega^k(X, \mathbb{C})$.

Topological action

We know that the *Euclidean action*, S_{Eucl}, of a physical system is the time integral of a Lagrangian, $\mathcal{L}[\Phi]$, from an initial time t' to a final time t, and is a functional of all the fields, Φ, (Goldstein, 1965):

$$S_{Eucl}[\Phi] = \int_{t'}^{t} d\tau \mathcal{L}[\Phi].$$

(11.15)

From this Euclidean action we can construct the evolution operator in the path-integral formulation:

$$\hat{M}_{tt'}(\mathbf{x}\chi, \mathbf{x}'\chi') = \iint_{\substack{\mathbf{x}(t)=\mathbf{x}, \mathbf{x}(t')=\mathbf{x}' \\ \chi(t)=\chi, \chi(t')=\chi'}} D\Phi \, e^{-S_{Eucl}[\Phi]}.$$

(11.16)

Comparing with eqn (11.11) we can then define the Euclidean action, S_{Eucl}, of the system as:

$$S_{Eucl}[\Phi] \equiv \{Q, \Psi\} = \int_{t'}^{t} d\tau \sum_{i} \left(\chi^i(\tau) \frac{\delta}{\delta x^i(\tau)} + B_i(\tau) \frac{\delta}{\delta \bar{\chi}_i(\tau)} \right) \Psi.$$

(11.17)

We then say that this action is *Q-exact*, namely it is of *topological* character (it does not depend on any metric).

Topological (de Rham) supersymmetry

In this language, the exterior derivative, $\hat{d}$, is

$$\hat{d} = \sum_i dx^i \wedge \frac{\partial}{\partial x^i} \equiv \sum_i \chi^i \frac{\partial}{\partial x^i}.$$

(11.18)

In view of the fact that the wedge product $\wedge$ is *antisymmetric*, eqn (11.18) shows that $\hat{d}$ 'creates' *fermionic* or anti-commuting variables ($dx^i \wedge \equiv \chi^i$), and 'destroys' *bosonic* or commuting variables (x^i).

Supersymmetry of the ground state

If supersymmetry is preserved, the *ground state*, $|\psi_{GS}\rangle$, of the evolution operator, $\hat{H}$, is *supersymmetric* (or $\hat{d}$-symmetric *singlet*): $\hat{d}|\psi_{GS}\rangle = 0$, but $|\psi_{GS}\rangle \neq \hat{d}|\psi_k\rangle$ (with ψ_k some differential form). This implies that its eigenvalue is zero: $E_{GS} = \langle \psi_{GS} | \hat{H} | \psi_{GS} \rangle = 0$.

Chaos and spontaneous breakdown of supersymmetry

It was discussed in (Ovchinnikov and Di Ventra, 2019) that *all* features of chaotic dynamics can be interpreted as the *spontaneous breakdown* of supersymmetry (cf. Section 3.9), namely the *ground state*, $|\psi_{GS}\rangle$, of the system is not annihilated by $\hat{d}$: $\hat{d}|\psi_{GS}\rangle \neq 0$, while $\hat{d}$ still commutes with the evolution operator $\hat{H}$ (eqn (11.19)). Following the discussion in Section 4.3.3, this means that chaos is actually the *low-symmetry*, *ordered phase* of the dynamical systems that support it. That is the reason why in (Ovchinnikov and Di Ventra, 2019) the word *chronotaxis* (rather than 'chaos') was suggested to describe this phenomenon. DMMs do *not* spontaneously break supersymmetry as they are *integrable* dynamical systems (see Sections 3.10 and 8.2.5). However, in the *numerical simulation* of DMMs one necessarily discretizes time (generating a *discrete map*). In this case, supersymmetry may be broken *explicitly*: $[\hat{H}, \hat{d}] \neq 0$ (this is sometimes called *soft supersymmetry breaking*). Although it was never observed in practice (for the DMMs studied so far), due to this possibility, chaos may indeed appear during dynamics described by a map (Ovchinnikov and Di Ventra, 2019).

In addition, the form of $\hat{H}$ in eqn (11.13) is $\hat{d}$-*exact*. Since the exterior derivative $\hat{d}$ is *nilpotent*, $\hat{d}^2 = 0$, it commutes with any $\hat{d}$-exact operator, so that $[\hat{d}, [\hat{d}, \cdot]] = 0$. This means that the evolution operator $\hat{H}$ *commutes* with $\hat{d}$:

$$[\hat{H}, \hat{d}] = 0. \tag{11.19}$$

In Physics, we know that such a relation means that $\hat{d}$ is a *symmetry* of the system, and since $\hat{d}$ transforms bosonic variables into fermionic ones, it is the *Noether (super-)charge* of the *topological supersymmetry* of the dynamics (and *cannot* be broken *perturbatively*; cf. Section 12.7.).

Finally, any state of the system in the exterior algebra language can be represented as the linear combination

$$|\psi(\mathbf{x})\rangle = \sum_{k=0}^{D} |\psi^k(\mathbf{x})\rangle. \tag{11.20}$$

This completes the program of studying the dynamics (3.1) by means of state vectors $|\psi\rangle$ in, and operators on a linear Hilbert space.

After this (very) brief detour into the TFT of *non-quantum* dynamical systems, we can go back to the main features of the dynamics of DMMs.

11.4   Instantons

The concept of instantons is often introduced in High-Energy Physics as an alternative way to calculate the *tunneling* matrix elements (the *amplitudes* of the tunneling effect) of quantum systems (see, e.g., (Coleman, 1977; Rajaraman, 1982)). In fact, instantons can be viewed as the classical counterpart of 'tunneling' in Euclidean (classical) space-time.

An important remark
Instantons can be viewed as quantum tunneling in *imaginary time* by doing a *Wick rotation*, $t \to -i\tau$ ($\tau \in \mathbb{R}$), that turns Minkowski space-time into Euclidean space-time. After this rotation, the potential barriers of the original quantum problem transform into potential wells, and vice versa (Coleman, 1977).

I will not go into too much mathematical details about these topological objects, and I will exploit their relation to quantum tunneling only in Section 11.6 for a better understanding of their role in the dynamics. Here, I will only enumerate their properties, which give rise to the efficient phase-space search of DMMs. Also, in this book, I will only consider instantons for *dissipative* systems (see definition in Section 3.8). The interested reader can delve deeper into this topic (also for conservative systems) by starting from the books (Coleman, 1977; Rajaraman, 1982).

Instantons are *non-quantum trajectories* that connect *critical points* (also called *classical vacua* in Quantum Field Theory) in the phase space with *decreasing* indexes (see schematic in Fig. 11.3).[13] Recall from Section 3.7 that the index of a critical point is the number of its *unstable* directions.

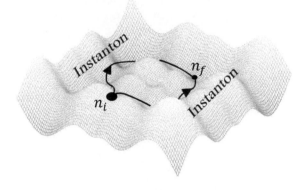

Therefore, an instanton in dissipative systems is a trajectory that always 'goes downhill' in the phase space, in the sense that it connects a critical point with a given degree of stability to a more stable one.[14] This is very similar to the ball that is driven by gravity to the exit point of a pinball machine, or the *directed percolation* of water flowing down a hill (see Examples 1.3 and 1.4 in Chapter 1).[15]

The name 'instanton' originates from the fact that it appears 'in an instant of time', although, as it is also clear from Fig. 11.2, this does not mean that an instanton is truly an instantaneous event. In fact, from the discussion in Section 11.2, for the physical DMMs we consider in this book the instantons have a 'time width' of order RC, where RC is the resistance-capacitance characteristic time of the circuit.[16]

An important remark

It was shown in (Bearden *et al.*, 2018) that the amount of memory in the system (as quantified, e.g., by the ratio between R_{off} and

[13]This is true for instantons in *dissipative* systems. For *conservative* systems (Section 3.8), instantons only need to connect *inequivalent* vacua (ground states), even if they have the same index. In this case, an instanton is half of a loop trajectory between two different vacua, with the other half being an *anti-instanton* (its time-reversed counterpart) (Coleman, 1977).

Fig. 11.3 The two solid lines indicate two 'low-energy', 'long-wavelength' trajectories (in a *non-convex* phase space) connecting two critical points (indicated by filled circles) with *different* indexes, namely from a critical point with $n_i \in \mathbb{N}$ unstable directions to one with *less* unstable directions, $n_f < n_i$. These trajectories are known in field theory as *instantons* (Coleman, 1977).

[14]In the language of supersymmetric TFT, the index of a critical point is the number of missing (unmatched) fermions. An instanton in dissipative systems then always *increases* the number of fermions.

[15]As I will show in Section 12.7, these analogies are indeed not far-fetched.

[16]This result can be obtained rigorously from the Lagrangian in eqn (11.16), by taking the second derivative of this Lagrangian with respect to the memory variables of the DMM. This gives the frequency of the instanton, and its inverse is its time width (Coleman, 1977).

On-shell and off-shell fields
Borrowing a terminology mainly used in (Quantum) Field Theory (Peskin and Schroeder, 1995), we may say that instantons are *on-shell* fields, because they obey the equations of motion (3.1). Instead, anti-instantons in dissipative systems do not obey those equations. Therefore, they are *off-shell* fields.

[17]From definition (11.21) and Fig. 11.4, one immediately sees that an instanton is a *topologically stable* object. It would cost an infinite amount of energy to remove the trajectory in between the initial and final critical points (that crosses zero in Fig. 11.4), while keeping the latter ones fixed. In other words, the very fact that instantons connect inequivalent vacua makes them topologically stable: they cannot be created or destroyed by *local operators* (local perturbations). Hence, the name *topological excitations* often used in Quantum Field Theory.

[18]Instantons are *on-shell* solutions of the equations of motion (3.1).

Fig. 11.4 An instanton is a non-quantum (classical) solution, $\mathbf{x}_{cl}(t,\sigma)$, of the dynamics (3.1) that connects two critical points, $\mathbf{x}_{cl}(-\infty,\sigma)$ (with index n_i) and $\mathbf{x}_{cl}(+\infty,\sigma)$ (with index n_f), with different indexes: $n_f < n_i$. The *modulus*, σ, is the center of the instanton ($\sigma = 0$ in the figure). The fluctuations in the modulus give rise to a fluctuational zero mode, $\delta\mathbf{x}_0(t) = \partial_t \mathbf{x}_{cl}(t,\sigma)$. In supersymmetric theories, an associated *fermionic zero mode* emerges. Zero modes express the *collective* behavior of the dynamics.

[19]An arbitrary translation of the instanton center.

[20]The dimension ($\dim \sigma$) of the space of parameters σ (the *moduli space*) is equal to the index of the initial critical point of the instanton (the number of missing fermions in the initial critical point).

R_{on} in the resistive memories of the circuit, see Section 7.3) determines the degree of curvature of the unstable directions, hence the time spent by the system at the critical points before an instantonic jump occurs. Very little memory (small curvature of the unstable directions) leads to an increased time spent by the system at the critical points (*classical vacua*) as also demonstrated by Fig. 7.14. However, Fig. 7.14 also shows that too much memory (hence too much curvature of the unstable directions), strongly repels the dynamics from critical points, thus slowing down the approach to equilibrium (solution of the problem).

Let's make these points a bit more rigorous. If $\mathbf{x}_{cr}^i$ is some 'initial' critical point (with $n_i \in \mathbb{N}$ unstable directions), and $\mathbf{x}_{cr}^f$ some 'final' critical point (with $n_f \in \mathbb{N}$ unstable directions) of the flow vector field $F(\mathbf{x})$ of an appropriate eqn (3.1) representing a DMM, namely $F(\mathbf{x}_{cr}^i) = F(\mathbf{x}_{cr}^f) = 0$, then the instantons are the *deterministic* trajectories, $\mathbf{x}_{cl}$, satisfying:[17]

$$\text{Instanton} := \begin{cases} \dot{\mathbf{x}}_{cl}(t,\sigma) = F(\mathbf{x}_{cl}(t,\sigma)), \\ \mathbf{x}_{cl}(-\infty,\sigma) = \mathbf{x}_{cr}^i, \\ \mathbf{x}_{cl}(+\infty,\sigma) = \mathbf{x}_{cr}^f, \\ n_f < n_i, \end{cases} \quad (11.21)$$

that connect the two critical points of F (Fig. 11.4).[18]

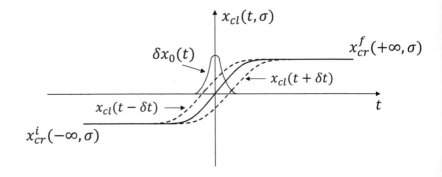

Since the *time-translation* operation $\mathbf{x}_{cl}(t) \to \mathbf{x}_{cl}(t \pm \delta t)$[19] still produces a state of the dynamics, instantons are highly *non-local* objects in time. This *time* non-locality is encoded in the *parameters* σ in eqn (11.21) that are called *moduli* (or *collective coordinates*) of the instantons.[20] As I have discussed in Section 4.3.5, and I will show explicitly in Section 11.7.1, this is accompanied by *spatial* non-locality in DMMs.

An important remark

I want to stress further the important fact that to each modulus, σ, is associated a *zero mode*. Zero modes in *any* physical system are an indication of *collective* behavior (Rajaraman, 1982). This non-local character allows instantons to connect critical points *anywhere* in the phase space. In other words, they represent 'tunneling' of the dynamics across arbitrary distances of the phase space. The reader interested in learning the more formal aspects of these notions applied to DMMs should read (Di Ventra and Ovchinnikov, 2019).

Instantons as manifolds
Consider the *unstable* manifold $M_u(\mathbf{x}_{cr}^i)$ of the initial critical point $\mathbf{x}_{cr}^i$, and $M_s(\mathbf{x}_{cr}^f)$, the *stable* manifold of the final critical point $\mathbf{x}_{cr}^i$. Properly speaking, an instanton between these two critical points is the *manifold* $I_{i,f} = M_u(\mathbf{x}_{cr}^i) \cap M_s(\mathbf{x}_{cr}^f)$, with dimension $\dim I_{i,f} = \text{ind}(\mathbf{x}_{cr}^i) - \text{ind}(\mathbf{x}_{cr}^f)$ ('ind' stands for 'index'). It is then the *family* of all trajectories between these two points.

Since critical points, $\mathbf{x}_{cr}$, in DMMs comprise voltages/currents and memory variables, and in our scheme voltages represent the variables of the original combinatorial optimization problem, instantons can connect variables's assignments differing by *many* literals, in fact a *macroscopic* number of them with respect to the size of the problem.

In other words, if DMMs employ instantons to compute (as I will show), then they are a highly *non-local* type of solver, as opposed to the 'local-search solvers' that are often employed for such problems (cf. Section 2.6). As I have discussed in Section 2.6, this is a very desirable feature to solve hard problems efficiently.

In addition, it is worth stressing once more that instantons in dissipative systems can *only* connect a critical point with a given index (number of unstable directions, see Section 3.7) with another one with *lower* index (a smaller number of unstable directions).[21] It *cannot* connect critical points with the same number of unstable directions.

Therefore, instantons are *topologically nontrivial* deterministic trajectories, namely, functions that have different limits (vacua) for $t \to -\infty$ and $t \to +\infty$ (Schwarz, 1993).

We can then summarize this important notion as:

Heteroclinic and homoclinic orbits
A trajectory in phase space connecting two arbitrary and distinct critical points is called a *heteroclinic orbit*. Instantons in dissipative systems are specific heteroclinic orbits connecting two distinct (in index) critical points. A trajectory that starts from a critical point and returns to the *same* critical point is called a *homoclinic orbit* (Hilborn, 2001).

[21] It is for this reason that one should say, more precisely, that the instantonic trajectories connect a *neighborhood* of one critical point to the *neighborhood* of another. The presence of unstable directions makes the critical points always somewhat repulsive. Only equilibrium points, with all stable directions (or at most with some center manifolds), are fully attractive to the dynamics, and the system can 'fall' into them exactly. However, to avoid overburdening the reader, I will not emphasize this point in the following, and simply say that instantons connect critical points of different stability.

Instantons (in dissipative systems)
A family of topologically nontrivial deterministic trajectories smoothly connecting pairs of critical points of increasing stability in the phase space.

Note that I wrote a 'family' of trajectories, because, given a pair of critical points with different indexes, there may be *many* trajectories that connect them (like in Fig. 11.3).[22]

Finally, let me point out that instantons, can only emerge because the equations of motion of DMMs are *non-linear*. A linear system does not

[22] In fact, since the initial state of the instanton has some unstable directions, a tiny perturbation to this state may lead to a totally different trajectory, but the *same* final state (see an example of this in Fig. 7.13).

support them (Rajaraman, 1982). This is yet another desirable feature to compute efficiently, as I have anticipated in Section 6.4.

11.4.1   Anti-instantons (in dissipative systems)

The (time-)reversed process,[23] from a critical point with a certain number of unstable directions to one with *more* unstable directions, is called an *anti-instanton*.[24] However, it requires *noise* to occur.[25] And, even in the presence of noise (whether physical or numerical), the probability of an anti-instanton occurring is *exponentially* smaller than that of the corresponding instantonic path (Birmingham *et al.*, 1991). This is because the 'tunneling' matrix element (namely the amplitude of the transition between the two critical points) of an anti-instanton contains an exponential factor of the type $e^{-\Delta V_{if}/\theta}$, where $V_{if} = V(\mathbf{x}_{cr}^i) - V(\mathbf{x}_{cr}^f) > 0$ can be interpreted as the 'potential difference' between the initial and final critical points (Fig. 11.5), and θ measures the noise strength.

[23] That goes *against* the flow defined by eqns (3.1).

[24] In the language of supersymmetric TFT, an anti-instanton in dissipative systems always *decreases* the number of fermions.

[25] In the absence of noise there are *no deterministic paths* that can lead from a more stable critical point to one less stable. Unlike their *conservative* counterpart, anti-instantons in dissipative systems are then *off-shell* solutions of eqns (3.1). In actual physical systems and in numerical simulations, we then expect anti-instantons to occur.

Fig. 11.5 In the presence of noise, in addition to instantons, *anti-instantons* may occur. These are trajectories in the phase space that connect two critical points (indicated by filled circles) with *increasing* indexes, namely *from* the critical point with number of unstable directions n_f *to* the critical point with number of unstable directions n_i, with $n_f < n_i$. However, the probability amplitude that these trajectories occur is suppressed by an exponential factor $e^{-\Delta V_{if}/\theta}$, where $V_{if} > 0$ is the 'potential difference' between the two critical points, and θ is the measure of the noise strength (Birmingham *et al.*, 1991).

[26] Note that this term must be a part of the flow vector field of DMMs, because these are dissipative systems (see Section 3.8 and Chapter 8).

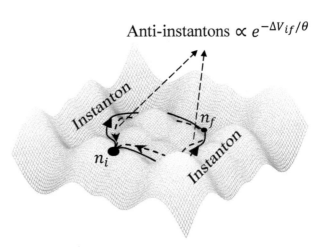

Anti-instantons $\propto e^{-\Delta V_{if}/\theta}$

The function $V(\mathbf{x})$ contributes to the flow vector field, F, in eqn (3.1), as $-\nabla_{\mathbf{x}} V(\mathbf{x})$.[26] We may then say that the reverse process, namely the occurrence of anti-instantons, is *gapped*.

Gapped and gapless systems
In Physics, we say that a system is *gapped* (in the thermodynamic limit) when there is a finite energy separation (gap) between its ground state (possibly degenerate) and its first excited state(s) (*excitations*). Namely, it takes a finite energy to excite the system. If that is not the case, we say the physical system is *gapless*.

An important remark
In the presence of noise, the system may also experience a *condensation* of instantons and anti-instantons (similar to the *Bose-Einstein condensation* of bosons; see Section 13.7). The condensation of (anti-)instanton pairs can break supersymmetry *dynamically* (cf. Section 11.3). In turn, this supersymmetry breakdown, can produce a *noise-induced chaotic phase*, even if the flow vector field,

$F(\mathbf{x})$, in eqns (3.1) remains *integrable* (Ovchinnikov, 2016). The 'solvable-unsolvable' transition I discussed in Section 8.6.1 may be a manifestation of this phase.

11.5 The principle of stationary action

Why would a DMM employ instantons to compute? The simple answer is that *it has no other choice*.

In Chapter 1, I provided the example of a ball driven (*directed*) by gravity to the exit point of a pinball machine (or the water flowing down a hill), and argued that the ball (or the water) has *no choice* other than to do what it has to do when it is driven by gravity and repelled by the pop bumpers (or the water flowing through the only paths available to it).

As I have anticipated in Section 1.5, this 'lack of choice' of physical systems is a manifestation of what we call the *principle of stationary action* (Lanczos, 1986).

To make this principle more precise, let's recall that in Physics, we typically associate a *functional* (which maps a function into a real number), we call *action*, S, to *non-dissipative* mechanical systems (Section 3.8) that may also be subject to some constraints (Arnold, 1989).[27]

And this procedure is not limited to *particle* systems. In fact, one can associate an action to *fields* as well, whether quantum or non-quantum (Peskin and Schroeder, 1995).

Furthermore, from Section 11.3 we have seen that to *any* equation of the type (3.1) (whether describing a non-dissipative or a dissipative system, and, indeed, with or without noise) we can associate a *topological* action S_{Eucl}, eqn (11.17).

Irrespective of the type of action we consider, if we want to find the *actual dynamics* of the physical system we then rely on the following:

> **Principle of stationary action**
> The *actual* motion of a physical system from time t_0 till time t_1 is the one that renders its action S *stationary*:
>
> $$\delta S = 0, \qquad (11.22)$$
>
> subject to the boundary condition that the variations of the configuration states at t_0 and t_1 are zero.

Here, the symbol δ indicates precisely the operation of *variation* of the action, compatible with any constraints on the system (Lanczos, 1986).

Anti-instantons
A *time-reversed* instantonic trajectory is called an *anti-instanton* (Coleman, 1977). In *dissipative systems*, an anti-instanton connects two critical points with *increasing* indexes, namely from the critical point with the smallest number of unstable directions to the critical point with the largest number of unstable directions. Their matrix elements are *exponentially* suppressed compared to the corresponding instantonic matrix elements.

[27] The word 'stationary' applied to a functional, S, means that, given a function in the domain of the functional, its variation in any possible direction from that function (we indicate with the symbol δS) vanishes.

Action in Mechanics
Consider a mechanical system subject to some constraints and with n degrees of freedom. Its dynamics can be described by introducing a function $\mathcal{L}(\mathbf{q}, d\mathbf{q}/dt, t)$ (called *Lagrangian*), where $\mathbf{q}$ are the n *generalized coordinates* (which define the *configuration space*), and $d\mathbf{q}/dt$ are the n *generalized velocities*. The *action* functional of such a system is then:

$$S[\mathbf{q}] = \int \mathcal{L}(\mathbf{q}, d\mathbf{q}/dt, t) dt,$$

which, given a configuration, $\mathbf{q}$, gives a real number.
The reader not very familiar with the concept of actions and action principles in dynamical systems may want to look at the book (Lanczos, 1986).

Extremum of a functional
As in the case of a function, $g(x)$, that reaches an *extremum* at x_0, when its differential at that point is zero, $dg(x_0) = 0$, we say that a function, f, is the *extremum of a functional*, $S[f]$, if, for any small variation, δf, of the function $f(x)$, the variation of the functional, $\delta S[f] \equiv S[f + \delta f] - S[f]$, is zero:

$$\delta S[f] = \int dx \, \frac{\delta S}{\delta f(x)} \delta f(x) = 0.$$

See, e.g., (Arnold, 1989).

[28]Including the critical points.

[29]We then say that instantons are the topologically non-trivial extrema (*global minima*) of the functional S_{DMM} (Schwarz, 1993). In supersymmetric TFTs, the stationarity of the action on instantons, eqn (11.23), can also be viewed as a consequence of the *localization principle*.

Localization principle
In supersymmetric (Quantum) Field Theories, due to the fact that fermionic fluctuations cancel exactly bosonic fluctuations (cf. Section 11.9), the path integral (e.g., eqn (11.16)) is always *localized* on the paths that render the action stationary. This is known as the *localization principle* of supersymmetric field theories (Hori *et al.*, 2000). Instantons are such stationary paths. It is also the reason why the matrix elements (11.24) can be computed exactly (Di Ventra and Ovchinnikov, 2019).

An important remark
In Mechanics, by 'configuration state' (or simply 'configuration') we mean the collection of generalized coordinates, $\mathbf{q}$ ('generalized' because they take into account the *constraints* on the system). Therefore, in that case, the boundary condition of the principle means $\delta\mathbf{q}(t_0) = \delta\mathbf{q}(t_1) = 0$ (Lanczos, 1986).

In the supersymmetric TFT of Section 11.3, it means that the variations of both the bosonic and fermionic variables vanish at the boundaries: $\delta\mathbf{x}(t_0) = \delta\mathbf{x}(t_1) = \delta\chi(t_0) = \delta\chi(t_1) = 0$. In order to obtain the equations of motion (3.1), the variation is done with respect to the conjugate momenta B_i, with the condition $\delta B_i(t_0) = \delta B_i(t_1) = 0$.

Even though DMMs are *dissipative* systems, they satisfy this principle with an appropriate action. If we call S_{DMM} the (topological) action of a given DMM, eqn (11.17), then the stationary principle (11.22) leads to

$$\delta S_{\text{DMM}} \bigg|_{\dot{\mathbf{x}}_{cl}(t,\sigma) = F(\mathbf{x}_{cl}(t,\sigma))} = 0, \tag{11.23}$$

where I have highlighted the fact that the action is stationary (in fact, minimal) *precisely* on the instantonic trajectories,[28] namely the ones that represent the *actual* dynamics of the system.[29]

In fact, the action has to be stationary also on the trajectory that connects the arbitrary initial state (not necessarily a critical point) to its 'closest' critical point in phase space (cf. discussion in Section 12.1). In other words, the principle of stationary action is satisfied over the *entire* trajectory, from the initial point of the dynamics until its last.

An important remark
The previous discussion—which is summarized in eqn (11.23)—also shows why a formulation of the dynamics in terms of an action, S, is very powerful:

- The action contains the *whole* information on the actual trajectory of a physical system (both in time and space). In this sense it is a *global* quantity, from which differential equations (which instead contain *local* information) can be derived by setting $\delta S = 0$.

- Of particular note is that the action contains information about the whole *history* of the system. This means that once the action is known, the whole dynamics of the system can be determined 'in one shot'.

- Unlike the equations of motion, the action can be evaluated also *off shell* (cf. Section 11.4), namely even for those trajectories that *do not* satisfy the equations of motion (e.g., anti-instantons in dissipative systems). This is the starting point for the description of *virtual processes* in Quantum Field Theory (Peskin and Schroeder, 1995).

- The action contains all information on the *symmetries* of the system. In turn, the symmetries determine the *conservation laws* of the dynamics (cf. discussion on *Noether's theorem* in Section 11.3). This means that once we know the symmetries of the action, we know how the dynamics are constrained, namely in which *sector* of the phase space the system is constrained to evolve.

These are the main reasons why a formulation of dynamics in terms of an action is often much more convenient than the same formulation in terms of the equations of motion. This is valid for both classical and quantum systems.

Sector of a physical theory
In Physics, the word *sector* is often used to indicate a particular subset of the general (Lagrangian or Hamiltonian) description of a physical system. It can also indicate the subset of the phase space (or Hilbert space for quantum systems) in which the system is constrained to evolve, with the constraints dictated by symmetries and/or external forces (Peskin and Schroeder, 1995).

Before discussing the dynamical long-range order in DMMs, let me first exploit the analogy between instantons and quantum tunneling to reinforce further the notion that instantons are the *only* low-energy/long-wavelength trajectories in the phase space a DMM can follow. This discussion should provide the reader with an intuitive understanding of this fact, in addition to the more formal result embodied in eqn (11.23).

11.6 Instantons and quantum tunneling

Let's recall the fact that instantons are the Euclidean (classical) analog of quantum tunneling events (Rajaraman, 1982). This analogy will help us understand why they happen suddenly after periods of relative 'quiet' time, as it is evident from, e.g., Fig. 11.2 (see also discussion in Section 12.1). In fact, from Fig. 11.2, we could say that they are 'rare events' compared to the overall time evolution of the system.

For quantum mechanical systems, we know that there exists an uncertainty principle between time, t, and its *conjugate variable*, the energy, E.[30] This means that the interval of time, Δt, it takes the system to vary the expectation value of all its observables by a standard deviation is related to its corresponding change of energy, ΔE, via the *uncertainty relation*: $\Delta E \Delta t \sim \hbar$, with $\hbar$ the (reduced) Planck constant (Messiah,

[30] Time and energy are indeed *conjugate variables* (like position and conjugate momentum of a particle) in the non-quantum case as well (Goldstein, 1965).

[31]To be precise, this uncertainty relation should read $\Delta E \Delta t \geq \hbar/2$, but the inequality (and the factor of 2) are irrelevant for the discussion that follows. Since the energy E is proportional to the frequency, ω, (e.g., for photons $E = \hbar\omega$) the uncertainty between frequency bandwidth and time reads $\Delta\omega\Delta t \geq 1/2$. This relation also tells us that zero (energy) modes are *slow modes*.

[32]Although, note that even the definition (and interpretation) of such a 'tunneling time' is not at all trivial in Quantum Mechanics (Messiah, 1958).

[33]And, as I have discussed in Section 11.2 this time width is of order RC.

[34]This change of energy is related to how many voltages vary during a time interval of order RC.

[35]Also, it is worth recalling that, instantons, like quantum tunneling, are *non-perturbative phenomena*, namely perturbation theory is inadequate to treat them (Coleman, 1977). As I discussed in Section 2.7, this is a desirable (if not necessary) feature to solve difficult computational problems. See also Section 11.10.

[36]For instance, for a sinusoidal wave moving with constant phase velocity v, its wavelength, λ, is related to its frequency, ω, as $\lambda = v/\omega$.

[37]As a typical example, the energy of a photon of frequency, ω, is related to its energy, E, as $E = \hbar\omega$ (Messiah, 1958).

[38]Note that I put the words 'lowest energy' between quotes, because by 'energy' here I do not mean the one associated only to the voltages, but the whole energy of the dynamics, *including* the one associated with the memory degrees of freedom.

1958).[31]

A tunneling event between two regions of space separated by an energetic barrier is indeed a 'rare event', in the sense that although the time it takes a particle to cross the barrier is extremely short,[32] the time the particle spends *outside* the barrier is long. This means that the time separation, Δt, between tunneling events is long.

Therefore, in view of the uncertainty principle, the change of energy between two successive tunneling events is very small. If the system were, say, at some energy E, it will essentially stay at that energy in between tunneling events.

If we translate this argument to the Euclidean case, where instantons represent the tunneling events, we then conclude that instantons are 'rare events' and, in between two instantonic events, the system is in some quasi-stationary configuration (in fact, in the neighborhood of a critical point) where its energy changes little (see also Fig. 12.8 in Section 12.1).

Along similar lines, we could also argue that a *single* tunneling event is very short, and conclude that the instantonic event is also very short,[33], meaning that during that short time a large variation of system energy, ΔE, has to occur (see again Fig. 12.8).[34] (Note that this analogy does *not* imply that during a single quantum tunneling event the total energy is not conserved, see, e.g., (Messiah, 1958).)

We can push this analogy between instantons and quantum tunneling even further to better understand the dynamics of DMMs.[35]

As I have discussed in Section 11.4, instantons may connect critical points anywhere in the phase space, hence they are highly *non-local* (collective) objects. In Physics language, we would say they have a 'long-wavelength character'. In some sense, being of long-wavelength character, instantons do not 'see' the microscopic details of the system, only its 'coarse-grained' features (cf. Section 13.7.5).

We also know that long wavelengths correspond to small frequencies.[36] And, by employing again the analogy with quantum-mechanical phenomena, frequencies are proportional to the energy, E.[37] Following this line of thought, we then see that the dynamics embodied by instantons are the ones corresponding to the 'lowest energy' possible.[38]

In view of the result (11.23), that instantons are the actual dynamical trajectories of a DMM, and the arguments I have outlined previously, I can then conclude that:

'Low-energy', 'long-wavelength' dynamics of DMMs
Instantons (avalanches) are the *only* 'low-energy', 'long-wavelength' (*collective*) dynamics of DMMs.

In the next Section 11.7, I want to spend a bit more time discussing what I mean by 'long-wavelength' (collective) dynamics. I will also provide an explicit example using a DMM that solves prime factorization (Di Ventra, Traversa and Ovchinnikov, 2017).

11.7 Dynamical long-range order (DLRO)

In the following I will discuss only DMMs that solve Boolean problems, but the same arguments can be applied to algebraic ones. To be specific, consider again the circuit for prime factorization of Fig. 7.4, which is a collection of SOLGs.

Now, suppose we had a device (like a *voltmeter*) that detects when an arbitrary voltage at the terminal gates, transitions from the logical 1 to the logical 0, and vice versa.

In fact, let's assume we had as many of these devices as needed to check on all the terminal voltages in the circuit. These devices define an 'operation' (or *observable*), namely provide us (the observers) with an *experimental procedure* to monitor when voltages anywhere (at the gates's terminals) cross from one logical value to another.

If we had such obervables we could then look at how different voltages *correlate* anywhere in the circuit, at a fixed time (*spatial correlations*). Or, if we focused on a representative terminal voltage in the circuit, we could determine how the correlations of a single voltage decay in time (*temporal correlations*).

We want to do these measurements during dynamics, when the DMM goes through a succession of instantonic trajectories, each connecting two arbitrary critical points. Following the supersymmetric TFT of Section 11.3, this procedure amounts to calculating the *matrix elements*, $\mathcal{I}$, of the chosen observables on instantons (the *amplitudes* of the transition from one critical point to the next).

Let's then use the operator language of supersymmetric TFT I discussed in Section 11.3, and denote with $\hat{O}_j$ the chosen observables for each voltage j in the circuit, with $|i\rangle$ the state vector (in the exterior algebra of the phase space) of the *initial* critical point of the instanton, and with $|f\rangle$ the state vector of the *final* critical point. By means of the same theory, and carrying out the calculation exactly as done in (Di Ventra *et al.*, 2017; Di Ventra and Ovchinnikov, 2019), the matrix elements, $\mathcal{I}$, of the chosen observables on instantons, are[39]

> **Instantons vs. fluctuations**
> Instantons are *not fluctuations*. The latter ones are not 'rare events'. Fluctuations tend to localize around critical points (while instantons connect them), and may occur also in linear systems, while instantons are *nonlinear* phenomena.

> **Topological sector of the theory**
> The observables $\hat{O}_j$ have to be chosen appropriately in order to reveal the topological character of the theory (Di Ventra *et al.*, 2017). The observables that give rise to topological invariants, as in eqn (11.24), are called Bogomol'nyi–Prasad–Sommerfeld (BPS) observables, and define the *topological sector* of the theory. The matrix elements $\mathcal{I} = \langle f| \prod_{j=1}^{l} \hat{O}_j(t_j)|i\rangle$ on *non-BPS* observables would instead give rise to correlations that decay *exponentially* with spatial distance and time. They define the *non-topological* sector of the theory (Hori *et al.*, 2000). As BPS observables relevant to DMMs, we can choose the operators $\hat{O}_\alpha(\hat{\Phi}) = \delta(\hat{x}^\alpha - 1/2)\hat{\chi}^\alpha$ (Di Ventra *et al.*, 2017). They can be interpreted as 'detecting' when the voltages on the terminals of the gates 'cross' the value 1/2, either toward the logical 0 or the logical 1. The missing ghost (fermion) in each unstable direction of the initial state is compensated by a fermion χ^α.

> **Matrix elements on instantons**
>
> $$\mathcal{I} = \langle f|\mathcal{T}\prod_{j=1}^{l}\hat{O}_j(t_j)|i\rangle \in \mathbb{Z} \Rightarrow \textbf{topological invariant,} \quad (11.24)$$

[39]See Section 11.9 for a derivation of eqn (11.24).

where l is the number of voltages considered, and the obervables are *time*

Time ordering
Intuitively, the fact that the observables in eqn (11.24) are time ordered is to take into account that the matrix elements (11.24) describe a process from a given initial time to a final time in a succession of events, as they appear from right to left in eqn (11.24). The path-integral formulation of the theory takes into account time-ordering naturally, namely, unlike for the matrix elements of operators, we do not need to worry about time ordering in the path integral (Peskin and Schroeder, 1995).

Correlation length
The spatial distance over which the *fluctuations* of some microscopic quantities of a system (e.g., its spins or voltages) are correlated at any given time is called *correlation length* (cf. Section 4.3.2). Unlike *discontinuous* (first-order) phase transitions, where the correlation length is typically *finite*, at a *continuous* phase transition the correlation length *diverges* in the thermodynamic limit (Goldenfeld, 1992).

Correlation time
The length of time a given perturbation acting at a certain time (and typically at a specific spatial location) on the system propagates in the future is called *correlation time* (Henkel *et al.*, 2008). At a *dynamical* continuous phase transition (e.g., directed percolation, Section 12.7.5) the correlation time *diverges* in the thermodynamic limit. This is called *critical slowing down*: it takes a very long (infinite) time for the system to 'forget' its initial state.

[40] In the *thermodynamic limit*, namely when their size goes to infinity (see Section 4.3.2 for a precise definition of thermodynamic limit).

ordered (indicated by the symbol $\mathcal{T}$), namely the product of operators is ordered, so that 'past' times appear to the right.

The words 'topological invariant' mean that the matrix elements (11.24) are *independent* of how far the voltages are located in the circuit, and on the time at which the measurements on voltages are done. The matrix elements can only depend on the *topology* of the phase space.

An important remark
Note that I have assumed all along that the instantons of DMMs *do not* interact, namely they form a *diluted gas* (Coleman, 1977). That this is indeed the case can be observed directly from the numerical simulations of DMMs, which show distinct voltage spikes at different times (cf., e.g., Fig. 11.2). This assumption is also the starting point of the *mean-field theory* of instantons/avalanches I will discuss in Section 12.5.3, which is also corroborated by numerical results.

11.7.1   Spatial DLRO

Since the matrix elements, $\mathcal{I}$, essentially quantify how voltages *correlate* in the circuit during the instantonic dynamics, their invariance with respect to the spatial location of the voltages in the circuit implies that the correlations among them *do not decay* with distance. As I have discussed in Section 4.3.2, the system then develops a spatial *dynamical long-range order* (DLRO). The origin of this DLRO is precisely due to the *collective* behavior of the instantons, which, in turn, is due to the *non-linearities* in the equations of motion of these machines.

In fact, in Section 4.3.2, I have discussed that *scale-free* correlations are observed in many natural phenomena, such as in the vicinity of continuous phase transitions. Away from the transition point, the correlations between spatially separated elements of the system, a distance **r** apart, decay exponentially.

At the transition, instead, such correlations decay with a power law, $\sim 1/|\mathbf{r}|^\alpha$ (see eqn (4.7)), with α some number dependent on some general properties of the system, such as its symmetries and dimensionality, and not on its microscopic details (Goldenfeld, 1992).

It turns out that $\alpha = 0$ for DMMs, so they are *ideally scale-free*[40] on instantons. The system is (dynamically) *rigid* on instantons.

This is similar to the (long-range) entanglement in Quantum Mechanics, I discussed in Section 4.3.1. The voltage correlations of DMMs on instantons decay with a peculiar power law: they do not decay at all.

However, unlike entanglement, this rigidity is non-quantum, hence does not require any cryogenic temperatures to be maintained, and develops *dynamically* (and *spontaneously*, namely without fine-tuning of parameters) during the solution search. It is an *emergent property* of the system (cf. discussion in Section 1.6: Criterion IV of MemComputing; and Section 12.6).

This is a very useful property because if a voltage needs to change its logical value to satisfy the logical proposition of its own gate, it will first attempt to do so by first interacting with the voltages at the terminals of the nearby gates. This is because the connectivity of the self-organizing circuit is *local*.

However, despite this local connectivity, any given voltage, at any given time, can change its value by correlating with voltages *anywhere* in the circuit, thus allowing the *simultaneous* 'flipping' of large chunks of literals (an instanton/avalanche) in the computational problem the DMM is attempting to solve (cf. discussion in Section 4.3.4).

The *correlation length*[41] of the voltages of DMMs then spans the entire spatial dimension of the machines. Hence, in the thermodynamic limit of machine size (Section 4.3.2), the *correlation length* of DMMs *diverges*.

Furthermore, as I have discussed in Section 11.5, a DMM cannot do anything else, since it has to always satisfy the stationarity of the action, eqn (11.23). This is the reason DMMs are able to solve hard problems efficiently, as I have shown explicitly with the numerical examples in Chapters 8, 9, and 10.

11.7.2 Temporal DLRO

From the topological invariance of the matrix elements (11.24), we also infer a *temporal* DLRO of DMMs.[42] This means that the instantonic trajectory is *sensitive* to the unstable directions of the initial state vector $\mathbf{x}_{cl}^i$. In other words, a small perturbation on the initial state $\mathbf{x}_{cl}^i$, may lead to a *different* intantonic trajectory connecting $\mathbf{x}_{cl}^i$ to the critical point $\mathbf{x}_{cl}^f$.

I showed this explicitly in Fig. 7.13, and as I mentioned in Section 11.4, it is not surprising in view of the fact that the initial critical point of an instanton has some unstable directions.

[41]Recall that the correlation length is the distance over which the microscopic degrees of freedom of the system (e.g., spins or voltages) correlate with each other (more precisely, their *fluctuations* are correlated). Beyond such a length, the fluctuations of such degrees of freedom are effectively independent. Therefore, a correlation length that spans the *entire* spatial dimension of the system implies that the fluctuations of its (microscopic) degrees of freedom are correlated at *all* length scales.

[42]By employing TFT (Section 11.3), we can understand that this temporal DLRO is due to the fact that if the time argument of the observables in eqn (11.24) changes, pairs of solutions with positive and negative Jacobian determinants appear, canceling each other; cf. eqn (11.4) and see, e.g., (Di Ventra *et al.*, 2017).

An important remark
Although this sensitivity may be confused with chaotic behavior (or even with *turbulent dynamics*), it is definitely *not chaos*! The reason is because chaotic behavior would lead to a completely different

(strange) attractor in the phase space, while the attractors of the DMMs, if properly designed, consist only of fixed points.

Turbulence
A fluid flow is said to be *turbulent* when it is characterized by irregular, chaotic patters in its velocity field. Fully developed turbulence is characterized by power-law correlations in both space and time. See, e.g., (Landau and Lifshitz, 1959).

Similar to the spatial component, we can then conclude that, in the thermodynamic limit of machine size, the *correlation time* of the voltages of DMMs *diverges*.

Example 11.1

Spatial and temporal correlations

To illustrate the spatial and temporal DLRO of DMMs, I show in Fig. 11.6 the spatial and temporal correlations of voltages (literals) in a SOLC solving the factorization of a 20-bit number (see also Fig. 11.2).

The normalized spatial correlations (at a time t corresponding to one instantonic transition) are defined as (cf. eqn (4.7)):

$$C(d) = \max_{(i,j) \in \Sigma_d} \frac{\langle\langle v_i(t) v_j(t) \rangle\rangle}{\sqrt{\langle\langle v_i^2(t) \rangle\rangle \langle\langle v_j^2(t) \rangle\rangle}}, \qquad (11.25)$$

where $\langle\langle ab \rangle\rangle = \langle ab \rangle - \langle a \rangle \langle b \rangle$ and $\langle\langle a^2 \rangle\rangle = \langle a^2 \rangle - \langle a \rangle^2$, with $\langle a \rangle$ the *ensemble average* of a function a; $v_{i,j}$ are the voltages at the positions i and j in the circuit, and Σ_d is the set of all pairs (i,j) such that $\mathrm{dis}(v_i, v_j) = d = |\mathbf{r}|$, where $\mathrm{dis}(\cdot)$ is the *graph distance* defined as the shortest path between two nodes (literals/voltages) in the SOLC.

The normalized temporal correlations, at a fixed literal position indicated by α, are instead:

$$C(\tau) = \frac{\langle\langle v_\alpha(t) v_\alpha(t+\tau) \rangle\rangle}{\langle\langle v_\alpha^2(t) \rangle\rangle}, \qquad (11.26)$$

for a time t in the middle of the transient dynamics of the solution search, and τ an arbitrary interval of time.

As anticipated, the correlations plotted in Fig. 11.6 do not decay at all in space or time other than at the boundaries of the system (in the case of time correlations, this happens when $t + \tau$ is comparable to the time to reach the solution). The decay at the boundary is to be expected and is typical of any finite-size system that shows scale-free correlations: it is a *finite-size effect* (Goldenfeld, 1992).

The boundary effects would of course be of lesser importance as the machine grows in size. More precisely we say that, in the *thermodynamic limit* (see definition in Section 4.3.2), the boundary effects are of measure zero.

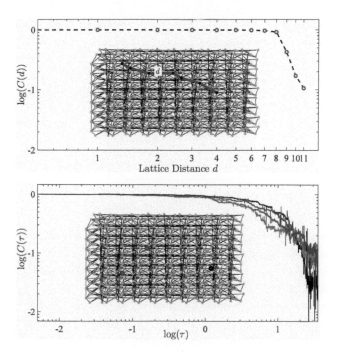

Fig. 11.6 Top panel: normalized *spatial* correlation $C(d)$ between voltages (literals) where d is the *graph distance* between voltages in the SOLC sketched in the inset of the same figure. **Bottom panel:** normalized *temporal* correlations $C(\tau)$ of four random selected voltages (literals) indicated in the inset of the same figure. The DMM in this example solves the factorization of the 20-bit number $497503 = 499 \times 997$. Reproduced with permission from (Di Ventra *et al.*, 2017).

A similar calculation of spatial correlations was done in (Sheldon *et al.*, 2019) for a completely different problem: finding the ground state of Ising spin-glass with *planted solutions* (see Section 9.2.2). In that case, the spatial correlations take a finite value all the way to the edge of the system, and, as the system size increases, they appear to saturate to a value that depends on the dimension of the system (see Fig. 6 in (Sheldon *et al.*, 2019)).

11.8 Boundary effects and spread in solution time

These results indicate that the *bigger* the machine (problem to solve) the *more rigid* the system should be in both spatial and temporal domain (because the *boundary* effects become negligible compared to the 'bulk' properties of the machine). In practice, this means the following.

Consider a problem the DMM is designed to solve, and generate an *ensemble* of instances for different sizes of such a problem (say 10^3 instances per size). Since the instances are different, for *each* size of the problem, a *spread* in the time to solution should be observed. This is indeed the case, as we have seen in Chapter 9 (see, e.g., Figs. 9.6 and 9.7).

However, this spread is not only due to the differences in the instances themselves, but also in how the machine solves them, namely how their *physical* properties change with the size of the problem.

In the thermodynamic limit, the boundary effects are of measure zero, meaning that the machine would not distinguish between the instances, in terms of *how* the problems are solved. As I have explained in Section 11.6, the fact that instantons have a 'long-wavelength' character means that they do not 'see' the microscopic details of the system, only its 'coarse-grained' features.

At *fixed* problem size, the instances differ in the *local* connectivity between the variables (voltages), which reflects into the local properties of the critical points in the phase space, not in the global structure of the latter. Therefore, at fixed (but very large) problem size, we indeed expect that the DMM would essentially 'see' the global features of the instances, not their coarse-grained properties.

This means that the *spread* in the solution time should *decrease* with increasing problem size. I have anticipated this phenomenon in Section 9.2.3, and it does not seem to be shared by other algorithms applied to the same instances.

Although we do not have a rigorous proof of this phenomenon,[43] its empirical evidence for some of the problems studied so far, further reinforces the long-range (non-local) character of DMMs.

[43] And, indeed, if it is universal across all combinatorial optimization problems.

An important remark

It is also not obvious if the spread in solution time would tend, in the thermodynamic limit, to zero or some finite value.

11.9 Matrix elements on instantons

The goal of this Section (which can be skipped at first read) is twofold.

First, I will provide a simple two-dimensional example of phase space in which the dynamics of a hypothetical DMM occurs (Ovchinnikov and Di Ventra, unpublished). This should help the reader to better understand how the (supersymmetric) TFT applies to these machines. Of course, this example is not quite realistic since the phase space of even the most elementary SOLG has more than two dimensions (see, e.g., Section 7.3). However, this is the simplest model that allows us to 'visualize' the various aspects of the theory.

Second, for the reader already familiar with (supersymmetric) TFT (or interested in learning more about it), I will show the explicit calculation of the matrix elements on instantons, eqn (11.24).

This calculation is standard in TFT, and should clarify some of the concepts I have briefly mentioned in this Chapter. It is adapted from (Di Ventra and Ovchinnikov, 2019).

A DMM's phase-space model

Let us assume we have a DMM with four equilibria (a_1, a_2, a_3, a_4) corresponding to four possible logically consistent states. These could be, e.g., the four states of an SO-AND gate (Chapter 7). Let's further assume these are states of a two-dimensional phase space, $X \subset \mathbb{R}^2$, (see Fig. 11.7). In addition, the phase space has saddle points (with one unstable direction) and unstable critical points (with two unstable directions).

Although this is very unlikely (see Section 12.1), let us assume the dynamics start from one of these unstable critical points, u_1, and proceed via a saddle point, s_1, to reach one of the equilibria, say a_1. Each of the critical points is a *local* (supersymmetric) vacuum.

> **Poincaré duality**
> Consider a closed n-dimensional oriented manifold X (which contains all its limit points; cf. Section 3.10). *Poincaré duality* (or *geometric duality*) then states that the k-th *cohomology* group of X is isomorphic to the $(n-k)$-th *homology* group of X, for all integers k (Fomenko, 1994). This means that a k-form in X is dual to a $(n-k)$-form.

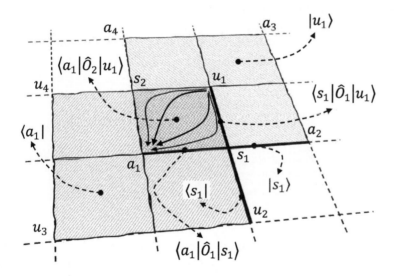

Fig. 11.7 Schematic of a two-dimensional phase space of a hypothetical DMM with four isolated equilibrium points (a_1, a_2, a_3, and a_4), unstable critical points (maxima: u_1, u_2, u_3, and u_4 in the figure), and saddle points (s_1 and s_2 in the figure). The dynamics start from u_1 and end at, e.g., a_1 directly or via the saddle point s_1. The perturbative supersymmetric states and instantonic matrix elements are explicitly indicated in the plot. The operator $\hat{O}_2$ has two fermions, while the operator $\hat{O}_1$ only one. The fermionic content of these operators compensates for the missing fermions in u_1 and s_1, respectively.

Following the TFT I have outlined in Section 11.9, the *kets* of these states are the so-called *Poincaré duals* of the local unstable manifolds of the corresponding critical point (cf. eqn (11.14)). They are:

$$|a_1\rangle = \delta^2(x - x_{a_1})dx^1 \wedge dx^2 \in \Omega^2(X, \mathbb{C}), \quad (11.27)$$

$$|s_1\rangle = \delta(x^\perp - x_{s_1}^\perp)dx^\perp \theta_{a_1 a_2}^{(1)}(x^\|) \in \Omega^1(X, \mathbb{C}), \quad (11.28)$$

$$|u_1\rangle = \theta_{a_1 a_2 a_3 a_4}^{(2)}(x) \in \Omega^0(X, \mathbb{C}). \quad (11.29)$$

Here, $x^\parallel$ and $x^\perp$ are coordinates *along* and *perpendicular*, respectively, to the curve a_1a_2 (Fig. 11.7). The functions $\theta^{(1,2)}$ are equal to 1 if the argument is within the domain of the corresponding shape, which is either the curve a_1a_2 or the area $a_1a_2a_3a_4$, and it is 0 otherwise. Similarly, the *bras* of the local supersymmetric states are the Poincaré duals of the local stable manifolds:

$$\langle a_1| = \theta^{(2)}_{u_1u_2u_3u_4}(x) \in \Omega^0(X,\mathbb{C}), \tag{11.30}$$

$$\langle s_1| = \delta(x^\perp - x^\perp_{s_1})dx^\perp\theta^{(1)}_{u_1u_2}(x^\parallel) \in \Omega^1(X,\mathbb{C}), \tag{11.31}$$

$$\langle u_1| = \delta^2(x - x_{u_1})dx^1 \wedge dx^2 \in \Omega^2(X,\mathbb{C}). \tag{11.32}$$

The local supersymmetric states are not supersymmetric in the *global* sense. In fact, the exterior derivative does not annihilate them. Rather,

$$\hat{d}|s_1\rangle = |a_1\rangle - |a_2\rangle, \tag{11.33}$$

which shows that the exterior derivative, $\hat{d}$, is the operator version in *cohomology* of the boundary operator, $\hat{\partial}$, in *homology* (Fomenko, 1994), because it provides the 'boundary' of the ket $|s_1\rangle$.

Nevertheless, the states are *locally* supersymmetric in the sense that the diagonal matrix elements on all $\hat{d}$-exact operators vanish. For example, for a 'reasonable' operator $\hat{O}$ (namely, one that is physically plausible):

$$\langle s_1|[\hat{d},\hat{O}]|s_1\rangle = 0. \tag{11.34}$$

The instantonic matrix elements are defined via these local supersymmetric states. Suppose we want to compute a matrix element for one such family of instantons, say the one connecting a less stable critical point, e.g., u_1, and a more stable critical point, such as a_1.

In the path-integral language, the overlap between any pair of local supersymmetric states is zero (they are *orthogonal*). For instance,

$$\int_{x(-\infty)=x_{u_1}, x(+\infty)=x_{a_1}} D\Phi e^{-S[\Phi]} = \langle a_1|u_1\rangle = 0. \tag{11.35}$$

In eqn (11.35), Φ is the collection of all the fields in the path-integral representation of the theory; $S[\Phi]$ is the corresponding action, eqn (11.17); and the path integral is over the fluctuations around the family of deterministic trajectories connecting the two critical points.

It is easy to see why the overlap (11.35) is zero. In fact, the trajectories that start at u_1 and end at a_1 form a two-parameter family, because the state u_1 has two unstable directions. This means that the instanton has two *moduli* (global collective variables), which provide two *fermionic zero modes* (Section 11.4).

Therefore, the fermionic path integral in eqn (11.35) misses two fermionic variables. According to *Berezin rules* of integration (see margin note), the whole path integral is thus zero.

In order to render the path integral non-zero, one must introduce into it an operator with two fermionic variables. An example of such a $\hat{d}$-exact operator is:

$$\hat{O}_2 = [\hat{d}, \omega^{(1)}], \qquad \omega^{(1)} \in \Omega^1(X, \mathbb{C}), \tag{11.36}$$

where $\omega^{(1)}$ is some 1-form on the exterior algebra of the phase space. (Note that $\hat{d}$ has fermionic degree 1; see eqn (11.18).) The instantonic matrix element of $\hat{O}_2$ is then

$$\langle a_1 | \hat{O}_2 | u_1 \rangle = \int_{A_{a_1 s_1 u_1 s_2}} \hat{d}\omega^{(1)} = \oint_{\partial A_{a_1 s_1 u_1 s_2}} \omega^{(1)}. \tag{11.37}$$

In the previous eqn (11.37), $A_{a_1 s_1 u_1 s_2}$ denotes the area formed by these critical points (see Fig. 11.7), whereas $\partial A_{a_1 s_1 u_1 s_2}$ is the boundary of this area. In the last equality of eqn (11.37), I have used *Stokes' theorem*.

If, instead, the instanton family connects the saddle point, s_1, to the equilibrium, a_1, its moduli space has dimension 1 (the saddle point s_1 has only one unstable direction), and the corresponding instanton matrix element between these local vacua is non-zero only if we choose an operator, $\hat{O}_1$ with one fermionic variable.

One such operator could be, e.g.,

$$\hat{O}_1 = [\hat{d}, f], \qquad f \in \Omega^0(X, \mathbb{C}), \tag{11.38}$$

with f a 0-form (a smooth function on the manifold). The matrix element of $\hat{O}_1$ between the local supersymmetric states of a_1 and s_1 is then

$$\langle a_1 | \hat{O}_1 | s_1 \rangle = \int_{s_1 a_1} \hat{d}f = f(s_1) - f(a_1). \tag{11.39}$$

Similarly, the matrix element on instantons between the critical points u_1 and s_1 is

$$\langle s_1 | \hat{O}_1 | u_1 \rangle = \int_{u_1 s_1} \hat{d}f = f(u_1) - f(s_1), \tag{11.40}$$

Integral of fermionic (Grassmann) variables
Berezin rules of integration of a Grassmann variable, χ, are:

$$\int d\chi = 0,$$

$$\int d\chi \chi = 1,$$

$$\int \chi d\chi = -1.$$

This can be extended to many fermionic variables, and means that an integral over Grassmann variables that contains *unmatched* fermions is always zero (Wegner, 2016).

Stokes' theorem
Consider a differential form, ω, and an orientable manifold, Ω. *Stokes' theorem* then states that

$$\int_\Omega \hat{d}\omega = \oint_{\partial\Omega} \omega,$$

with $\partial\Omega$ the boundary of Ω (Göckeler and Schücker, 1989). Note that if the volume Ω is n-dimensional, its boundary $\partial\Omega$ has $n-1$ dimensions.

because, even though the unstable point u_1 provides two fermionic zero modes, one of them is compensated by the fermion in s_1. Therefore, only one zero mode is truly unmatched.

Note that both $\hat{O}_2$ and $\hat{O}_1$ are $\hat{d}$-exact operators, and the expectation value (diagonal matrix elements) of such operators over the (*global*) supersymmetric states would be zero.

Matrix elements on instantons

After these preliminaries, I can now show the explicit calculation of the matrix elements (*correlators*) on instantons (between two critical points, $\mathbf{x}_i$ and $\mathbf{x}_f$) relevant to actual DMMs. Namely, I can prove the result in eqn (11.24) that these matrix elements are *topological invariants* (Di Ventra *et al.*, 2017; Di Ventra and Ovchinnikov, 2019).

Given a set of l observables $O_{\alpha_j}(\hat{\Phi})$ (with $\hat{\Phi}$ the operators of the corresponding fields) we then want to compute the following matrix elements:

$$\mathcal{I} = \langle f | \mathcal{T} \prod_{j=1}^{l} O_{\alpha_j}(\hat{\Phi}(t_j)) | i \rangle, \tag{11.41}$$

where the operators are in the *Heisenberg representation*, $\hat{\Phi}(t) = \hat{M}_{0t}\hat{\Phi}\hat{M}_{t0}$, with $\hat{\Phi}$ being *Schroedinger operators*, and $\hat{M}_{t0}$ is the evolution operator (11.16). Since the choice of the reference time instant is irrelevant, it can be taken to be zero, and $\mathcal{T}$ denotes the operator of chronological (time) ordering (which orders products of operators so that 'past' times appear to the right of the expression).

The states $\langle f |$ and $| i \rangle$ are the bra and ket of the f- and i-vacua, respectively, i.e., supersymmetric perturbative ground states, associated with the respective critical points, $\mathbf{x}_f$ and $\mathbf{x}_i$.

As in the previous two-dimensional example, they are the Poincaré duals of the local stable and unstable manifolds of the critical point for the bra and ket of the vacuum, respectively. Namely, they are differential forms that are constant functions (*without* fermions) along the manifolds, and δ-function distributions *with* fermions in the transverse directions (cf. eqns (11.27), (11.28), and (11.29)).

As I have anticipated in Section 11.7, not all observables are good to reveal the *topological sector* of the theory. The Bogomol'nyi-Prasad-Sommerfield (BPS) observables relevant to DMMs, that do reveal such a sector, can be chosen as:

$$O_\alpha(\hat{\Phi}) = \delta(\hat{x}^\alpha - 1/2)\hat{\chi}^\alpha. \tag{11.42}$$

These observables 'detect' (e.g., by means of a *voltmeter*) when the

---

Schroedinger vs. Heisenberg representations

The operator version of supersymmetric TFT (as discussed in Section 11.3) parallels the algebraic description of Quantum Mechanics. Therefore, as in Quantum Mechanics, we can work in the *Schroedinger picture* in which the states are dynamical quantities, but the observables are not (except for an explicit time dependence). Or we can work in the *Heisenberg picture* in which the states are time-independent, while the operators vary in time (Messiah, 1958). The two representations provide the *same* physical results.

voltages on the terminals of the gates 'cross' the value $1/2$, either toward the logical 0 or the logical 1. The missing ghost (fermion) in each unstable direction of the initial state is compensated by a fermion χ^α.

Also, the observable $O_\alpha(\hat{\Phi})$ is Q-exact,

$$O_\alpha(\hat{\Phi}(t)) = \{Q, \theta(\hat{x}^\alpha(t) - 1/2)\}, \tag{11.43}$$

with $\theta(x)$ being the Heaviside step function. And the product of any number of Q-exact operators is also Q-exact: $O_{\alpha_1}O_{\alpha_2}\ldots = \{Q, \theta(\hat{x}^{\alpha_1}(t) - 1/2)O_{\alpha_2}\ldots\}$, due to the nilpotency of the differentiation by Q.

The calculation of $\mathcal{I}$ is then done in the standard manner (Hori *et al.*, 2000). First, we take advantage of the fact that the supersymmetric description is *coordinate-free* (since it can be written in terms of differentials).

We can then choose as coordinates the instanton moduli, σ, and fluctuations around them,

$$x^i(t) = x^i_{cl}(t, \sigma) + \overbrace{\ldots}^{\text{fluctuational modes}}, \tag{11.44}$$

where, $x^i_{cl}(t, \sigma)$ is the instanton solution, and the dots represent all the other fluctuational modes, which vanish at the critical points (see Fig. 11.4).

As discussed in Section 11.4, each modulus provides one *fermionic zero mode* (related to the translation of the instanton center) to the deterministic equations of motion for the fermions,

$$(\hat{\partial}_t - TF)\lambda_j(t, \sigma) = 0; \quad \lambda^i_j(t, \sigma) = \frac{\partial x^i_{cl}(t, \sigma)}{\partial \sigma^j}, \tag{11.45}$$

where $TF^i_k = \partial F^i/\partial x^k$. Equation (11.45) is obtained by differentiating eqn (11.21) over σ^j once.

We then introduce the supersymmetric partners of the moduli, ν^j. The associated zero mode (*tangent* to the moduli space) is

$$\chi^i(t) = \lambda^i_j(t, \sigma)\nu^j + \ldots, \tag{11.46}$$

where once again the dots represent all the other modes.

The matrix elements (11.41) are then

$$\mathcal{I} = \int \prod_{i=1}^l d\sigma^i d\nu^i \delta(x^{\alpha_i}_{cl}(t_i, \sigma) - 1/2)$$

$$\times \left(\lambda^{\alpha_i}_j(t_i, \sigma)\nu^j + \ldots\right) \iint D\Phi' e^{-\{Q, \Psi^{(2)}(\Phi')\}}. \tag{11.47}$$

Bosonic and fermionic zero modes

In supersymmetric theories, to each *bosonic zero mode* is associated a *fermionic zero mode*. This can be seen from the fact that the action in eqn (11.11) is Q-exact. In fact, from eqn (11.7) we have

$$\{Q, x^j(\tau)\} = \chi^j(\tau).$$

This means that the (bosonic) field $x^j(\tau)$ is (supersymmetrically) transformed into the (fermionic) field $\chi^j(\tau)$. If the former has a bosonic modulus, σ^j, then the latter has a supersymmetric partner, ν^j, as well, by setting

$$\{Q, \sigma^j\} = \nu^j; \quad \{Q, \nu^j\} = 0.$$

The fermionic zero mode is then given by eqn (11.46). See also (Birmingham *et al.*, 1991).

Path integrals as integrals over the moduli space and its tangent

The result in eqn (11.47) expresses one of the main features of TFT: the path integral of a Q-exact observable, like (11.43), over a functional space of fields, Φ, reduces to an integral over the *moduli space* and its *tangent*.

Gaussian integrals

Integrals over a *quadratic* form of the integration functions, e.g., for one function $g(x)$, of the type

$$\int Dg\, e^{-\int dx\, g^2(x)},$$

are called *Gaussian integrals*. They can be solved exactly.

Cancellation of determinants

Consider an n-dimensional matrix A. The following is true for *fermionic* Gaussian integrals:

$$\int d\bar{\chi}d\chi\, e^{-\sum_{i,j}\bar{\chi}_i A^i_j \chi^j} = \det(A),$$

where $d\bar{\chi}d\chi \equiv \prod_{k=1}^{n} d\bar{\chi}_k d\chi_k$ is the (ordered) measure of Grassmann variables. For a *bosonic* Gaussian integral instead:

$$\int \frac{d\phi^* d\phi}{\pi^n} e^{-\sum_{i,j}\phi_i^* A^i_j \phi^j} = \frac{1}{\det(A)}$$

with ϕ a complex n-dimensional vector. (Note that the Gaussian integral for complex variables exists only if the real part of all the eigenvalues of A are positive.) We then see that if we perform a Gaussian integral that contains *both* fermionic *and* bosonic variables, the determinant of the fermionic integration *cancels exactly* (up to a sign) the determinant of the bosonic integral. This is the origin of the cancellation of bosonic and fermionic (non-zero) modes in supersymmetric theories (cf. discussion in Section 11.5). See, e.g., (Wegner, 2016) for the demonstration of these statements.

Intersection number

In *Algebraic Geometry*, the *intersection number* quantifies the number of times two or more high-dimensional curves intersect (Hori et al., 2000).

The path integral in the second line of eqn (11.47) is over all the other modes, and only the *Gaussian* part of the action (*one loop*) is left in the exponent (the term $\{Q, \Psi^{(2)}(\Phi')\}$ provides a *Gaussian integral*).

Such integrals are always unity due to the supersymmetric cancellation of the fermionic and bosonic determinants (the *localization principle* of supersymmetric theories; cf. Section 11.5) (Hori et al., 2000). This is the reason why the *one-loop approximation* (quadratic in the fields) is *exact* in the present case.

Using the property of the δ-function:

$$\int g(\mathbf{x})\delta(f(\mathbf{x}))d\mathbf{x} = \sum_{\mathbf{x}_\alpha, f(\mathbf{x}_\alpha)=0} \frac{g(\mathbf{x}_\alpha)}{|f'(\mathbf{x}_\alpha)|}, \tag{11.48}$$

one is left with

$$\mathcal{I} = \sum_{\sigma_0, x_{cl}^{\alpha_i}(t_i,\sigma_0)=1/2} \text{sign Det} \left.\frac{\partial x_{cl}^{\alpha_i}(t_i,\sigma)}{\partial \sigma^j}\right|_{\sigma=\sigma_0}, \tag{11.49}$$

which are indeed *topological invariants*, $\mathcal{I} \in \mathbb{Z}$, hence they cannot change by any continuous deformation.

Note that this result holds both in the limit of zero noise (*deterministic limit*), as well in the presence of perturbative noise. This is the (topological) reason why DMMs are robust against noise and perturbations, as I will explicitly show in Section 12.7.

Correlators as intersection numbers on instantons (moduli space)

The final result on the *correlators* in eqn (11.49) allows us to see, in yet another way, how the (spatial and temporal) DLRO—which I have explicitly shown in the example of Fig. 11.6—emerges in DMMs.

First of all, the BPS observables (11.42) I have chosen are the *Poincaré duals* of the hyperplane $x^\alpha = 1/2$. Therefore, the matrix elements $\mathcal{I}$ can be interpreted as an *intersection* of a collection of such hyperplanes on the instanton, with σ_0 the point of the *moduli space* where all variables $x_{cl}^{\alpha_i}$ acquire the value $1/2$.

This intersection, however, is *independent* of how far the terminal voltages in the DMM are separated from each other spatially. This *spatial* DLRO then originates directly from the spatial nonlocality of the collective instantonic variables (the instanton moduli).

Second, the matrix elements (11.49) *cannot change* by changing the time argument of the observables in eqn (11.41). For instance, for two observables, the 'on-time' instantonic matrix element

$$\langle f|\hat{O}_1(0)\hat{O}_2(0)|i\rangle = 1, \qquad (11.50)$$

is the *intersection number* of the two hyperplanes, $x^{1,2} = 1/2$, and cannot change even if the time argument of, say, $\hat{O}_1(t)$ changes. This is because pairs of solutions with positive and negative Jacobian determinants appear, canceling each other in eqn (11.49).

As I have discussed in Section 11.7.2, this invariance with respect to time arguments reflects the *temporal* DLRO of the transient dynamics of DMMs. It is a consequence of the fact that the initial state of the instanton has unstable variables (missing fermions), and thus the trajectory is highly sensitive to initial conditions on the unstable manifold of the initial critical point.

11.10 The non-perturbative character of the solution search by DMMs

Let me conclude this Chapter by highlighting a point I have briefly anticipated in Section 2.7: the *non-perturbative* dynamics of DMMs during solution search.[44] We can now understand that this fact is a consequence of the *instantonic* dynamics of DMMs.

This is because instantons, like their quantum tunneling counterpart, cannot be treated using perturbation theory (Coleman, 1977). This means that there is no 'small parameter' that can be used to describe the low-energy, long-wavelength dynamics of DMMs starting from a 'non-interacting' system to which a 'small perturbation' is added.[45]

In other words, as soon as we switch on the dynamics of DMMs (by setting the voltages at the appropriate 'input terminals'; see Chapter 7), the memory degrees of freedom strongly couple the voltages representing the logical variables of the combinatorial optimization problem. The voltage-memory system then becomes a *strongly coupled system*, for which perturbation theory is inadequate.

An important remark
Intuitively, the memory degrees of freedom 'mediate' (and participate in) the interactions among the voltages/logical variables (cf. discussion in Section 4.3.5 and see Fig. 4.5) in such a way that the whole system becomes *strongly correlated*.

This is reminiscent of *Quantum Chromodynamics*, where the *gluons* mediate and participate in (act as 'exchange particles' for) the

Correlators on instantons and 'tunneling' amplitudes
The result expressed by eqn (11.49), or eqn (11.50), is quite remarkable. It means that the *transition amplitudes* between two critical points (mediated by the appropriate observables) are *independent* of the *local* structure of the phase space, and depend only on its *global* shape (topology). Since instantons are the Euclidean analogue of quantum tunneling (Section 11.6), this means that the 'tunneling' amplitudes between the critical points are topological invariants. One can then locally perturb the trajectories of a DMM within the instanton manifold, and such amplitudes would not change (cf. Section 12.7).

[44]I have indeed argued in that Section, that such a feature is very desirable (and quite possibly necessary) for the solution of difficult combinatorial problems.

[45]Mathematically, I cannot write the Lagrangian in eqn (11.15) as $\mathcal{L} = \mathcal{L}_{\text{free}} + g\mathcal{L}_{\text{int}}$, where $\mathcal{L}_{\text{free}}$ is a Lagrangian describing, say, non-interacting voltages ('free' Lagrangian), while $\mathcal{L}_{\text{int}}$ describes their interaction via memory variables, with g some 'small' parameter, namely much smaller than the lowest characteristic energy of the system.

strong interactions between *quarks* to form *hadron bound states* (such as protons and neutrons) (Peskin and Schroeder, 1995).

In the case of DMMs, the new (dynamical) state—formed by the interaction of voltages mediated by the memory degrees of freedom—is a succession of instantons (a *composite instanton*; Section 12.1). See also Section 13.7.

[46]As I will show in the next Chapter 12, the strongly coupled nature of the degrees of freedom of DMMs gives rise to a (non-equilibrium and long-range) *ordered state*.

I can then summarize this point by stating that:[46]

DMMs and strong correlations
The dynamics of DMMs showcase features of *strongly coupled systems.*[a]

[a]In fact, the memory variables (like gluons) *self-interact*, thus generating strong *non-linearities* in the dynamics (cf. discussion in Section 13.7).

It is this physical property that allows them to solve difficult combinatorial optimization problems efficiently, as I have shown explicitly in Chapters 8, 9, and 10.

After this discussion on the topological properties of DMMs, I am now ready to discuss their approach to equilibrium (how they proceed to solve a problem), and explicitly show interesting features of the transient state of their dynamics, such as the underlying *critical branching process*. We will finally be able to understand why they are robust against noise and perturbations.

11.11   Chapter summary

- In this Chapter, I have discussed that DMMs employ *topological* properties of the associated dynamical system to compute: *critical points* in the phase space and trajectories connecting them (*instantons*).
- In fact, instantons (avalanches) connect critical points of *increasing* stability (with lower index), and are the *only* 'low-energy', 'long-wavelength' (*collective*) dynamics of DMMs.
- The reverse processes (*anti*-instantons) that connect critical points of *decreasing* stability (with higher index) require noise to occur, and are *gapped*, namely they are *exponentially* suppressed compared to the instantonic trajectories.
- The matrix elements on instantons, namely the 'transition ampli-

tudes' between critical points with decreasing indexes are *topological invariants*, namely they depend only on the topology of the phase space.

- This implies an ideal *dynamical long-range order* in *both* the spatial *and* temporal domains.

- The instantonic dynamics of DMMs then share some resemblance with the spatial correlations of entangled quantum systems.

Approach to equilibrium and the ordered state of DMMs

Out of intense complexities, intense simplicities emerge.
Winston Churchill (1874–1965)

Takeaways from this Chapter

- Digital MemComputing machines (DMMs) find the solution of a problem via a *composite instanton*.

- The number of instantonic steps, and the *physical* time to reach the solution (if it exists), scale at most *polynomially* with problem size.

- The state DMMs *self-tune* into is characterized by a *critical branching process*.

- The topological features of DMMs make them robust against noise, both physical and numerical.

- DMM's dynamics in the presence of noise can be modeled as *directed percolation* of the state trajectory in the phase space.

- MemComputing machines share some physical features observed also in the animal brain.

In Chapter 11, we have seen that DMMs employ *instantons* to connect critical points of increasing stability in the phase space. In fact, we have seen (Section 11.6) that instantons are the *only* 'low-energy', 'long-wavelength' dynamics in the phase space of DMMs. However, given an initial condition, how does the physical system end up into such a state?

We have also seen that instantons are *avalanches* of voltages at the terminals of the self-organizing gates (see discussion in Section 11.2). What is the size of these avalanches? And what is their distribution?

Does it tell us anything about the dynamics of DMMs?

Finally, we know that the topological features of the dynamics (critical points/instantons) should make them robust against noise and perturbations. However, what is the actual time evolution of the system when noise is present? And when would the machines stop working with increasing noise levels?

In this Chapter I will answer all these questions. I hope these answers will provide an even deeper understanding of how DMMs work.

12.1 Approach to equilibrium and composite instanton

For the sake of discussion, let's consider again a search problem (Section 2.3.4), like the satisfiability (SAT) problem I discussed in Section 9.1.1 or the factorization of an integer (Section 9.1.2). We would then design a DMM to solve these problems as I have shown in those Sections. We further provide an arbitrary initial condition to the state dynamics, and let these DMMs self-organize till a solution is found.[1]

The initial condition of the dynamics

The initial state, $\mathbf{x}(t = 0)$, of the DMM is an *arbitrary* state in the whole phase space X. As I have already anticipated in Section 11.2, since the phase space, X, is vast (even for relatively 'small' problem sizes), it is very unlikely that we start the dynamics at a critical point of the flow vector field F of the equations of motion of the DMM (see margin note). However, we know that the phase space is interspersed with critical points, whether saddle points, or the equilibrium points representing the solution of the given problem.[2]

Any of these critical points, whether saddle points or equilibria, have *stable manifolds*, namely manifolds on which the dynamics are *attracted* toward the critical point (see Section 3.7). Since the DMM is a dissipative system, the initial state will then be attracted by the stable directions of the 'closest' critical point in the phase space (Fig. 12.1).[3]

Compact phase space and critical points
Since the phase space, X, of DMMs is *compact* (see Section 3.10 for the definition of 'compact'), every point in it is on the stable manifold of some critical point, and on the unstable manifold of some critical point. The phase space then partitions into a collection of stable manifolds and a collection of unstable manifolds of critical points (Milnor, 1963).

---

[1] If a solution exists.

[2] Of course, I am assuming that the DMMs are designed without any equilibria, other than the solutions, and with no periodic orbits. Otherwise, one has to satisfy condition (iii) of Section 6.3.2, and the analysis relates to an *ensemble* of initial conditions.

[3] Following the discussion I presented in Section 2.6, during this phase, DMMs explore/discover the *short* length scales of the problem they are attempting to solve.

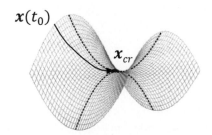

$\mathbf{x}(t_0)$

$\mathbf{x}_{cr}$

Fig. 12.1 The initial state of the dynamics, $\mathbf{x}(t = 0) = \mathbf{x}(t_0)$, is attracted by the *stable manifold* of the 'closest' critical point, $\mathbf{x}_{cr}$ (in the figure, a saddle point).

Composite instanton
It is the collection of all *elementary* instantons a DMM needs to find the solution of a given problem (Di Ventra and Ovchinnikov, 2019).

Instanton condensation and log-conformal theories
It has been shown that some models in which instantons condense are *log-conformal* (Frenkel *et al.*, 2007). This means that the correlations in those models have *logarithmic corrections*, and are invariant under *conformal transformations* (namely transformations that preserve angles locally).

An important remark
Note that the previous statement is definitely true for a *point-dissipative system* (see Section 7.8), in view of the fact that, in such as case, any point in the phase space is attracted to the global attractor (Hale, 2010). If the DMM does not have a global attractor, the basin of attraction of its equilibria must be 'large enough' (it does not decay exponentially with problem size, as discussed in Section 6.6). Then, the initial state falls on the stable manifold of some critical point (due to the compactness of the space), and from there, with high probability, it will be attracted to the equilibria.

However, there is an important caveat to the previous statement. The initial state can be attracted to the 'closest' critical point, provided the *unstable* directions are not 'strong enough' to *repel* the dynamics. As I have discussed in Section 11.1, this is indeed the case: critical points of DMMs have unstable directions that are not unstable enough to repel the dynamics (Bearden *et al.*, 2018).

More precisely, the *magnitude* of the real part of the eigenvalues, $\lambda_{i,u}$, of their unstable directions (see definition in Section 3.7), is *much smaller* than the magnitude of the real part of the eigenvalues, $\lambda_{j,s}$, of their stable directions: $|\text{Re } \lambda_{i,u}| \ll |\text{Re } \lambda_{j,s}|, \forall i, j$.[4]

---

[4]Even if this were not true for all stable and unstable directions of a critical point, it is enough that there exists at least one stable direction whose real part of its eigenvalue is much larger (in absolute value) than the largest real part of the eigenvalue of the unstable directions: $\exists j : \max_i |\text{Re } \lambda_{i,u}| \ll |\text{Re } \lambda_{j,s}|$.

[5]We can also replace the word 'hop' with 'percolate', in view of the fact that in the presence of noise—physical or numerical—the (anti-)instantonic dynamics can be described as a *directed percolation* process (see Section 12.7).

[6]During this phase, DMMs explore/discover *also* the *long* length scales of the problem they are attempting to solve.

Dynamics after the first critical point is reached
After the system reaches the first critical point, it can only go through instantonic trajectories, which are the only 'low-energy, 'long-wavelength' dynamics of the system (cf. discussion in Section 11.6).

From that point on, the dynamics 'hop'[5] from one critical point of a given stability to another one more stable, and so on, until the system reaches the equilibrium point (if it exists, or one of the equilibrium points if there are more than one).

A schematic of the dynamics of a DMM in its phase space is sketched in Fig. 12.2.

Since the whole dynamics are a collection of individual instantonic 'jumps', we can then conclude the following:[6]

Transient dynamics of digital MemComputing machines
The transient dynamics of DMMs is characterized by a *composite instanton*.

DMM's trajectory in phase space

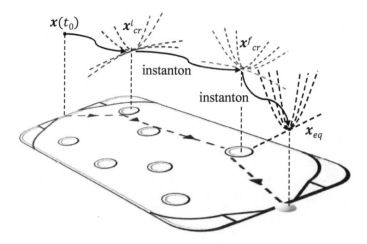

Fig. 12.2 The trajectory of a DMM solving a problem starts from an arbitrary initial state, $\mathbf{x}(t_0)$, and 'falls' into the 'closest' critical point with some unstable directions (negative-curvature parabolas) and stable directions (positive-curvature parabolas): a saddle point, $\mathbf{x}^i_{cr}$. The system then 'hops' to another saddle point with *less* unstable directions, $\mathbf{x}^f_{cr}$, along an instantonic trajectory. This is repeated till the system ends up into a fully stable equilibrium point, $\mathbf{x}_{eq}$, (if it exists) with also center manifolds (straight line) representing the memory degrees of freedom. Notice the analogy with the ball traveling down a pinball machine (Example 1.3). The pop bumpers are the critical points, and the instantons, the trajectories connecting them.

Critical points, excitations, and relaxation processes

I can also make the following analogy. The critical points in phase space can be interpreted as (topological[7]) *excitations* (excited states) of the system above its ground state (the equilibrium state; cf. Fig. 10.11).[8] How 'far' such excitations are from the ground state is characterized by the number of their unstable directions (the *index* of the critical points; Section 3.7).[9]

A physical system that is prepared in an excited state will tend to *relax* toward its ground state.[10] The instantonic processes then correspond to the *relaxation* of these excitations toward the steady state(s) of the DMM. Therefore, we can also provide the following interpretation:

Transient dynamics of DMMs

The transient dynamics of DMMs is characterized by *relaxation processes* (instantons) of *excitations* (critical points) toward the ground state(s) of the system.

Note, however, that due the *temporal* dynamical long-range order (DLRO) (Section 11.7.2) the instantonic trajectories to reach equilibrium may be *substantially* different even for initial conditions that differ very little.

If the equilibrium point is truly the solution of the problem (for a search problem), then all the self-organizing gates of the DMM are satisfied, meaning that the terminal voltages have reached the logically consistent solution. In that case, the internal memory variables *decouple* from the voltage dynamics, and they may acquire any arbitrary value

[7]They are topological because they cannot be created or destroyed by *local perturbations*. Note, however, that instantons are also called *topological excitations* in Quantum Field Theory (cf. Section 11.4).

[8]Possibly more than one, if there are more steady states.

[9]The index of critical points is then their 'excitation number'.

[10]This is another manifestation of the principle of stationary action (Section 11.5).

Anti-instantons and time-reversed processes
Since anti-instantons connect critical points of *decreasing* stability (Section 11.4), namely states with *increasing* excitation numbers, they can be viewed as *time-reversed* relaxation processes.

[11]Note that for some SAT problems, e.g., those that contain 3-literal clauses, one could also have some center manifolds associated with the voltages, not just the memory variables. This is because the clauses could be satisfied for some literals/voltages, irrespective of what value some other voltages acquire.

Defects
Any violation of local equilibrium in a condensed medium (as, e.g., in Fig. 12.3) is called a *defect* of the medium. It could be a point or a high-dimensional object, and typically showcases nontrivial topological properties (Schwarz, 1993).

[12]We then say that the $\mathbb{Z}_2$ symmetry is *spontaneously broken* in the ground state (cf. discussion in Section 3.9).

Fig. 12.3 Schematic of a ferromagnet on a finite 2-dimensional lattice with two ground states: all spins down or all spins up. Starting from one ground state (say, all spins down), we can flip all spins at the far left of the lattice, generating a *soliton* or *kink*. This soliton then moves undisturbed toward the far right of the lattice, leaving the system in the other ground state: all spins up.

Solitons
'Solitary waves', or *solitons*, are localized wave packets that propagate in a medium at constant velocity without changing their profile. Instantons can be viewed as solitons in one lower dimension (Rajaraman, 1982).

Logical defects
These are *domain walls* in DMMs that separate regions of space between logically consistent and logically inconsistent sets of self-organizing gates.

(they are now *independent* of the dynamics of the voltages).

As a consequence, the equilibrium points of DMMs have, in addition to stable directions, also *center manifolds* (namely, with Re $\lambda_i = 0$, see Section 3.7). These are 'flat directions' in the phase space that do not affect the equilibrium.[11]

12.2   Instantons, solitons, and logical defects

The previous discussion should help 'visualize' the actual dynamics of DMMs through phase space. However, we can rely on yet another physical analogy that may further clarify the workings of DMMs.

As a way of introduction, let me examine first the following experiment. Consider the ferromagnet on a *finite* 2-dimensional square lattice I introduced in Section 4.3.2, Fig. 4.3.

That ferromagnet has two ground states (*vacua*): one with all spins down, the other with all spins up (see Fig. 12.3). Initially, the system is in either one of the two, say all spins down.[12]

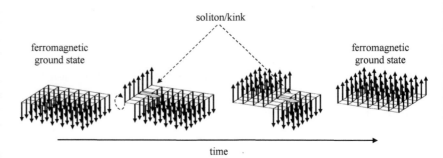

We then take all the spins at one edge of such a ferromagnet, and we flip them all, from spin down to spin up (Fig. 12.3).

There is then a 'wall' (a *defect*) that separates all spins up from all spins down. This *domain wall* (which is of a *topological* nature: it is *protected* by the lattice boundary conditions) then travels undisturbed from the far left of the lattice to its far right. This traveling domain wall is what we call a *soliton* or *kink* (Rajaraman, 1982).

This soliton then separates regions of the system that correspond to different ground states (vacua). In order for the system to transition from one ground state to the other it has to 'rid itself' of this soliton, by letting it travel all the way from the far left to the far right.

What does all this have to do with DMMs?

The reason is that instantons are the analogue of solitons in one lower dimension (Rajaraman, 1982).[13] As I have exemplified previously, solitons emerge in many non-linear physical phenomena, and are *topological defects* in a medium, that propagate undisturbed from one spatial location to another.

The 'medium' in a DMM is its physical realization with an actual self-organizing circuit (whether logic or algebraic), and its voltages at the gates of this circuit. The voltages represent logical variables, which at steady state satisfy some logical or algebraic relation.

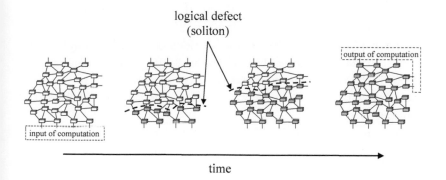

logical defect
(soliton)

output of computation

input of computation

time

Away from a solution, the voltages at the gates do not satisfy the correct relations. Therefore, during the solution search a DMM supports a certain number of *logical defects*. Solitons in DMMs are precisely these logical defects.[14]

This is illustrated in Fig. 12.4 which represents a 2-dimensional self-organizing circuit. At the beginning of the dynamics the DMM has a certain number of logical defects, which 'travel' toward the 'exit' points of the circuit (representing the output of the computation).

The DMM dynamics can then be viewed as an instantonic process during which the logical defects (solitons) are being *pushed out* of the circuit. In other words, during the transient solution search a DMM attempts to 'rid itself' of these defects till the solution is found.

This is similar to the ferromagnetic example in the preceding Section 12.1, where the system needs to get rid of the soliton to transition from the state with all spins down to the one with all spins up. In the DMM case, the solitons (instantons) occur *after* the immediate transient, namely after the system has evolved (in phase space) from the initial state $\mathbf{x}(t_0)$ to the 'closest' critical point.

[13]To be precise, instantons are the processes that *destroy* solitons (kinks), in the transition from one vacuum to another (as in Fig. 12.3). Anti-instantons, instead, *create* them (anti-solitons/anti-kinks). Solitons can then also be destroyed by annihilating with anti-solitons. This process, however, requires noise in dissipative systems like DMMs.

Fig. 12.4 The DMM is represented as a collection of elementary gates (rectangular boxes) connected together to represent the original combinatorial optimization problem. At the initial time, an input representing the problem instance is fed at the appropriate terminals. The DMM then generates a number of 'logical defects' or *solitons*: 'domain walls' that separate the regions of empty (unsatisfied) and filled (satisfied) boxes (gates). The DMM then attempts to rid itself of logical defects (solitons) till one of the equilibria (solutions) is found. Adapted with permission from (Di Ventra and Ovchinnikov, 2019).

[14]For those readers familiar with the notion of solitons, it is also important to stress that, strictly speaking, solitons are defined in *continuous* space models or on lattices that allow a 'coarse-graining' procedure (one in which the system is scaled so that microscopic details are smoothed out), and with interactions between nearest neighbors only (Rajaraman, 1982). This is the reason solitons have typically *finite* width. DMMs, instead, when implemented physically, represent circuits, and these are almost never structured lattices. Therefore, a 'coarse-graining' procedure is not appropriate for these systems. It is for this reason that in certain situations a DMM's solitonic configuration (a logical defect) may occupy the bulk of the circuit (Sheldon *et al.*, 2019).

This critical point is then the initial (perturbative) 'ground state' (or 'classical vacuum') from which the soliton 'travels' to the 'exit point' (the output terminals) of the machine and leaves the system in the next critical point (perturbative ground state/classical vacuum). And the process repeats itself till an equilibrium is found (if the latter exists).

12.3   Number of instantonic steps to solution

Having discussed the approach to equilibrium of a DMM in terms of instantons, or equivalently, solitons, we can now count how many instantonic steps, $\mathcal{N}$, (the size of the *composite instanton*) a DMM needs to reach the solution of a given problem. It turns out this calculation is quite simple (Di Ventra and Ovchinnikov, 2019).

As I have discussed in Section 6.3.2, ideally, DMMs should be designed as point-dissipative systems (Hale, 2010). This means that all trajectories of the system will eventually end up into one of the equilibria representing the solution of the problem, *irrespective* of the initial conditions.

In Section 11.4 I also pointed out that instantons can *only* connect critical points of a given stability to critical points that are *more stable* (with a smaller index, or smaller number of unstable directions; Fig. 12.5).

Algebraic Topology
The branch of Mathematics that uses tools of abstract *Algebra* to study and characterize *topological spaces* is known as *Algebraic Topology* (Hatcher, 2002). I have made use of it in several parts of Chapters 11 and 12, in particular, in Section 11.3.

Fig. 12.5 Number of instantonic steps to reach solution. After an initial condition, $\mathbf{x}(t_0)$, the system is attracted to the 'closest' critical point whose index, $\mathrm{ind}(\mathbf{x}_{cr}^1) = i_1$, can be *at most* equal to the dimension D of the phase space. An instanton connects critical points differing by at least one unstable direction, and the dimension of the phase space, D, grows at most *polynomially* with problem size. Therefore, the approach to equilibrium, $\mathbf{x}_{eq}$ (with zero index: $i_{eq} = 0$), requires a number, $\mathcal{N}$, of instantonic steps that can only grow at most *polynomially* with problem size.

Number of instantonic steps to solution

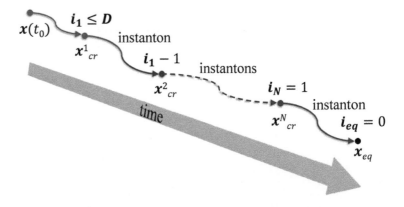

However, we know that the number of unstable directions is *at most* equal to D (the dimensionality of the phase space, X). The latter, in turn, can only grow *linearly* with the number of degrees of freedom (see Section 3.2).

Irrespective of the type of DMM chosen (e.g., a linear or a quadratic circuit), the dimension D grows at most *polynomially* with problem size (Traversa and Di Ventra, 2017; Di Ventra and Traversa, 2018*a*). Therefore, even if all instantons in the system connect two critical points differing by only one unstable direction, we can conclude that the total number of instantonic steps to reach equilibrium can only grow at most *polynomially* with system size:

Number of instantonic steps to solution

$$\mathcal{N} \sim O(\mathcal{P}(D)), \qquad (12.1)$$

with $\mathcal{P}(D)$ some polynomial of D whose degree may change with the given problem considered and/or the DMM chosen to solve it.

This result is *independent* of the particular paths the system chooses to reach equilibrium. In fact, as I have already mentioned in Section 11.7.2, due to the *temporal* DLRO, two initial conditions that differ infinitesimally from each other, may lead to very different paths. However, since instantons (in dissipative systems) can only connect critical points of increasing stability, they have to 'shed' unstable directions till they reach equilibrium, no matter the path chosen.

The previous argument holds also in the presence of moderate noise, namely noise not so intense as to destroy the topology of the phase space. In that case, *anti-instantons* (which connect critical points of given stability to critical points that are *less* stable) would appear in the dynamics. Anti-instantons would then try to drive the system *away* from equilibrium. However, as I have discussed in Section 11.4, their probability to occur is exponentially suppressed (they are *gapped*) compared to the instantonic processes (Birmingham *et al.*, 1991). This means that, even in the presence of (moderate) noise, instantons are still the 'low-energy', 'long-wavelength' dynamics of a DMM, so that eqn (12.1) still holds. As a consequence, DMMs break *ergodicity*.

An important remark

Anti-instantons, being gapped, can *at most* change the degree of the polynomial $\mathcal{P}(D)$ in eqn (12.1). They cannot transform a polynomial scaling into an exponential one, unless the noise is strong enough to induce a *condensation* of (anti-)instantons, with consequent *noise-induced chaotic phase* (see Section 11.4).

As I have discussed in the previous Section 12.2, instantonic steps correspond to the elimination of solitonic configurations of *logical defects*. This means that the number of logical defects a DMM needs to eliminate

---

Duality between Computational Complexity Theory and Algebraic Topology
The calculation that led me to eqn (12.1) can be classified as an application of *Algebraic Topology* (previous margin note) to *Computational Complexity Theory* (Chapter 2), since it transforms the task of finding the number of steps to solution of a given combinatorial optimization problem to the counting of topological objects (instantons of a *Cohomological Field Theory*) necessary to reach it. In fact, in Section 12.8 I will argue that the critical points (connected by instantons) form a topological space known as *CW complex*. All this suggests an interesting *duality* between the two fields of Mathematics worth exploring further (cf. discussion in Section 13.8.5).

Ergodicity breaking in DMMs
A dynamical system is *ergodic* if its trajectory in the phase space traverses, during time evolution, any neighborhood of any point of the phase space. In that case, *time averages* of observables are equal to their *ensemble averages* (Kubo, 1957). Instantons are the ultimate low-dimensional manifolds of the phase space of DMMs: the instanton manifolds *foliate* the phase space (cf. Section 3.10). As a consequence, DMMs explore a substantially smaller portion of the whole phase space (cf. Section 12.9). Therefore, DMMs are *not* ergodic systems: they *break ergodicity*.

to reach the solution of a given problem can only grow *polynomially* with problem size.

12.3.1 Continuous time to solution

I can push the previous analysis a step further and determine how the total *physical* (continuous) time, T_{phys}, to solution[15] scales with increasing size of the problem.

For this, consider a Boolean problem written in *conjunctive normal form* (CNF) (see Section 2.4). In addition, assume as I have done in Chapter 8, that for each OR gate there is only a fixed number of memory variables, say two memory variables per clause. As it can be easily checked, the argument that follows is *independent* of how many memory variables per gate one employs, since their number contributes at most a prefactor to the degree of scalability.

Let's further assume that when we increase the number of variables, N, in the Boolean problem (corresponding to the number of voltages in the self-organizing circuit), we keep the *ratio*, $\alpha_r = M/N$, between number of clauses, M, and the number of variables, constant. This is what I mean by 'problem size'.

As I have already discussed, the number of unstable directions is at most equal to the dimensionality of the phase space, D. For a CNF circuit with two memory variables per clause, this dimensionality is $D = N + 2M = N(1 + 2\alpha_r)$. In this case, then, the dimensionality of phase space grows *linearly* with problem size, N.

Assuming that each instanton connects two critical points whose indexes differ by 1, we then conclude that the total number of instantonic steps (in the composite instanton) to reach equilibrium can only grow at most *linearly* with system size, which is a particular case of eqn (12.1).[16]

We also know from Section 11.4, that the 'time width' of an elementary instanton depends *only* on *fixed* physical parameters of the circuit, like its RC time,[17] and *not* on its size. Let's call this time $T_{inst,j}$ for each instanton j.

Recall from Section 11.4 that the system spends some time at the initial critical point before making an instantonic jump to the next (distinct) critical point.[18] Call this time, $T_{cr,j}$. It is typically longer than the instantonic time width: $T_{cr,j} \gg T_{inst,j}$.

From Section 11.4, we know that also this time is *independent* of the size of the problem, and depends only on the amount of memory in the system (Bearden *et al.*, 2018). It is essentially the time the system spends 'exploring' the stable directions before 'realizing' there are unstable directions it can 'escape' from (Fig. 12.6). Therefore, this time will be larger the smaller the curvature of the unstable directions compared to the curvature of the stable directions.

---

[15]This is the time it takes the system state to be within a *fixed distance* from the equilibrium point (cf. property (ii) in Section 6.3.2).

[16]However, note that, in principle, *any* Boolean problem can be written in CNF (Section 2.4).

[17]With R the characteristic resistance, and C the capacitance of the circuit.

[18]There is also the initial time from the initial state of the dynamics to the 'closest' critical point. This time is also *independent* of the size of the problem.

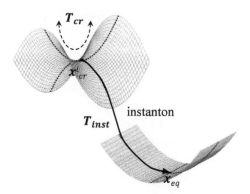

Fig. 12.6 Schematic of an instanton connecting a saddle point and an equilibrium point (with a center manifold). The time it takes the system to jump from the initial critical point to the equilibrium point T_{inst} is typically much shorter than the time the system spends at the initial critical point, T_{cr}: $T_{cr} \gg T_{inst}$. This is because the latter involves the time it takes the system to 'explore' the stable directions, before 'escaping' from the unstable ones.

This, however, does not mean that this time is exactly the same for all critical points (due to possible variations in the curvature of their unstable directions). Therefore, let's call $T_{max} = \max_j(T_{cr,j} + T_{inst,j})$ the *maximum* time required to do an instantonic jump in the phase space, including the time the system spends on initial critical points. The important thing is that this time does *not* depend on the size of the phase space, hence on the problem to solve.

From the previous discussion we then see that the maximum total *physical* time, T_{phys}, required by the system to reach the solution of a given Boolean problem of size N, expressed in conjunctive normal form (CNF), is $T_{phys} \leq N(1 + 2\alpha_r)T_{max}$.

However, an instanton can connect critical points that differ by *more* than one unstable direction (the initial and final critical points ought to have only different indexes). This means that the number of steps to reach the solution could scale *(sub-)linearly* with the system size.

Putting all this together we then conclude that:[19]

> ### Continuous time to solution for a Boolean problem in CNF
>
> $$T_{phys} \sim O(N^\theta), \quad \theta \leq 1. \tag{12.2}$$

We have seen numerical examples of this sub-linear scalability in Chapter 9.

[19] A similar analysis on the circuit in Example 7.2 for prime factorization would show that the continuous time to solution in the case of prime factorization is $O(N^2)$, with N the number of bits of the number to factor. This is because that circuit contains $O(N^2)$ self-organizing logic gates (Traversa and Di Ventra, 2017).

> ### An important remark
> I want to stress again that the result (12.2) does *not* imply NP = P! In this Section, I have discussed the scalability of *continuous* (physical) time of a DMM as a function of the size of the problem. The famous conjecture, instead, relates to Turing machines operating in *discrete* time.

12.4   Instantons as 'oracles' of DMMs

At this point, we can make an interesting analogy between DMMs and *non-deterministic* Turing machines (see definition in Section 2.3.1).[20] Although the dynamical system describing a DMM is *deterministic* (assuming any type of noise is negligible), its actual trajectory in the phase space is *not* easy to predict *a priori*.

This is because of the sensitivity of the trajectories to initial conditions (due to temporal DLRO, Section 11.7.2), as well as the gigantic size of the phase space for typical combinatorial optimization problems. In fact, even for 'small' problems, say of tens of variables and comparable number of constraints, the corresponding phase space would be so large as to make any attempt to predict the instantonic dynamics hopeless. For such phase spaces it would be even challenging to determine *all* of their critical points, let alone the 'low-energy' trajectories connecting them!

As I have anticipated in Section 1.5, due to these reasons, in Physics, we would call DMMs *complex systems*: those for which inferring their collective dynamics from their elementary constituents is not an easy task. This means that even if the machines are deterministic, knowing which specific instantonic path the system will take after each critical point cannot be determined *before* we observe it (or simulate it numerically). And yet, the machine converges to the correct solution employing a number of instantonic steps that can grow at most *polynomially* with problem size, eqn (12.1).

In Section 2.3.1, we have seen that *non-deterministic* Turing machines can be interpreted as operating assisted by an *oracle* (Garey and Johnson, 1990). The latter is an 'agent' that via some internal mechanism *unknown* to the user, is able to *correctly* 'guess' each step of the computation done by the machine, and guide it to the correct solution in polynomial time, even if the solution tree of the given problem grows exponentially with its size (Fig. 12.7).

Instantons seem to do precisely that: they are able to 'guess' correctly the path they have to travel in the phase space, a feat that eludes an external observer. However, the *fundamental* difference with the oracle of the non-deterministic Turing machine, is that instantons do not actually 'guess' their path: they have *no other choice* but to render the action of the physical system stationary (eqn (11.22) and Fig. 12.7), with that action determined by the physical laws underlying the dynamics of DMMs, namely their equations of motion.

In other words, it is the Physics of the problem that dictates the correct path traveled by the system in the phase space, not some 'magical', know-it-all agent that can predict the future. With this important difference in mind, we can nonetheless interpret the whole set of instantons

[20] Recall from Section 2.3.1, that 'non-deterministic', in the context of Turing machines, does *not* mean 'stochastic' or 'noisy'.

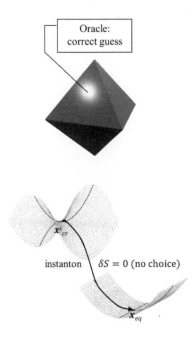

Fig. 12.7 An 'oracle' is a hypothetical mechanism (a black box) that, at each step of the computation, correctly guesses the next step to take (Section 2.3.1). Instantons, instead, 'guess' the correct path to follow according to well-defined physical laws. They have *no choice* other than to render the action, S, of the system stationary: $\delta S = 0$.

during the transient phase of the dynamics of DMMs (the *composite instanton* of DMMs) as a *physical* realization of an oracle (Di Ventra and Traversa, 2018*a*).

12.5 Critical branching processes in DMMs

We have seen that instantons are the field-theory analogue of *avalanches* (Section 11.2). During an avalanche (or instantonic jump) only a certain number, S, of voltages/literals transitions, say, from logical 0 to logical 1, or vice versa (see, e.g., Fig. 11.2).[21]

This behavior is easily seen by considering the example of the Ising spin glass I introduced in Section 9.2.2. In that case, we can monitor the number of spins (represented by voltages at the appropriate self-organizing gates), that flip (change from +1 to −1) at any given time.

If we plot the energy of the system,[22] then we see what I have anticipated in Section 11.6: the energy of the *sub-space* of voltages changes *only* in between sudden jumps, the latter indicating the occurrence of instantons/avalanches (Fig. 12.8).

[21]Do not confuse the size of the avalanches, S, with the action of the dynamics.

[22]Here, I consider only the energy of the voltages/spins subspace.

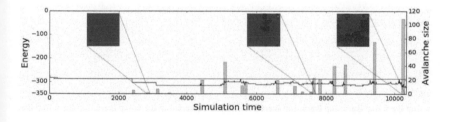

Fig. 12.8 Left axis: energy of the voltage-variable subspace of a DMM solving the 2-dimensional spin-glass problem (Section 9.2.2) as a function of simulation time. The red line is the energy for the same machine but *without* memory variables. **Right axis**: the sizes of the avalanches/instantons (the number of voltages/spins that change sign) are plotted as gray bars. Some avalanches are depicted in the inset in red. Reprinted with permission from (Sheldon *et al.*, 2019).

From the same figure we also see that the energy will settle into a *local minimum* in the *absence* of memory variables. This is again a confirmation that we need memory variables to transform local minima of the original problem into saddle points (cf. discussion in Section 7.2.1).

12.5.1 MemComputing vs. Monte Carlo sampling

I want take this opportunity to make a short detour that will provide a better understanding of how differently DMMs explore the space of 'energies' compared to stochastic methods, such as Monte Carlo, used in optimization problems (see, e.g., Chapters 9 and 10, or (Sauer, 2018),

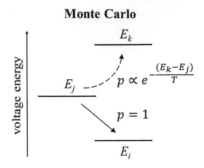

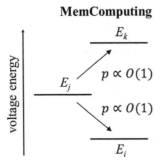

Fig. 12.9 Comparison between the energy exploration of the voltage (or spin) sub-space using a traditional Monte Carlo method vs. MemComputing. In the Monte Carlo method, higher energy states have a probability of being explored that is *exponentially* suppressed. A MemComputing machine, instead, can 'jump' from a given state E_j to one higher, E_k, or lower, E_i, with equal and finite 'probability'. In fact, the 'jumps' are *not* stochastic. Rather, the machine 'chooses' the appropriate physical path as I have explained in Section 12.1.

---

[23]At least after the very short initial transient of the dynamics.

for a quick introduction to some of these methods). Here, by 'energy' I mean the energy associated with the *original* variables (e.g., voltages or spins), not the total energy that includes also the energy associated with the memory variables.

In Section 10.4.3, I have shown explicitly that DMMs are considerably better at finding a good approximation of the lowest (ground-state) energy (or the ground state itself) than Gibbs sampling (a form of Monte Carlo method; see Section 10.2.1) in an energy landscape defined by a neural network (in that Section, a restricted Boltzmann machine). That energy landscape is indeed not much different than the one of the Ising spin glass of Fig. 12.8.

For instance, in Fig. 10.5 I have shown that the energies found by the MemComputing dynamics are substantially lower (sometimes showing relative differences of more than 1000%) than those found by Gibbs sampling within the same amount of time. We can now understand better why these machines show advantages in such a task compared to traditional stochastic methods.

It is clear from Fig. 12.8, that the energy of the *sub-space* of voltages can *both* decrease *and* increase as the machine goes through its solution search. This is because the actual dynamics are carried out in the *full* phase space of voltages plus memory variables, *not* in the reduced phase space of voltages only. It is only in this full phase space that the system explores the 'low-energy' dynamics I have discussed in Section 11.6.

In addition, from Fig. 12.8 we see that the system can make substantial jumps, even to *higher* energy states of the voltage sub-space. In fact, from that figure it seems that there are almost as many energy transitions to higher energy states as to lower energy states.[23]

In other words, the DMM easily *samples* states that may be energetically far from each other in the positive (higher energy) direction, in addition to the negative (lower energy) direction. Typically, this is *not* the case in stochastic methods (like the Monte Carlo one) where sampling energetically distant states, in the higher-energy direction, has an *exponentially low probability* of happening (see Fig. 12.9).

For instance, if we consider again *simulating annealing* to solve the Ising spin glass (Section 9.2.2), the acceptance probability for a transition from a spin state with energy E to another with energy $E' > E$ is suppressed *exponentially* by a factor $e^{-(E'-E)/T}$ at fixed temperature T (setting the Boltzmann constant $k_B = 1$). In other words, transitions to higher-energy states in the spin configuration are exponentially unlikely.

However, there is no reason to assume that those high-energy states should not be explored in the search for the global minimum. In fact, transition to those states may be the *only* way to reach the global minimum, by starting from an arbitrary initial state. The memory variables of DMMs then expand the phase space of the original problem to 'open up' the paths that connect all these states[24] in the phase space. The results in Fig. 12.8 seem to support these conclusions.

12.5.2 Avalanche sizes during solution search

After this detour, let me get back to the discussion of avalanche sizes. From Fig. 12.8, we see that the avalanches come in different sizes: some small, some large. In fact, some avalanches even span the *entire* dimension of the system. It is then natural to ask what the *distribution* of the *size*, S, of these avalanches is, during the solution search.

As I will now argue, this distribution is also *scale-free*, namely of the type $S^{-\tau}$ with $\tau = 3/2$ (Bearden *et al.*, 2019). This means that DMMs *self-tune* to a state in which avalanches are characterized by a *critical branching process*, and that this state persists during their transient dynamics before reaching equilibrium.[25]

[25]Again, this is somewhat reminiscent of what is known in some literature as *self-organized criticality* (Section 11.2). However, as before, this phenomenon can be understood from the more rigorously grounded (supersymmetric) Topological Field Theory of dynamical systems (Ovchinnikov, 2016).

An important remark

The fact that the distribution of the size of avalanches has to be *scale-free* is a consequence of the fact that *instantons* are *collective* objects, with this collective behavior encoded in their *moduli* (Section 11.4). In fact, in the language of supersymmetric Topological Field Theory (TFT), the *topological supersymmetry* is *effectively* (although *not* globally) broken on instantons (i.e., the matrix element of the exterior derivative, $\langle f|\hat{d}|i\rangle_{\text{instanton}} \neq 0$ on instantons), giving rise to the dynamical long-range order I have discussed in Section 11.7. This order can be interpreted as due to the 'release' of *goldstinos* (the *fermionic Goldstone modes* associated with the moduli) every time the system transitions from one local vacuum (critical point) to another more stable (Di Ventra and Ovchinnikov, 2019). See also Section 12.9.

Goldstinos

Fermions that emerge due to the *spontaneous* breakdown of *supersymmetry* are called *goldstinos*, or fermionic Nambu-Goldstone modes (Hori *et al.*, 2000). They are the analogue of the massless (zero-energy) Goldstone modes of spontaneously broken (bosonic) continuous symmetries (see discussion in Section 3.9).

In other words, DMMs go through what appears to be a *dynamical phase transition* during the solution search.[26] As I have discussed in Section 4.3.4, tuning a system to a continuous phase transition (or 'edge of chaos'), and utilizing its critical state, has been identified as a desirable feature to compute efficiently (Langton, 1990).

DMMs, properly designed, behave similarly: they *self-tune* into a (long-range) ordered state during dynamics for the purpose of *computing efficiently*.[27]

Like the DLRO I have discussed in Section 11.7, the word 'self-tuning' means that the system does not need any external influence to enter this state: it will *spontaneously* 'fall' into it[28] once the dynamics are initiated

[26]See, however, Section 12.9 where I discuss more in depth the nature of this state.

[27]This 'self-tuning' is precisely the initial transient in which the initial state of the system, $\mathbf{x}(t_0)$, is attracted to the 'closest' critical point in the phase space.

[28]It is *attracted* by it.

from some arbitrary initial condition.

Let me now quantify the previous discussion with what we call *mean-field theory*, as done in (Bearden *et al.*, 2019). I will provide later numerical examples.

12.5.3   Mean-field theory of avalanches

I first recall that the memory variables in DMMs have a much *slower dynamics* than the voltage variables (Section 11.2), and the system spends some time at a critical point before initiating the next instanton/avalanche (see again Figs. 12.6 and 12.8). This implies that each avalanche is *independent* of the others generated.

I will then assume this condition to be valid throughout the solution search. Using a Physics terminology, this assumption may be called *mean-field approximation*, since the behavior of a single avalanche is *representative* of the behavior of all other avalanches, and these do not 'interact' with each other.

Now, let's take an arbitrary voltage in one of those avalanches. Every time that voltage flips, say, from +1V to −1V, or vice versa, so that its corresponding Boolean variable changes from logical 1 to logical 0, or the reverse, *on average*, it will only have enough strength (power) to affect one other voltage in the circuit (which I call its 'offspring'); see Fig. 12.10 and margin note.[29]

In turn, this 'offspring' voltage, on average, will have enough strength to only affect at most one other voltage at the next step, and so on. This creates a *branching process* (Athreya and Ney, 2004) in which one voltage 'generates' (or better flips) some other voltage in the circuit, which, in turn, flips another one, and so forth.

Now, all voltages in the circuit are equally important. This means that the *distribution* of the offspring voltages must be the same for each individual voltage at every 'generation' step. It must also be *independent* of both the number of voltages that flip in the circuit at that step, and

---

Mean-field theory

A many-body system is well described by a *mean-field theory* when its evolution can be effectively represented by the dynamics of just one of its elements interacting with the average (mean) effect of all the other elements in the system. In other words, the dynamics of a single element, in the *mean field* of all the others, are representative of the whole many-body dynamics (Messiah, 1958). In (Quantum) Field Theory it is often called *tree-level approximation*. It is a reasonable approximation when *fluctuations* with respect to the 'average dynamics' can be neglected.

---

[29]Of course, since the dynamics of DMMs are continuous, and voltages/literals at some gates may also appear at other gates, this power may be distributed to a few voltages, not just one. However, I am making here an argument on the *average* power one voltage can provide to another, for the latter to transition from +1V to −1V, or vice versa.

Fig. 12.10 Graphical representation of a *branching process*. The filled circles represent 'individuals' (e.g., voltages), generating a certain random number of 'offsprings' (according to some probability distribution). Some of these offsprings will not generate any further individuals, and are indicated with an open circle. The process can continue *ad infinitum* or go to an extinction state (of the population of individuals) after a finite number of generations.

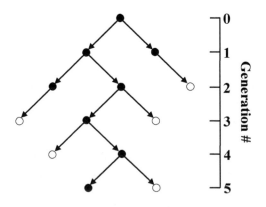

the number of voltages that are offspring of any given voltage.

Since the average number of offspring voltages, $\mu \to 1$, is the same at any given step, these conditions are such that the offspring distribution (namely the distribution of the number k of offspring voltages at a given step) must be a *Poisson-distributed process*:

$$p_{\text{k-offspring}} = e^{-\mu} \frac{\mu^k}{k!}, \tag{12.3}$$

where the average number of affected variables is $\mu \to 1$.

Under these conditions, it can be demonstrated that the number of 'descendants' of a flipped voltage, namely the size, S, of the avalanches is an integer random variable described by the *Borel distribution* (see, e.g., (Tanner, 1961) for a derivation of this result):

$$p_S = \frac{(\mu S)^{S-1} e^{-\mu S}}{S!}. \tag{12.4}$$

The expectation value of S is given by $\langle S \rangle = 1/(1-\mu)$. Therefore, due to $\mu \to 1$, the average avalanche size *diverges* (in the thermodynamic limit). In fact, using the Stirling approximation (Arfken, Weber and Harris, 1989) for the factorial in eqn (12.4), we have

Distribution of the size, S, of the avalanches in DMMs

$$\lim_{\mu \to 1} p_S \to S^{-3/2}, \tag{12.5}$$

namely, a *scale-free distribution*, $S^{-\tau}$, with $\tau = 3/2$.

Therefore, DMMs showcase a *critical branching process* in the voltages involved in the instantonic jumps during their transient dynamics. In the thermodynamic limit this supports an *infinite* average cluster size.

These results then further reinforce what I have anticipated in Section 1.6 (criterion IV of MemComputing): DMMs operate at a *dynamical long-range ordered state* without, however, the need to tune any external parameter to reach such a state (see also Section 12.8).

12.5.4   Numerical examples

To corroborate this theoretical prediction, numerical simulations of DMMs have been done using equations of motion similar to the ones I have introduced in Chapter 9. Random satisfiable 3-SAT instances of different variable sizes have been taken from the annual SAT competition (www.satcompetition.org).

The choice of random instances is to insure that any feature found is a feature of the dynamics of DMMs, rather than a feature of the SAT instances solved.

Branching process

Consider a population of organisms in which each individual produces a random number n of individuals (first generation) with given probability. Each of these individuals, independently, generates its own random number of individuals (second generation) with given probability (possibly different from the probability of the previous generation). And the process can go on for generations. The collection of random numbers of individuals per generation defines a stochastic process known as a *branching process* (Tanner, 1961).

Poisson distribution

Consider the number of events that occur *independently* of each other, and at a constant average rate, in a time interval. The *discrete* probability distribution of this number is called *Poisson distribution*, eqn (12.3) (van Kampen, 1992).

Borel distribution

Consider a population of organisms that reproduce and die. Assume, further, that the number of 'offsprings' of a single organism follows a Poisson distribution, with a mean, μ, that is less than 1. The *number of descendants* that a single organism in this population can generate is distributed according to a *Borel distribution*, eqn (12.4) (Tanner, 1961).

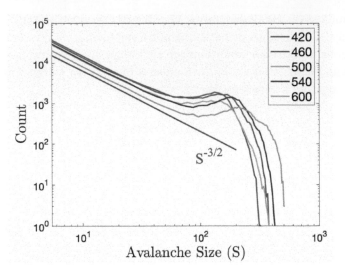

Fig. 12.11 Distribution of avalanche sizes, S, for different variable sizes of random 3-SAT instances. The red line is proportional to $S^{-3/2}$ (a *critical Borel distribution*), and is shown for comparison. The curve for each variable size originates from 100 solutions. Reprinted with permission from (Bearden *et al.*, 2019).

Finite-size effects

Strictly speaking, a scale-free distribution has a rigorous meaning only in the *thermodynamic limit* (see Section 4.3.2), namely in the limit of *infinite* system size. In actual physical systems, and their numerical simulations, such a limit is not reached. Therefore, *finite-size effects* emerge when considering long-wavelength features (e.g., avalanches whose size is comparable to the finite size of the system). The 'bump' appearing in Fig. 12.11 at the end of the scale-free part of the distribution is due to such an effect. Also, there is a natural cutoff for the minimal size of the avalanches, which is 1.

Scale-free distributions and extreme events

A scale-free distribution is characterized by events (e.g., size of avalanches) that do not have a typical scale, and their average (when it exists) is not a good measure of 'likelihood'. Rather, the events are distributed over scales differing by many orders of magnitude. This means that *extreme events* (e.g., those corresponding to very large avalanche sizes) are not as rare as in the Gaussian distribution (van Kampen, 1992).

I refer the reader to (Bearden, Sheldon and Di Ventra, 2019) for the details of how the avalanches have been detected numerically. Here, I just report the results in Fig. 12.11.

The scale-free part of the distributions in that figure is proportional to $S^{-\tau}$, with $\tau = 1.51 \pm 0.02$ which is consistent with the mean-field analysis I have presented in the previous Section 12.5.3. The 'bump' appearing in the distributions of Fig. 12.11 is typical of many *finite-size* power-law distributions (Pruessner, 2012).

As I have previously remarked, the scale-free properties of the distribution of the size of the avalanches could have been deduced using the supersymmetric TFT of dynamical systems of Section 11.3, as a consequence of the *collective* ('long-wavelength') nature of instantons.

However, while the general form, $S^{-\tau}$, can be obtained relatively easily using TFT, the coefficient τ is not so easy to determine, and the mean-field analysis I discussed previously seems the most straightforward (and simple) way to extract it.

12.6 Dynamical long-range order and critical branching as emergent phenomena

At this stage, it is worth stressing a very important point. The dynamical long-range order (DLRO) that I have anticipated in Section 11.7, and the critical branching behavior I have just discussed are *emergent phenomena* of DMMs.

We could also say they are *epiphenomena* of time non-locality. They are indeed very desirable properties of any MemComputing machine (criterion IV of MemComputing in Section 1.6).

'Emergent phenomena' means that despite the connectivity of the self-organizing gates is *local*, the dynamical behavior of a *collection* of them gives rise to these observed features.[30] Ultimately, we can attribute these features to *time non-locality*, the *non-linearity* of the equations of motion of DMMs, and the *coupling* of their self-organizing gates.[31]

The fact that these are emergent phenomena is very appealing. It means that we do not need to introduce them *ad hoc*, by connecting, say, each gate to *all* gates in the circuit. We do not need any 'global' coupling between the gates. In fact, given a problem to solve, the number of connections between gates scales at most *polynomially* with increasing number of gates.[32]

Since connections between gates ultimately requires energy expenditure (if the machines are built in *hardware*), this means that the *energy consumption* does *not* need to increase exponentially with problem size for the machines to develop these features. We then confirm what I have anticipated in Section 4.3.4: a machine that computes at some sort of 'edge of chaos' can solve hard problems efficiently.[33]

12.6.1 Quenching and information flow

Finally, before concluding this Section and discussing the robustness of DMMs against noise and perturbations, let me make a point that I think may further help understand their dynamics.

When we switch on the input voltages to initiate the DMM dynamics we set these machines *very far* from a steady state (equilibrium point of the dynamics). In fact, we set them into an *unstable* state.[34] We then let the system evolve on its own, until it finds an equilibrium point/solution of a given problem (if it exists). However, as it is evident from, e.g., Fig. 11.2, the system 'quickly' attempts to reach a steady state (an equilibrium point). It is as if the system 'cools' some degrees of freedom (the voltages at the terminals of the self-organizing gates) very fast in an attempt to reach the equilibrium point, while essentially 'freezing out' the other ones (the memory degrees of freedom). This is similar to what we call *quenches*.[35] In fact, 'quenched dynamics' are characteristics of many phenomena that typically showcase *scale-free* properties in the probability distribution of their avalanche sizes (Pruessner, 2012).

During the DMM 'quench', the 'fast' degrees of freedom (the voltages in the case of DMMs) transfer (syntactic) *information* to the 'slow' degrees of freedom (memory in DMMs).[36] The latter ones then act as a *reservoir* of information, providing *feedback* to the voltages themselves, and promoting the spatial long-range order I have discussed in Section 11.7. In a language that is most familiar in Quantum Mechanics,

[30] These features cannot be inferred from the properties of the single units only: they emerge from their interaction. They are proper of *complex systems* (see Section 1.6).

[31] Similar to the coupling between spins in a spin system like the one in Fig. 4.3 of Chapter 4. The coupling between the spins may be *local*, as in eqn (9.1), but the system undergoes a *phase transition* at a critical temperature, with consequent long-range order/criticality (cf. discussion in Section 4.3.2).

[32] Indeed, for certain problem specifications, this number may not scale at all.

[33] This also shows that such a machine explores/discovers the structure of the problem it is attempting to solve at *all length scales* (cf. discussion in Section 2.6).

[34] Or an *excited* state, if it is a saddle point (cf. Section 12.1).

[35] Note that this process is the opposite of *annealing*, where the system is allowed to 'cool' slowly.

[36] See Section 1.3 for a distinction between *syntactic* and *semantic* information.

[37]In fact, the interaction between two physical systems can always be interpreted as one system 'measuring' the state of the other, even if such a state is never ultimately recorded by an external observer; see, e.g., (Di Ventra, 2008).

Physical vs. numerical noise
Physical noise is any type of stochastic process that influences the otherwise deterministic dynamics of a physical system. It can be caused by a variety of sources, e.g., temperature, charge, or spin fluctuations, etc. (van Kampen, 1992).
Numerical noise (or *truncation error*) is due to the discretization of time in the integration of the ODEs (3.1), and hence the transformation of *continuous* dynamics into a *discrete* map (cf. Section 8.4). This type of noise is 'worse' than the physical noise, for two reasons. (i) The numerical noise/error *accumulates* during integration, thus degrading the numerical solution of eqns (3.1), as the simulation progresses. The physical noise, instead, typically acts at disjoint (and usually very narrow) time intervals during the actual dynamics of the system. (ii) Numerical integration schemes may introduce *extra* critical points in the phase space, thus changing its topology *explicitly*, even for small integration time steps (see Section 12.7.3). This cannot happen with moderate *physical* noise.

[38]Stability against perturbations is also a typical feature of *point-dissipative systems* (Section 7.8). This implies that, for these dynamical systems, the flow on the global attractor is stable under perturbations of the flow vector field in eqns (3.1) (Hale, 2010).

we could also say that this process is analogous to the memory variables 'measuring' (recording) the values of the voltages at any given time.[37]

12.7 Robustness against noise and perturbations

Armed with all these results, I can now address the point I made in Section 1.6 regarding the *robustness* of MemComputing machines against small imperfections and noise (criterion VI of MemComputing).

Note that this robustness refers to *both* the physical (*hardware*) implementation of MemComputing *and* its numerical (*software*) simulations. In fact, as I have shown in Chapters 8, 9, and 10, even the *simulations* of DMMs, despite suffering from (substantial) numerical errors, reach equilibria in times that typically outperform traditional algorithms.

To emphasize this point even further, it is worth stressing that the majority of the simulations I have reported in this book employed an *explicit* method (Euler's method) to integrate forward the ordinary differential equations (ODEs) (Chapter 8). And explicit methods are notorious for generating substantial numerical errors during simulations (Press *et al.*, 2007). Numerical errors are akin to *noise* introduced in the physical system, although of a different nature, and arguably more deleterious (see the margin note and discussion in Section 12.7.3).

Therefore, there has to be some fundamental reason why these errors (noise) do not seem to affect the solution search significantly. The reason is as follows.

12.7.1 Topological considerations

We have seen that DMMs employ objects of a *topological* nature to compute: *critical points* and trajectories between them (*instantons*). The topological character of these objects means that they are constrained by the *global* structure of the phase space (its topology), *not* by its local (geometric) properties.

The dynamics of DMMs in the presence of noise was briefly introduced in Section 11.3, see discussion after eqn (11.1). In that Section, I have anticipated that also the action (S_{Eucl} of eqn (11.15)) of the *stochastic* dynamics is of a topological nature, implying that the matrix elements on instantons, eqn (11.24), are *topological invariants* even in the presence of moderate noise.

All this implies that in order for the machine *not* to operate as designed, one needs to make *global* changes to the structure of the phase space. In practice, such global changes are not likely to occur with relatively small noise or perturbations.[38]

The DMMs' dynamics are then *stable* ('topologically protected') against these effects. Instead, to considerably affect the dynamics would require *dramatic* changes of the system.

In simple words, one really needs to change substantially the physical topology (architecture) of the DMM circuit (by, e.g., destroying connections between gates), or modify the functioning of the single gates themselves to see substantial changes of the phase space global structure.

We then say that DMMs are *not fault-tolerant* (see also Section 12.10).

12.7.2 Physical noise

As discussed in Section 11.3, some types of physical noise can be modeled by adding stochastic terms to eqns (3.1):

$$\dot{\mathbf{x}}(t) = F(\mathbf{x}(t)) + \sum_{a=1}^{k} e_a(x)\xi^a(t); \quad \mathbf{x}(t = t_0) = \mathbf{x}_0, \quad (12.6)$$

where, again, $\{e_a(x) \in TX_{\mathbf{x}}, a = 1...k\}$ is a collection of vector fields that couple the noise to the system, and $TX_{\mathbf{x}}$ is the tangent space of X at point $\mathbf{x}$.

The noise is characterized by the *moments* of its distribution. For instance, for *Gaussian white noise*, we have the following first two moments, $\langle \xi^a(t) \rangle = 0$, and $\langle \xi^a(t)\xi^b(t') \rangle = \Gamma \delta(t - t')\delta_{ab}$, where Γ is the noise strength (van Kampen, 1992).

An example of the effect of such a noise on the operation of a single self-organizing (SO) AND gate, as the one I introduced in Example 7.3 of Chapter 7, is shown in Fig. 12.12. One can see from that figure that even for large noise levels, the SO-AND gate converges to the correct logical proposition of the gate.

In fact, by relating the noise strength to temperature, as done in (Bearden *et al.*, 2018), one finds that the largest value of the noise intensity in Fig. 12.12 corresponds to a temperature that would, most likely, induce physical instabilities of the resistive memories of the gate.

However, note that, due to the temporal DLRO (Section 11.7.2), the instantonic trajectory itself, that connects one critical point to another, may indeed change (and even dramatically) with local perturbations or noise. On the other hand, the critical points (hence the final equilibrium point), as well as the ('tunneling') amplitudes (11.24), cannot.

I showed this explicitly in Fig. 7.13, and it can also be inferred from Fig. 12.12, which reports two standard deviations from the ensemble averages of the voltage trajectories of the SO-AND gate of Fig. 7.12, with the largest deviations occurring, as expected, in between critical points (along the 'bulk' of the instantonic trajectory). The noise in Fig. 12.12 is applied to the resistive memory dynamics of the SO-AND, eqn (7.11). It is then *additive* for the memory variables, but *multiplicative* for the voltages.

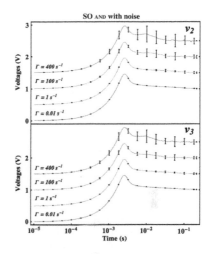

Fig. 12.12 Dynamics of the SO-AND gate of Fig. 7.12 with added white noise: eqn (12.6) with just one noise term of varying strength, Γ. The terminal voltage v_1 is fixed at $+1$V. The **top panel** shows the terminal voltage v_2, while the **bottom panel** shows the voltage v_3 of the SO-AND. Each curve is the ensemble average over 100 simulations, with curves shifted upward for clarity. Error bars indicate 2 standard deviations. The final voltage equilibrium point is $(1, 1, 1)$V, which is one of the correct logical propositions of the gate. Reproduced with permission from (Bearden *et al.*, 2018).

Additive and multiplicative noise

If the noise term in eqn (12.6) does *not* depend on the state of the system, $\mathbf{x}(t)$, it represents *additive noise*. If it *does* depend on the state $\mathbf{x}(t)$, we call it *multiplicative noise* (van Kampen, 1992).

12.7.3 Numerical noise

The topological arguments I made in Section 12.7.1 apply to *numerical noise/errors* as well. In this case, however, we need to make a further, important consideration.

I have said that we need to change the topology of phase space in order to change the number of critical points, and vice versa (Fomenko, 1994). Since (moderate) physical noise cannot change the topology of phase space, the number of critical points is unaffected by this type of noise. In view of the fact that the 'low-energy' dynamics are instantonic (connecting distinct critical points), physical noise (in moderation) should then not affect the functioning of DMMs, namely their ability to solve the problems they have been designed for, which is what we see in, e.g., Fig. 12.12.

On the other hand, integration schemes *may* change (in fact, increase) the number of critical points, hence change the topology of phase space, even for relatively small integration steps (Stuart, 1994).

Regular integration schemes
To see this explicitly, consider the ODEs (3.1), and let's apply first the *forward Euler* integration scheme (a *discrete map*) at a time step, t, over an interval of time Δt (cf. eqn (8.11) in Section 8.4.1):

$$\mathbf{x}(t + \Delta t) - \mathbf{x}(t) = F(\mathbf{x}(t))\Delta t, \quad \mathbf{x}(0) = \mathbf{x}(t_0), \quad \textit{forward Euler.} \quad (12.7)$$

It is obvious from eqns (12.7) that whenever the *original* ODEs (3.1) have a critical point, $F(\mathbf{x}(t)) = 0$, *also* their approximate representations have a critical point: $F(\mathbf{x}(t)) = 0 \implies \mathbf{x}(t + \Delta t) = \mathbf{x}(t)$.

The reverse is clearly true as well: whenever the forward Euler eqns (12.7) have a critical point, $\mathbf{x}(t + \Delta t) = \mathbf{x}(t)$, the original ODEs (3.1) have the *same* critical point: $\mathbf{x}(t + \Delta t) = \mathbf{x}(t) \implies F(\mathbf{x}(t)) = 0$.

The critical points of the original equations are then in a *one-to-one* correspondence with the critical points of the (forward Euler) maps, eqns (12.7). We call numerical methods that have this property, *regular integration schemes* (Stuart, 1994).

Irregular integration schemes
Consider now the *explicit trapezoid* method (Section 8.5), which approximates the ODEs (3.1) as (Sauer, 2018)

$$\mathbf{x}(t + \Delta t) - \mathbf{x}(t) = [F(\mathbf{x}(t)) + F(\mathbf{x}(t) + F(\mathbf{x}(t))\Delta t)] \frac{\Delta t}{2}, \quad \textit{trapezoid,} \quad (12.8)$$

with same initial conditions: $\mathbf{x}(0) = \mathbf{x}(t_0)$.

Now, if we have a critical point of the *original* ODEs (3.1), then it is clear that this is also a critical point of eqns (12.8), because all terms with the flow field, F, on the right-hand side of eqns (12.8) are zero, hence $\mathbf{x}(t + \Delta t) = \mathbf{x}(t)$.

Conversely, when $\mathbf{x}(t + \Delta t) = \mathbf{x}(t)$, then $F(\mathbf{x}(t)) = 0$ in eqns (12.8) is indeed a critical point of such equations, which, in turn, is a critical point of eqns (3.1). So far, so good.

However, the property $\mathbf{x}(t + \Delta t) = \mathbf{x}(t)$ may be satisfied by *additional* critical points, namely those corresponding to the relation

$$F(\mathbf{x}(t)) = -F(\mathbf{x}(t) + F(\mathbf{x}(t))\Delta t), \quad \textit{ghost critical points.} \quad (12.9)$$

We call these critical points *ghosts*, because they do *not exist* in the original, *continuous-time* ODEs (3.1).

Of course, condition (12.9), may *never* be satisfied if, say, Δt is small enough,[39] or the flow field, F, takes appropriate forms. Nonetheless, it is an unwanted possibility.

In other words, the critical points of the original ODEs *may be* only a *sub-set* of all critical points of the explicit trapezoid map. This issue is not limited to the trapezoid method.

In fact, it is true also for, e.g., higher-order Runge-Kutta methods (Section 8.5) as it can be easily checked as I just did. If that is the case, namely there is *no* one-to-one correspondence between the set of critical points of the continuous-time ODEs and the set of critical points of their approximate maps, we say that the latter ones have been generated by *irregular* integration methods.

Ghost critical points
Those critical points introduced by an *irregular* integration scheme, in addition to those of the original continuous-time dynamics, are called *ghosts* (Stuart, 1994).

[39] In that case, $F(\mathbf{x}(t)) \approx -F(\mathbf{x}(t))$, which is satisfied only by $F(\mathbf{x}(t)) = 0$, if $F(\mathbf{x}(t))$ does not vary dramatically in very short intervals of time.

An important remark
The proliferation of ghost critical points is even more dramatic for *non-autonomous* dynamical systems, namely those that depend *explicitly* on time: $\dot{\mathbf{x}}(t) = F(\mathbf{x}(t), t)$ (Section 3.2). For instance, the trapezoid method (as well as any higher-order Runge-Kutta method) for these systems does not even have the same critical points as the original ODEs, while the forward Euler is still regular.

Trajectories in the presence of noise
It is worth pointing out that in the presence of noise (physical or numerical) *all* trajectories are, in principle, available to a physical system. This situation is then somewhat reminiscent of Quantum Mechanics (Messiah, 1958).

Again, since a change in the number of critical points may imply a change of topology (Fomenko, 1994), we then see that for *irregular* integration methods the *topology* of phase space of the original, continuous dynamics may be changed *explicitly* by a bad choice of integration step and/or particular choices of approximate flow field. Therefore, particular care needs to be put into selecting the integration scheme and step.

In fact, in Chapter 8 I have shown that the explicit trapezoid method and the Runge-Kutta 4th order method provide a slightly worse integration scheme compared to the forward Euler one (in terms of scalability of the number of steps to solution vs. the number of variables) for selected SAT problems (see Fig. 8.7). This could be attributed to the presence of ghost critical points in the *map phase space* defined by these two integration schemes, while, as we have seen, the forward Euler method cannot introduce any ghost critical point.

Therefore, combined with the fact that the number of function evaluations is minimal for the forward Euler (see again Fig. 8.7), the latter provides a good enough integration scheme for the DMMs's ODEs.

12.7.4 Conditioning

To further understand the reason why explicit methods are reasonable for the ODEs of DMMs, let me discuss how the *form* of the ODEs (3.1) of a dynamical system affects the propagation of numerical errors.

To illustrate this, let's consider the forward Euler method, eqns (12.7). For simplicity, let us first assume we have only one variable x in eqns (3.1), namely that equation is just $\dot{x} = F(x)$.

It can be shown (Sauer, 2018) that the *global* truncation error, g_{i+1}, (Section 8.5) for such a method at iteration step $i + 1$, is related to the global truncation error, g_i, at the previous iteration i, as:

$$g_{i+1} = (1 + \Delta t \frac{\partial F}{\partial x})g_i + O((\Delta t)^2). \tag{12.10}$$

We see from eqn (12.10) that if $\partial_x F < 0$ then the global truncation error *decreases* at successive iterations. We call a dynamical system with such a property, *well conditioned*.

On the other hand, if $\partial_x F > 0$, then the error *increases*. We call such a dynamical system *ill conditioned*.

It is obvious that, if the ODEs are well-conditioned, a numerical method like the forward Euler will provide a reasonable approximation to the exact solution (of course, with small enough time interval). Otherwise, the method will quickly diverge. (Examples of this are *chaotic dynamics* which are ill-conditioned on a *global* scale; cf. discussion in Section 3.10.)

The situation is not so straightforward in the case of the ODEs for DMMs. Let me explain.

Condition number of linear equations

Consider a system of equations $Ax = b$ with A an $n \times n$ matrix, x an $n-$vector of unknowns, and b a constant $n-$vector. Suppose we want to solve this system of equations numerically. We define the *condition number* of the matrix A, cond(A), as follows:

$$\text{cond}(A) = ||A||_\infty \cdot ||A^{-1}||_\infty,$$

where $|| \cdot ||_\infty$ is the *norm* of the matrix (the maximum absolute row sum of the matrix). If cond(A) $= 10^k$, with k some integer, and p is the precision with which we represent numbers on a computer, then we should expect about $p - k$ correct digits in the solution of x, without considering the errors introduced by the numerical method itself (Sauer, 2018). Note, however, that cond(A) is only an upper bound. As a rule of thumb, a problem is said to be *well-conditioned* if $k \ll p$. It is *ill-conditioned* if k is a large fraction of (or comparable to) p ($k \lesssim p$) (Trefethen and Bau III, 1997).

In Section 12.1, I have shown that the dynamics of DMMs go through a *composite instanton*, namely a succession of disjoint instantonic jumps between distinct (also in index) *critical points*, namely those for which the flow vector field $F = 0$.

In fact, I have argued in Section 11.4 (see also Fig. 11.3) that there are *several* (a *family* of) trajectories[40] that connect a pair of critical points differing in index (number of unstable directions).

Let us then focus on the neighborhood of two distinct critical points of eqns (3.1). In Section 3.9, I have discussed how to expand the trajectory, $\mathbf{x}(t)$, in the neighborhood of any critical point (see eqn (3.8)).[41] I re-write this expansion here for convenience:

$$\mathbf{x}(t) \approx \mathbf{x}_{cr} + \sum_i \mathbf{u}_i e^{\lambda_i t}, \tag{12.11}$$

where the sum is over the eigenvalues λ_i and the associated eigenvectors $\mathbf{u}_i$ of the Jacobian matrix (3.9).

As I have already discussed in Section 3.9, in the neighborhood of a critical point there are some *attractive* (or stable) directions (Re $\lambda_i < 0$) and *repulsive* (unstable) directions (Re $\lambda_i > 0$). From (Bearden *et al.*, 2018), we also know that the *magnitude* of the real part of the eigenvalues, λ_i, of their unstable directions, is *much smaller* than the corresponding quantity of their stable directions (cf. discussion in Section 12.1).

Let us then consider just a pair of distinct critical points $\mathbf{x}_{cr}^i$ and $\mathbf{x}_{cr}^f$ (whose index differs by at least 1), with $\mathbf{x}_{cr}^i$ having a larger number of unstable directions than $\mathbf{x}_{cr}^f$. Let's also assume we know *all* the (instantonic) trajectories that connect them (see Fig. 12.13).

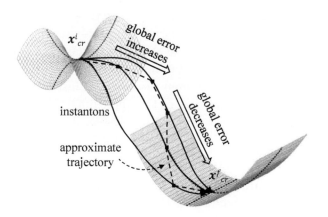

[40]For continuous dynamics, an infinite number of trajectories.

[41]This is a *linear* expansion, which allows us, in principle, to compute also the *condition number* of the corresponding system of equations in the neighborhood of critical points (Trefethen and Bau III, 1997).

Fig. 12.13 Schematic of a family of instantons (solid lines) connecting a saddle point, $\mathbf{x}_{cr}^i$, and an equilibrium point, $\mathbf{x}_{cr}^f$, with a center manifold. The dynamics in the proximity of the initial critical point, $\mathbf{x}_{cr}^i$, are *ill-conditioned* due to the *unstable* direction(s). Any numerical method will approximate the exact trajectory with a succession of discrete trajectories. If the approximate trajectory does not leave this family of instantons, it will converge to the next critical point $\mathbf{x}_{cr}^f$. This is because, the ODEs are *well-conditioned* in the neighborhood of $\mathbf{x}_{cr}^f$ due to the *stable* direction(s).

In the neighborhood of $\mathbf{x}_{cr}^i$

In the neighborhood of the initial critical point, $\mathbf{x}_{cr}^i$, the system has unstable directions, so that $\partial_{x_i} F > 0$ along those directions. From the previous discussion, and eqn (12.10), we then conclude that the ODEs of DMMs are *ill-conditioned* in the short period of time immediately after

Shadowing theorem
For certain chaotic systems it has been mathematically proved that even though the *numerical* trajectory diverges exponentially (due to numerical errors) from the true trajectory with the same initial conditions, there exists a 'true' (namely without error) trajectory, with a slightly different initial condition, that *shadows* (stays near) the numerical trajectory. This is called the *shadowing theorem*, and it is the reason the fractal structures of chaotic systems obtained numerically are assumed to be realistic, and not just numerical artifacts (Ott, 1993).

---

[42]Giving rise to a positive *time-averaged Lyapunov exponent*, namely a positive real part of the largest eigenvalue of the Jacobian matrix time-averaged over the *entire* dynamics. However, as in the case of chaotic dynamics, we expect that the numerical instantonic trajectory is *shadowed* by a true (noise-less) instantonic trajectory. This means that the approximate trajectory (as depicted in Fig. 12.13) would be shadowed by an actual trajectory obtained by slightly changing the initial condition in the neighborhood of the initial critical point.

[43]Within times of order RC.

the system is about to perform an instantonic jump to the next critical point $\mathbf{x}_{cr}^f$.

Being ill-conditioned in that period of time means that the global truncation error of successive time steps of the integration method increases, driving the *approximate* dynamics *away* from the original instantonic trajectory toward other instantonic trajectories (see Fig. 12.13). This is just a consequence of what I have discussed in Section 11.7: the *temporal* DLRO of the instantons is due to the presence of unstable directions at the initial critical point, with consequent *sensitivity* of the trajectory (but *not* of the critical points) to small perturbations.

However, in the same Section 11.7 (see also Section 12.9), I have stressed that this sensitivity to small perturbations in the neighborhood of the initial critical point is *not* chaos! In fact, for chaotic behavior to occur, that sensitivity should *persist* during dynamics (Hilborn, 2001).[42]

In the neighborhood of $\mathbf{x}_{cr}^f$

In the case of DMMs, instead, the instantonic dynamics quickly[43] converge toward the next critical point, $\mathbf{x}_{cr}^f$, so that *all* instantonic trajectories (within the family of those belonging to the specific pair of critical points) will be guided toward it (see Fig. 12.13). Therefore, even if, at the initial time, the numerical trajectory was moving 'away' from the correct trajectory, it quickly finds itself in a region of phase space where $\partial_{x_i} F < 0$, namely the ODEs in that region are *well-conditioned*.

This means that, at first (during the first time steps of the numerical integration from a critical point), the global truncation error may increase due to the *ill-conditioned* character of the ODEs in that time interval. However, so long as the initial global truncation error does *not* bring the dynamics away from the family of instantons connecting the two critical points, the rest of the approximate dynamics will *reduce* the global truncation error till the system dynamics converge to the neighborhood of $\mathbf{x}_{cr}^f$. This is illustrated in Fig. 12.13.

From global error increase to global error decrease

In fact, since the real part of the eigenvalues, λ_i, of the unstable directions of critical points, is much smaller (in magnitude) than the corresponding quantity of their stable directions (Bearden *et al.*, 2018), we also expect that during the 'initial' phase of an instantonic jump, the truncation error does not increase as fast as it decreases in the 'final' part of the jump.

In other words, the truncation error during the 'initial' part of the instantonic jump can be easily 'corrected' (in fact, canceled) by the 'final' phase of the instantonic jump (Fig. 12.13). This is a further indication of the robustness of the ODEs of DMMs to numerical noise.

All this implies that, even if the phase space trajectory is not faithfully represented by the numerical method, so long as the critical points of the flow vector field are preserved by the integration scheme, the solution search can sustain quite large numerical errors in the trajectories themselves (as I have shown in Chapter 8; see, e.g., Fig. 8.6).

12.7.5   Directed percolation in DMMs with noise

Numerical noise allows us to study yet another interesting phenomenon pertaining to DMMs, which I hope will clarify even more their dynamics, and help further understand their robustness against perturbations: *directed percolation* of the state, $\mathbf{x}(t)$, in phase space. Let me clarify what I mean by that.

Directed percolation revisited

As I have briefly mentioned in Example 1.4 of Chapter 1, directed percolation refers to the phenomenon by which a liquid, e.g., water, passes through a porous material along a given direction, as typically driven by gravity.

In that example I have discussed how water is directed down hill by gravity toward its 'exit' point (the estuary) at the bottom of the hill, and does so by following several paths (see Fig. 1.5 in Chapter 1).

Let's now perturb the hill by randomly introducing local sinks for the water to fall in, while going downhill (Fig. 12.14). If there are not many sinks, then water will still reach the estuary, because some paths toward the bottom exit are still available to it.

Instead, if there are too many sinks, we may even stop the flow of water completely, so that it will never reach its estuary: no path can join the top of the hill and the estuary. We say that the system ends up into an *absorbing state*, from which it can *never* get out.

In fact, directed percolation is an important example of *non-equilibrium phase transitions* (it is, in particular, a *continuous* one), namely phase transitions, like the ones I have discussed in Section 4.3.2, that occur *away* from thermodynamic equilibrium (Henkel *et al.*, 2008).[44]

Therefore, there has to be a 'critical' number of sinks, n_c, at which at least one path allows water to percolate from top to bottom of the hill.

An important remark

The phase transition of directed percolation from the macroscopically *permeable* (active/fluctuating) phase to the *absorbing* phase is typically characterized by counting the number $N(t)$ of 'active' sites (Fig. 12.15) at each 'time step' t. At criticality, $\langle N(t) \rangle \sim t^\theta$,

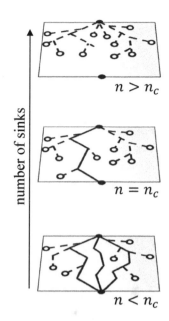

Fig. 12.14 Schematic of *directed percolation* of water from a source at the top of a hill (top filled circle) to the bottom exit (bottom filled circle). A certain number, n, of randomly distributed sinks (open circles), where water gets *absorbed*, is introduced. When that number is smaller than a critical value, n_c, the water has several paths (continuous lines) to percolate from top to bottom. When $n > n_c$ the water has no path to percolate and goes into an *absorbing state* (dashed lines). At the critical number, $n = n_c$, then there is at least one path (continuous line) for the water to percolate downhill.

[44]That directed percolation is an out-of-thermodynamic-equilibrium phenomenon can also be inferred from the presence of an *absorbing state*. This means that *detailed balance* (namely that the rate of transition *into* a state is the same as the one to transition *out* of it), typical of systems at thermodynamic equilibrium, is violated, because the system can never get out of the absorbing state.

with θ a critical exponent (Henkel *et al.*, 2008). This is *not* the same quantity I will report here, eqn (12.16). The reason why I do not attempt to compute $N(t)$ is because that quantity is not easily accessible in DMMs, in view of their vast phase space.

[45]The following analysis follows closely (Zhang and Di Ventra, 2021).

[46]Equivalently, we say that the bond is *open* (to the passage of water) with probability p.

[47]Written this way, the directed percolation can be interpreted as a (stochastic) *cellular automaton* (cf. Section 4.3.4).

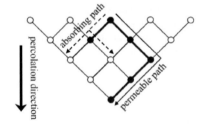

Fig. 12.15 Schematic of (1+1) *bond directed percolation* on a 'cone-shaped' lattice. Filled circles indicate that the probability to transition into that ('active') vertex is finite. Empty circles indicate that the probability to transition into those vertices is zero (they are disconnected). A path from the top (source) occupied vertex is *absorbing* if it ends up into a vertex with all disconnected sites beneath it. It is *permeable* if it connects the initial occupied vertex to the bottom of the lattice.

Let's now make these arguments a bit more rigorous.[45]

Consider again the schematic of water percolation downhill as in Fig. 12.14. This time, however, let's *discretize* the surface of the hill into a *lattice*, where the vertices are the pores (sinks) in which the water can get absorbed or not, and the edges are the *bonds* (channels) connecting nearest-neighbor pores.

The lattice then looks something like that in Fig. 12.15. Let's further assume that each bond connects neighboring pores with probability p.[46] Otherwise, the pores are disconnected (the corresponding bond is *closed* to the passage of water).[47]

To make direct contact with the dynamics of DMMs where the system starts from an arbitrary state in the phase space, and ends up into, say, one solution/equilibrium point (if it exists), we consider a lattice that has an arbitrary number of 'top' sites, and only one 'exit' site. The system then starts from one, randomly chosen, 'active' site at the top of the lattice, the initial state, and percolates downward toward the exit (solution of the problem).

We then say that a path is *permeable* if it connects the initial occupied (source) vertex to the bottom of the lattice. A path is *absorbing* if, instead, it ends somewhere in the lattice before the exit, with all sites beneath it disconnected.

It is easy to understand that by increasing p, the system goes through a *phase transition* (Section 4.3.2) from an *absorbing phase* into a *permeable phase* at a critical threshold p_c (Henkel *et al.*, 2008).

This is because if the probability to connect a vertex to the next is very small, then, most likely, the percolation will end up into absorbing paths. On the other hand, if that probability is close to one, then essentially all paths will be permeable. And, again, since the absorbing state cannot be escaped from, the phase transition is *non-equilibrium*.

Let's now calculate the percentage of permeable paths as a function of the probability p out of all possible paths (permeable plus absorbing). In anticipation of the relation to MemComputing machines, whose phase space has dimension $D \gg 2$, let's then consider a D-dimensional *cone-like* lattice which is the generalization in D dimensions of the two-dimensional 'inverted-triangle' lattice of Fig. 12.15.

Let's then start (first 'time' step, $T = 0$) from any top (source) vertex. Each bond from this vertex can permeate with probability p to the ones below it. We can then calculate the expectation of number of permeable paths as follows.

The *average* number of permeable paths at any given site is equal to the sum of permeable paths at all sites that precede (are 'above') it, multiplied by p. At the first time step ($T = 1$) this is simply Dp, because there are D sites preceding a given site (see Fig. 12.15).

Since the percolation events are independent going down the sites, if T is the number of time steps needed to reach the bottom exit point, then the expected number of permeable paths is

$$\langle n_{\mathrm{perm},T} \rangle = (Dp)^T. \tag{12.12}$$

We can now calculate the *average* number of absorbing paths at the i-th time step. This is the average number of permeable paths at the $(i-1)$-th time step times the probability that all bonds at that time step are absorbing: $(1-p)^D$. However, this is not fully correct at the 'boundaries' of our D-dimensional conical lattice, because a site at the boundary would have less than D connected bonds (see Fig. 12.15). Nevertheless, since we are interested in the limit of $D \to \infty$, I can assume that such a probability is valid also at the 'boundaries' of the lattice, as deviations from that probability are of measure zero in that limit.

Then, the total average number of absorbing paths is

$$\langle n_{\mathrm{abs},T} \rangle = (1-p)^D \sum_{i=0}^{T-1} \frac{(T+1-i)^{D-1}}{(D-1)!} (Dp)^i. \tag{12.13}$$

The percentage of permeable paths, r, is then

$$r = \frac{\langle n_{\mathrm{perm},T} \rangle}{\langle n_{\mathrm{perm},T} \rangle + \langle n_{\mathrm{abs},T} \rangle}. \tag{12.14}$$

Now, the exact calculation of this quantity is far from trivial. However, we are only interested in the neighborhood of the percolation transition, namely near the *critical probability* p_c. We can then analyze r in the limits $T \to \infty$ and $D \to \infty$.

I refer the reader to (Zhang and Di Ventra, 2021) for an explicit derivation of r under these conditions, and the various approximations to obtain the critical percolation threshold p_c. Here, I just provide the final result (valid for $p > 1/D$):

$$\langle n_{\mathrm{a},T} \rangle = \frac{(Dp)^{T+1}}{2} \left(\frac{1-p}{\ln Dp} \right)^D \mathrm{erfc}(\sqrt{D} \frac{1 - \ln Dp}{\sqrt{2 \ln Dp}}), \tag{12.15}$$

where 'erfc' is the 'complementary error function' (Arfken *et al.*, 1989). Using this result we obtain from eqn (12.14) the ratio of permeable paths:

$$r = \frac{1}{1 + \frac{1}{2} Dp (\frac{1-p}{\ln Dp})^D \mathrm{erfc}(\sqrt{D} \frac{1 - \ln Dp}{\sqrt{2 \ln Dp}})}. \tag{12.16}$$

Bond vs. site percolation
The percolation model in which bonds on a graph are open with probability p is called *bond percolation*. If that probability is assigned to the vertices of the graph, we call it *site percolation*. The critical probabilities of the two models are typically different. Every bond percolation model can be reformulated as a site percolation model on a different graph. The reverse, however, is not necessarily true (Henkel *et al.*, 2008). For DMMs in the presence of noise, it is natural to map the dynamics of their state trajectories to a bond percolation model (Zhang and Di Ventra, 2021).

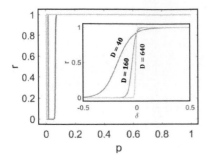

Fig. 12.16 The ratio, r, of permeable paths vs. total paths, eqn (12.16), on a D-dimensional lattice as a function of the percolation probability, p. The inset shows the same ratio as a function of $\delta = Dp - e$. A phase transition occurs near $p_c \sim e/D$. The fact that the r vs. δ curves all cross at the same point, irrespective of the size D, is an indication of *scale invariance*. Adapted with permission from (Zhang and Di Ventra, 2021).

[48]The observable r can be interpreted as the probability to have an infinite (in the thermodynamic limit) percolating path.

[49]Since Δt in DMMs is inversely related to the probability p in directed percolation, these plots are curving in opposite directions.

[50]Coupled with the constraints imposed by the self-organizing gates, the state dynamics are then *driven* toward the solution.

[51]The 'gradient' over the index of critical points can then be viewed as the 'direction' of percolation.

First of all, notice that the transition occurs at $p_c \sim \frac{e}{D}$. We can then define $p = \frac{e+\delta}{D}$, where $\delta \in \mathbb{R}$ is a small number. This way, $\ln Dp \approx 1 + \frac{\delta}{e}$, and using $\lim_{x \to \infty}(1 + \frac{1}{x})^x = e$, to order $O(\delta)$, eqn (12.16) becomes

$$r = \frac{1}{1 + \frac{1}{2}e^{1-e-\delta-D\delta/e}\text{erfc}\left(-\sqrt{\frac{D}{2}\frac{\delta}{e}}\right)}. \tag{12.17}$$

In the limit of $D \to \infty$, the divergence comes from D appearing in the exponent. Therefore, the transition happens exactly at $\delta = 0$. When $\delta > 0$, $r \to 1$; when $\delta < 0$, $r \to 0$.[48]

The behavior of eqn (12.16) is plotted in Fig. 12.16 for different dimensions D. I anticipate here that this ratio has a striking resemblance with the numerical results on the *solvable-unsolvable transition* I showed in Chapter 8 (cf. Fig. 8.5 in that Chapter[49]). Let's now see how all this relates to DMMs.

Directed percolation of the state trajectory in phase space

Let us then consider the numerical simulation of the ODEs of DMMs as I have discussed in Chapter 8. For the sake of discussion, let's employ the simplest integration method: forward Euler (Section 8.5.1 and eqns (12.7)). However, let's now *increase* the level of numerical noise/error by, e.g., increasing the time interval Δt in the forward Euler procedure.

For the MemComputing machines made of electrical circuits (as the ones I have considered in this book) the 'gravity' effect is the electrical potential that drives electrons in the circuit, once the input voltages have been switched on (see, e.g., Chapter 7). In phase space, the dynamics then initiate at some point, $\mathbf{x}(t = 0)$, and 'percolate' along instantonic trajectories till a solution is found (if it exists).[50]

In the absence of noise, the continuous-time dynamics proceed via instantons that connect critical points with a given index (number of unstable directions) to those with lower index (Section 11.4). In analogy to bond directed percolation, we may interpret critical points as 'pores' (or 'vertices'), while instantons are the 'channels' connecting 'neighboring' critical points (in the sense that their indexes are different). In the absence of noise then instantons connect 'adjacent' critical points with 'permeable probability' one.[51]

Physical noise

Now, let's introduce physical noise into the system. As I have discussed in Section 11.4, noise introduces *anti-instantons*, namely the time-reversed process from a critical point with a given index to one with *higher* index (Fig. 12.17).

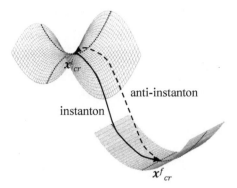

This process is *gapped*, in the sense that it is exponentially suppressed (see Fig. 11.5). Nevertheless, if an anti-instanton occurs right after an intanton for the *same* pair of critical points, we can conclude that, for all practical purposes, the critical point with *lower* index has not been reached (even if the anti-instanton trajectory is different from the instanton trajectory). In other words, it is an 'absorbing' critical point ('vertex').

Numerical noise

As I have discussed in Section 12.7.3, also the numerical integration (in time steps of size Δt) of the continuous-time dynamics of DMMs introduces some type of 'noise' (truncation error). Therefore, also in that case, if we consider an *ensemble* of trajectories,[52] there is a non-zero probability for anti-instantons to 'suppress' some instantonic jumps.[53]

In the language of directed percolation there is then a *finite* probability, p, for the state vector to 'percolate' from one critical point to another in the phase space, and a *finite* probability $1 - p$ for some critical points to be 'absorbing'.

In fact, we expect that the larger Δt, namely the less accurate the representation of the actual continuous dynamics is, the more numerical errors are introduced, hence the less probability for the state vector to 'percolate' and reach the equilibrium/solution of the given problem.[54]

In the opposite limit, when $\Delta t \to 0$, we then expect p to increase. Therefore, there must be some 'critical' Δt_c, at which one or a few paths are available to the state vector to reach the solution.

I have indeed discussed this in Section 8.6.1, where I have shown numerical evidence of a *solvable-unsolvable transition* as a function of Δt, with the latter scaling as a *power law* with problem size (see Fig. 8.4 in Chapter 8).

I can now interpret this transition in the language of directed percolation, by relating the *inverse* of the time step, $1/\Delta t$, to the percolation probability, p.

[52] For instance, by varying the initial conditions.

[53] Or the numerical noise is so large that the initial global truncation error does not allow the approximate trajectory to reach the next critical point, making the latter, for all practical purposes, *absorbing* (see discussion in Section 12.7.4).

[54] In the term 'numerical noise' we also take into account the fact that the integration scheme may introduce *additional* critical points (Section 12.7.3), which may further contribute to 'absorbing' the dynamics.

In *proximity* of Δt_c, we can then take as an *ansatz* of this relation the following expression:

$$p - p_c \propto \frac{f}{\Delta t}, \tag{12.18}$$

[55] In (Zhang and Di Ventra, 2021) this relation was chosen as $\delta \equiv Dp - e = a\left(\frac{1}{N\Delta t} - b\right)$, where a and b are parameters, and N is the number of variables of the combinatorial optimization problem, with $N \sim D$.

with f some function of D that has to be chosen appropriately.[55]

By using this relation, we can then relate the fraction of basin of attraction, A, I showed in Fig. 8.5 to the ratio of permeable paths in eqn (12.17), thus interpreting the numerical simulations I presented in Chapter 8 as a *bond directed percolation* of the state dynamics in phase space (cf. Fig. 12.16 and Fig. 8.5).

Robustness against numerical errors

Finally, we can use the correspondence between the dynamics of DMMs in the presence of noise and directed percolation, to qualitatively explain why the DMMs integrated numerically can still find the solution even in the presence of large numerical errors.

In DMMs, the dimensionality of the composite instanton, D, is usually very large. From the directed percolation analysis I have discussed previously we have also learned that the percolation threshold is $p_c \sim e/D$, see eqn (12.17).

In addition, the dimensionality D in directed percolation is proportional to the size of the problem a DMM attempts to solve, with this problem size characterized by the number of logical variables N: $D \sim N$.

Therefore, even if the majority of paths in the phase space are destroyed by truncation errors, the DMM can still reach equilibria (solutions to the problem) so long as the probability of an instantonic path is larger than $e/D \sim e/N$.

This result further reinforces the notion that, in simulations, one may *not* need a time step Δt to decrease exponentially with problem size, since DMMs operate at, or at the edge of, a *non-equilibrium ordered state* even in the presence of numerical noise.

Since numerical noise is in some sense 'worse' than *physical* noise (cf. discussion in Sections 12.7.2 and 12.7.3), we expect these conclusions to hold even in the presence of physical noise.

An important remark

The previous results do *not* imply that DMMs simulated numerically fall into the directed percolation *universality class* (see Section 4.3.2). In fact, it has been conjectured (Henkel *et al.*, 2008; Grassberger, 1982), that a given model belongs to such a universality class if:

(i) The model displays a continuous phase transition from a fluctuating active phase into a unique absorbing state.

(ii) The transition is characterized by a non-negative one-component *order parameter* (see margin note).

(iii) The dynamic rules are short-ranged.

(iv) The system has no special attributes such as unconventional symmetries, conservation laws, or *quenched* (frozen) randomness/disorder (see margin note).

DMMs solving a combinatorial optimization problem satisfy properties (iii) and (iv).[a] However, property (ii) is not so easy to check. In directed percolation, the order parameter is the average density of active sites (see Fig. 12.15). This would correspond to the average density of 'achievable' (permeable) critical points in a DMM. However, this quantity is not easy to access in a DMM due to the vast dimensionality of its phase space (cf. discussion in Section 12.4). As a consequence, property (i) is also not obvious.

[a]Here, the quenched randomness/disorder is not in the problem to solve, but in the dynamical equations of DMMs.

12.8 DMMs, non-equilibrium phase transitions, and the topological hypothesis

The discussion on the robustness of the solution search of DMMs to noise, and its analogy with directed percolation, somewhat completes the characterization of their dynamics occurring at an *ordered state*.[56] This happens *without* the need to tune any external parameter: the machines 'naturally' *self-tune* into such a state on their own, once the dynamics are initiated at some initial state (cf. Section 12.5).[57]

However, before summarizing this point, let me make a further interesting consideration. We can relate the previous results to what is known as *topological hypothesis* on the origin of *equilibrium* phase transitions (Pettini, 2007). This hypothesis posits that equilibrium phase transitions (continuous or not) are associated with *topological changes* in the *configuration space* of the system (cf. Section 11.5) as a function of some external parameter, e.g., temperature (Pettini, 2007).

In fact, various numerical results on model Hamiltonian systems support this hypothesis, further indicating that a phase transition corresponds to a 'combination of many simultaneous *elementary topological transitions*' (Pettini, 2007).

Order parameter
A system transitioning from one phase of matter to another may be characterized by an *order parameter* which quantifies the degree of 'order' of the different phases by varying non-trivially across the phase boundary (Goldenfeld, 1992). For instance, for the magnetic example of Fig. 4.3 in Chapter 4, it is the average spin polarization, which is zero in the (disordered) paramagnetic phase, and finite in the (ordered) ferromagnetic phase.

Quenched disorder
When the disorder of a physical system does not change in time, we say it is *quenched* (or 'frozen'). The spin glasses I introduced in Section 9.2.2 are an example of such a disorder (Henkel *et al.*, 2008).

Annealed disorder
When the stochastic parameters defining the disorder of a system vary in time in relation to its degrees of freedom, we say that the disorder is *annealed*.

[56]As if the machines were operating at the edge of some sort of (dynamical) *phase transition* (in a high-dimensional space), where the critical point separates a *quiescent* phase (in-between instantons) from an *active* phase (instantons).

[57]The ordered state is *attractive* for the dynamics.

Elementary topological transition
We call any change of topology associated with a *single* critical point of the flow an *elementary topological transition* (Pettini, 2007).

[58]Intuitively, if we visualize instantons as flow lines connecting different critical points (like in Fig. 12.17), an observer moving along those lines would see a change of 'shape' of the sub-manifolds of phase space whenever crossing a critical point. This change of shape occurs with the subtraction of *k-handles* (the handle of a cup is a 1-handle) to the sub-manifold across a critical point, with k some integer. This intuitive discussion can be made more rigorous with the help of *Morse theory*.

Morse theory
Morse theory refers to the body of statements in *Differential Topology* (the study of differentiable functions on differentiable manifolds) that establishes a connection between the topology of a manifold and the critical points of a real-valued smooth function on that manifold (a *Morse function*) (Milnor, 1963).

Composite instanton as a CW complex of the phase space
The *composite instanton* (Section 12.1) defines a finite (with dimension $\leq D$) *CW complex* of the phase space (C = 'closure-finite'; W = 'weak topology'). A CW complex is a topological space that can be constructed by inductively 'gluing together' elementary building blocks called *cells*. The critical points with index k define (via an isomorphism) k-cells of dimension k. The cells, from the equilibrium point (0-cell) to the first critical point of the dynamics, can be 'glued' together (intuitively 'via instantons') to form the CW complex. Since the complex is finite, it is also compact (Section 3.10). A DMM then 'discovers' the CW complex that leads to the solution. See, e.g., (Hatcher, 2002) for a precise definition of this concept.

In our DMM case, there is indeed an 'elementary topological transition' any time the system 'crosses' a critical point in between two consecutive instanton trajectories.[58] The collection of these elementary topological transitions finally leads the DMM to the equilibrium point, solution of the problem. Since DMMs are just one example of physical systems that showcase these properties (Henkel *et al.*, 2008), it is tempting to generalize the topological hypothesis to the non-equilibrium (and general) dynamical systems by stating that:

Topological hypothesis of non-equilibrium phase transitions
Non-equilibrium phase transitions originate from a collection of elementary topological transitions between appropriate sub-manifolds of the phase space.

However, whether this hypothesis is universally true or not requires further investigation, and I leave this for future work.

Irrespective, the dynamics of DMMs seem to have some features of what we call *criticality* (however, see Section 12.9), with the extra, important property that they *spontaneously* self-tune into their ordered state, without the need of adjusting external parameters. Let me then summarize these features here:

Long-range ordered state of DMMs:

(i) DMMs go through a succession of elementary instantons: a *composite instanton* (Section 12.1), i.e., a *CW complex*.

(ii) They develop both *spatial* and *temporal long-range order* (Section 11.7).

(iii) Their correlation time and length *diverge* on instantons (in the thermodynamic limit, Section 11.7).

(iv) The size of the avalanches (number of voltages) involved in the instantonic trajectories follows a *critical branching process*, with *diverging* average (again, in the thermodynamic limit, Section 12.5).

12.9   Effective breakdown of topological supersymmetry on instantons and the nature of the ordered state

For the reader familiar with supersymmetric TFT (Section 11.3), I can also argue that the dynamical state of DMMs is a sort of (*supersymmetry-protected*) *topological ordered state* as follows.

For concreteness, let us consider a search problem (Section 2.3.4), e.g., the 3-SAT problem with at least one solution, and the DMM designed to solve it, as described by the eqns (8.2), (8.3), and (8.4) in Section 8.2.5.

I have discussed in Section 7.2.1 that the *discrete* (logic) variables of the original combinatorial optimization problem are first transformed into *continuous* variables (voltages in a circuit realization of DMMs); see Fig. 7.7 in Section 7.2.1, and Fig. 12.18.

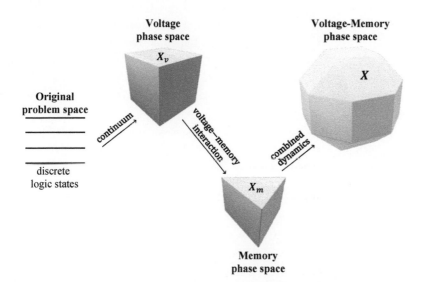

Fig. 12.18 The original *discrete space* of Boolean variables of a combinatorial optimization problem is first transformed into a *continuous phase space* X_v of variables ('voltages' in a circuit realization of DMMs). The coupling (interaction) of these variables with additional *memory* variables, spanning their own phase space X_m, brings about the *combined* phase space, X, on which the DMM solves the problem.

The phase space, X_v, of the continuous (voltage) variables has several local minima, in addition to the equilibria which are the solutions of the original problem. Without adding memory, the system would then, most likely, get stuck in some local minimum, not the solution of the problem.

(This can be easily seen if, e.g., in eqns (8.2) we set the short- and long-memory variables equal to 1. The rigidity term disappears and only the gradient remains. These gradient dynamics quickly settle into a local minimum; see Fig. 12.8.)

In this reduced phase space I can apply the supersymmetric TFT I have outlined in Section 11.3. This means that there is an exterior derivative, $\hat{d}_v$, associated with the exterior algebra of X_v: $\Omega(X_v, \mathbb{C})$.

Let us now consider the dynamics of the memory variables only. For instance, we can set the clause function in eqns (8.3) and (8.4) to some constant. The memory variables evolve in another phase

Excitations and Goldstone modes

In (Quantum) Field Theory, elementary particles can be viewed as *excitations* of the (global) ground state (*vacuum*) of the system (Schwarz, 1993). When a symmetry is broken, the system transitions into a new ground state (cf. Section 4.3.3). The broken-symmetry *spectrum* (see Fig. 10.11) is then different from the spectrum of the unbroken-symmetry phase. It is thus not surprising that new particles/modes can emerge due to symmetry breaking, such as the *Goldstone modes* (cf. Section 3.9). In the case of DMMs, *goldstinos* (fermionic zero modes associated with the instanton moduli; Section 12.5.2), are 'released' during dynamics, even if the supersymmetry is not spontaneously broken globally. This is because the instanton 'location' in time breaks 'spontaneously' time translation in Euclidean space, thus giving rise to a zero mode.

space, X_m (Fig. 12.18), and the exterior derivative on the exterior algebra of X_m, $\Omega(X_m, \mathbb{C})$, would be some $\hat{d}_m$.

When we switch on the interaction between the voltages and the memory variables at the initial time, their *strongly coupled dynamics* explore the combined phase space, X (Fig. 12.18). The associated Hilbert space is the tensor product of the Hilbert spaces of the two subsystems, $\Omega(X, \mathbb{C}) = \Omega(X_v, \mathbb{C}) \otimes \Omega(X_m, \mathbb{C})$, and the exterior derivative is $\hat{d} = \hat{d}_v + \hat{d}_m$, as defined in (11.18).

As I have discussed in Section 12.1, as soon as we switch on the interaction between voltage and memory variables, the low-energy dynamics quickly settle into a critical point (*perturbative vacuum*), call it $|i\rangle$, from which it will proceed, via an instantonic jump, to another perturbative vacuum, $|f\rangle$, which is more stable than $|i\rangle$.

However, 'more stable' means it has less unstable directions, or in the language of supersymmetric TFT, it has more *fermions*. Suppose $|f\rangle$ has one more fermion than $|i\rangle$, namely the indexes of these two critical points differ by 1.

(An instanton that connects two critical points differing by two or more unstable directions can be thought of as a succession of instantons between critical points whose indexes differ by 1.)

From eqn (11.18), we see that the operator $\hat{d}$ has a fermion (ghost) number of 1, which compensates for the missing fermion of $|i\rangle$. This means that, following *Berezin rules* of integration over Grassmann variables (see Section 11.9), the matrix element of this operator between $|i\rangle$ and $|f\rangle$ is *not* zero (see Fig. 12.19):

$$\langle f|\hat{d}|i\rangle_{\text{instanton}} \neq 0, \quad \textit{TS breaking on instantons.} \tag{12.19}$$

Fig. 12.19 On instantons the topological supersymmetry (TS) is *effectively broken*, namely the matrix element of the Noether charge $\hat{d}$ on the instanton is not zero, while it still commutes with the evolution operator $\hat{H}$: $[\hat{H}, \hat{d}] = 0$. This effective breakdown of TS on instantons is accompanied by the 'release' of *goldstinos*: fermionic zero modes expressing the long-range order of the dynamics. However, this effective breakdown of TS does *not* correspond to the *spontaneous* breakdown of the *global* vacuum. Hence, *no chaos* develops.

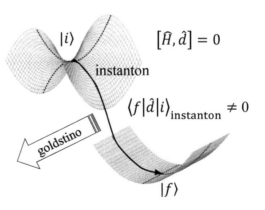

However, $\hat{d}$ is still a symmetry of the evolution operator, eqn (11.13), via eqn (11.19). This means that the *low-energy dynamics* (namely, the instantonic dynamics) do *not* share the same symmetry as the evolution operator. Following the discussion in Section 4.3.3, we can then say that the topological supersymmetry (TS) is *effectively broken on instantons* (in an 'instanton background').

Unbroken global TS symmetry and boundary effects

Note, though, that the TS is still *unbroken* globally. This means that the *true vacuum* or ground state of the Hilbert space (the exterior algebra of the phase space X), $|\psi_{GS}\rangle$ (which is *nontrivial* in the *de Rahm cohomology*, namely it is *closed* but not *exact*), is still *annihilated* by the Noether charge, i.e., it is $\hat{d}$-symmetric: $\hat{d}|\psi_{GS}\rangle = 0$. As a consequence, *no chaos* can develop.

As I have discussed in Section 11.9 (see, e.g., eqn. (11.33)), the exterior derivative, $\hat{d}$, is the operator version in *cohomology* of the boundary operator, $\hat{\partial}$, in *homology* (Fomenko, 1994). This means that the *global unstable manifolds* of DMMs have *no boundary*.

Instantons, instead, are *manifolds with boundary*. (The dimension of these manifolds is equal to the difference between the indexes of the initial and final critical points of the instanton; Section 11.4.)

This is why *any $\hat{d}$-exact operator* (of the type $[\hat{d}, \hat{O}]$, with $\hat{O}$ some physical operator) has non-zero matrix elements on them, but they do *not* showcase true chaos: the only contribution of the matrix element on instantons of a $\hat{d}$-exact operator comes from the *boundary* of the instanton manifold, *not* from its 'locally chaotic bulk' (see, e.g., eqns (11.37), (11.39), and (11.40)).

In other words, instantons are *not chaotic dynamics*! (In fact, in the presence of chaos—non-integrable dynamics—there are no distinguishable instantons, even without noise.)

Symmetry restoration by instantons and absence of chaos

I can make the previous point even more compelling by arguing that the presence of instantons is not conducive to chaos: it *prevents* chaos from happening (in the absence of condensation with anti-instantons; cf. Section 11.4.1).

This is because instantons may *restore* symmetries. In fact, restoration of (super-)symmetries due to instantons appears in various types of (Quantum) Field Theories (Polyakov, 1977).

We can intuitively understand this phenomenon by first analyzing a quantum-mechanical analogue. Consider then a single particle (for simplicity, without spin) in a *symmetric* double quantum

Bounded manifold vs. boundary of a manifold
Recall from Section 3.10 that a D-manifold, X, is *bounded* if all its points are within a fixed distance from each other. Instead, the *boundary* of X (∂X), is a $(D-1)$-manifold which is the *complement* (it does not include the elements) of the *interior* of X (Int$X$: the D-manifold with points in X having neighborhoods homeomorphic to an open subset of $\mathbb{R}^n$) (Fomenko, 1994).

Closed and exact differential forms
A differential form of degree k, ϕ_k (Section 11.3), is *closed* if $\hat{d}\phi_k = 0$, with $\hat{d}$ the exterior derivative. The form ϕ_k is *exact* if there exists a $(k-1)$-form, θ_{k-1}, such that $\phi_k = \hat{d}\theta_{k-1}$. Since $\hat{d}$ is *nilpotent* ($\hat{d}^2\alpha = 0$ for *any* form α), an exact form is also closed. The reverse is not necessarily true, and *nontrivial* forms, namely those that are closed but not exact, express the *nontrivial topology* of the manifold. This is because a differentiable manifold is *locally* homeomorphic to Euclidean space (Section 3.2). In Euclidean space *all* closed forms are exact. Therefore, any closed form on a manifold is *locally* exact. It is then the nontrivial topology of the manifold that renders a closed form not exact *globally* (Göckeler and Schücker, 1989).

Logical defects revisited
Since the topological supersymmetry is (effectively) broken in different parts of the DMM phase space differently, the dynamics generate *logical defects* (cf. Section 12.2). The phenomenon of generating (*topological*) defects, namely violations of local equilibrium in condensed media, happens also for other types of broken symmetries (Schwarz, 1993).

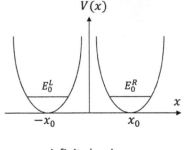

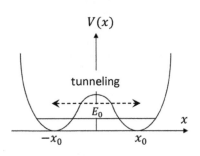

Fig. 12.20 Ground state of a single (spinless) particle in a symmetric double quantum-well potential $V(x) = V(-x)$. **Top panel:** the two quantum wells are separated by an infinite barrier. Each quantum well has a ground state, one on the left of the origin (with energy E_0^L) and one on the right (with energy $E_0^R = E_0^L$). The particle is either in the left quantum well or the other: the ground state does not possess the parity symmetry ($x \to -x$) of the potential. The original symmetry is then *broken* in the ground state. **Bottom panel:** the two quantum wells are separated by a finite barrier. Tunneling is then possible and the (unique) ground state (with energy E_0) is a linear combination of the ground states of the isolated wells. The new ground state has now the same parity symmetry as the potential: tunneling has *restored* the symmetry.

well potential $V(x)$, with x the one-dimensional spatial dimension (Fig. 12.20). The Hamiltonian of such a system is ($\mathbb{Z}_2$) invariant under *parity transformation* $\mathcal{P} : x \to -x$ (because $V(x) = V(-x)$).

If the barrier between the two quantum wells were *infinite*, they would be completely decoupled. This means that the state closest to the bottom of *each* quantum well is a ground state of the particle (these two ground states are orthogonal to each other). Those two vacua would be degenerate for the full Hamiltonian, and (depending on the initial conditions) the particle can be either in one or the other vacuum, but not both: the $\mathbb{Z}_2$ symmetry is broken in the ground state, similar to the spontaneously broken symmetry of the two-dimensional ferromagnet I discussed in Section 4.3.3.

However, if we have a *finite* barrier between the two quantum wells, tunneling occurs (Fig. 12.20). Tunneling couples these two vacua so that the *global* ground state of the system is now a linear combination of the two isolated vacua. As a consequence, this ground state has *again* the parity symmetry $x \to -x$. In a word, tunneling has *restored* (*non-perturbatively*) the original symmetry.

Let us see how this relates to DMMs. I have discussed in Section 11.9 that the local supersymmetric vacua (which are orthogonal to each other; see eqn (11.35)), are not annihilated by the TS charge, $\hat{d}$; see eqn (11.33).

If instantons were not present, this would suggest a spontaneous breakdown of TS. On the other hand, instantons are the Euclidean analogue of tunneling: they allow transitions ('tunneling') between the local supersymmetric vacua of the DMMs (cf. Section 11.6). Without them, there would be *no* transitions between the local supersymmetric vacua, like there are no transitions between the two vacua of a double quantum well when the barrier between the two wells is infinite.

Instead, as for the quantum case, in the presence of instantons, the new *global* supersymmetric state is a linear combination of these vacua, so that the *global* TS is *restored* (*non-perturbatively*) in the ground state. Since the global TS is restored, *no chaos* can develop in the system; see, e.g., (Ovchinnikov and Di Ventra, 2019) for a full discussion of the relation between chaos and (spontaneous) breakdown of TS. This is another way of demonstrating that DMMs are *integrable systems* (Section 3.10).

On the nature of the ordered state of DMMs

Now, since a broken (effectively or not) symmetry implies a more ordered/rigid phase (Section 4.3.3), we may conclude that the low-energy (instantonic) dynamics of DMMs correspond to the *low-symmetry, ordered phase* of the system.

And the long-range order of this phase is expressed by the instanton zero modes (Section 11.21), whose supersymmetric partners are *fermionic Nambu-Goldstone modes* or *goldstinos* (Section 12.5.2).

Therefore, every time a zero mode is 'released' ('liberated') during dynamics, the associated goldstino is liberated as well (Fig. 12.19). This is seemingly the nature of the *ordered state* in DMMs.

Supersymmetry-protected topological order?

Let me, however, make a further remark about this ordered state. As I have amply discussed previously, TS is *effectively* broken on instantons, but *not* globally. If we then consider the *ground state* (global vacuum) of DMMs, we would conclude that such a state is *not* the result of a true symmetry breaking phenomenon. A different classification of the ordered state would then be needed. A first attempt at such a classification is as follows.

Recall first that instantons are *topological excitations* (Section 11.4), namely they cannot be created or destroyed by *local* perturbations. In the presence of noise, anti-instantons may be created, but this process is exponentially suppressed, and no other 'excitation' can occur in the system.

The instantonic phase of DMMs is then *rigid* with respect to local perturbations. In addition, instantons showcase a sort of long-range order that is very much reminiscent of (long-range) quantum entanglement (Section 11.7). During the instantonic phase, different instantons (connecting different pairs of critical points) co-exist, but they are quite distinct: they cannot be continuously transformed one into another via some smooth deformation of the Lagrangian in eqn (11.15). The number of local supersymmetric vacua they connect is related to the topology of the phase space.

Since the global TS is not truly broken during the instantonic phase, as I have already mentioned, the *global* ground state (vacuum) of a DMM supports these (topological) excitations, and can be viewed as a *superposition* of the local supersymmetric vacua. This can be understood more rigorously by considering eqn (11.20), which shows that *any* state in the exterior algebra of the phase space (the Hilbert space) is a linear combination of states (differential forms) of *all* degrees.

As for the ground state, we know that each *de Rham cohomology class* provides one $\hat{d}$-symmetric state of the form $|\psi^k(\mathbf{x})\rangle =$

Topological order

The ground state of some gapped quantum many-body systems (cf. Section 11.4) is characterized by a *degeneracy* (number of states that share the same energy) which depends only on the topology of the space (the degeneracy cannot be lifted by *local* perturbations). This type of ground state cannot be classified according to the symmetry-breaking picture I have discussed in Section 4.3.3. We then say that such systems showcase *topological order* (Wen, 2013). The latter is not typically associated with any symmetry and is characterized by (long-range) quantum entanglement (Section 11.7).

de Rham cohomology classes

The kth *de Rham cohomology class* is the vector space of all k-forms (Section 11.3) whose exterior derivative, $\hat{d}$, is zero, modulo forms that are $\hat{d}$-exact, namely they are the exterior derivative of $(k-1)$-forms (Hatcher, 2002). Each (equivalence) class has exactly one *harmonic form*. These are differential forms, $|\phi^k(\mathbf{x})\rangle$, on which the (Hodge) Laplacian $\hat{\Delta} = [\hat{d}, \hat{d}^\dagger]$ (with $\hat{d}^\dagger$ the Hermitian conjugate (*codifferential*) of $\hat{d}$) evaluates to zero: $\hat{\Delta}|\phi^k(\mathbf{x})\rangle = 0$.

$|\phi^k(\mathbf{x})\rangle + \hat{d}|\cdot\rangle$ ($k = 0, \ldots, D$), where $|\phi^k(\mathbf{x})\rangle$ is a *harmonic differential form*, and we can add any $\hat{d}$-exact form $(\hat{d}|\cdot\rangle)$, since $\hat{d}$ is nilpotent (Ovchinnikov, 2016).

In other words, the global vacuum of DMMs is *degenerate*, and its degeneracy depends on the topology of the phase space (which determines the non-empty de Rham cohomology classes (Hatcher, 2002)). (And one ground state cannot be transformed into another by the Lie group of transformations $\hat{T}_g = e^{ig\hat{d}} = 1 + ig\hat{d}$.)

Apart from the underlying supersymmetry, all this points to an intriguing analogy with the *topological order* of the ground state of certain quantum many-body systems, which are good candidates for topological *quantum* computation (Freedman *et al.*, 2002). However, it is not yet fully clear how far this analogy can be pushed (e.g., if it holds in the thermodynamic limit).

Fermionic and bosonic entropy of the ordered state of DMMs

Finally, let me attempt to quantify the change of *entropy* (loss of information) in going from the initial instantonic jump till the system finds the equilibrium as follows.

We have two types of variables, bosonic and fermionic, which we expect to exchange information during the solution search. Let us then define the following *fermionic entropy* $S_\chi = \ln(i \sum_j \bar{\chi}_j \chi^j + 1)$, which quantifies the information present in a given critical point (the $+1$ is to avoid a logarithmic divergence in the absence of fermions). Starting from the initial critical point of the *composite instanton* at time t_0, this entropy is some $S_\chi(t_0) \leq \ln(D+1)$ (D is the dimension of the phase space). In fact, at the beginning of the composite instanton, we expect $S_\chi(t_0) \sim 0$, because the initial critical point is most likely far from the equilibrium point (if the latter exists).

The system then has little 'knowledge' of the equilibrium and of the (unstable) directions it has to take to reach the solution. This lack of knowledge can then be quantified with the dual *bosonic entropy* $S_x = \ln(D+1) - S_\chi$, so that at time t_0 the entropy of the bosonic sector is quite high: $S_x(t_0) \sim \ln D$.

However, as the dynamics progress, the system transitions to more stable critical points, with an increased loss of information in the fermionic sector, and a consequent increase of information in the bosonic sector. If the combinatorial optimization problem has a solution, its corresponding equilibrium point has D fermions (only stable directions, with at most center manifolds; cf. Section 3.7).

Entropy and loss of information

The concept of *entropy* is directly related to *loss of information* (Kubo, 1957). In fact, it quantifies how much (syntactic) information we have about a physical process. The more information we have of a physical process, the less its entropy (cf. Sections 1.3 and 5.3.2). During dynamics, a physical system interacting with a reservoir (an *open system*) can either increase or decrease its entropy by *transferring* information between its degrees of freedom and those of the reservoir. In the case of DMMs, we can think of the fermions χ^i as belonging to some 'fermionic reservoir' interacting with the bosonic variables x^i (memory variables and voltages in a circuit realization of DMMs). If the problem a DMM attempts to solve has a solution, then the final (fermionic) entropy would be maximal for such a problem $S_\chi(t \to \infty) = \ln(D+1)$ (D is the dimension of the phase space). If it doesn't, the DMM eventually ends up into a (quasi-)periodic orbit (cf. Section 7.9) with some lower (residual) entropy: $S_\chi(t \to \infty) = \bar{S}_\chi < \ln(D+1)$.

The final bosonic entropy is then $S_x(t \to \infty) = 0$, and the fermionic one is maximal: $S_\chi(t \to \infty) = \ln(D + 1)$. The bosonic entropy of the DMM *decreases* during the instantonic dynamics, at the expense of the fermionic one, leading to an 'increasingly ordered' state. We can then say that the DMM 'learns' the final/equilibrium state of the dynamics (solution of the original problem) at the expense of an increase in (fermionic) entropy (loss of information) of a coupled 'fermionic reservoir'.

12.10 Brain-like features of MemComputing machines

After an exhaustive discussion of the dynamics of DMMs, let me conclude this Chapter with some *analogies* (as well as *differences*) between the functioning of MemComputing machines (in general, not just the digital ones) and the biological brain. This discussion will allow us to come full circle and tie with the MemComputing criteria I have introduced in Section 1.6 of the Preamble of this book.

Also, recall from Section 5.1, that *artificial* neural networks form a specific *sub-class* of the more general concept of MemComputing machines (Traversa and Di Ventra, 2015). Although artificial neural networks are used as *models* of biological nervous systems, they are definitely *not* the same thing (Kohonen, 1989).

Arguably, it is not yet fully understood how the brain actually functions. However, some of its physical characteristics have been demonstrated experimentally, and these are the ones I am referring to here. In addition, let me stress again that what I will present here are only *analogies*. In fact, since the actual knowledge of the operation of the animal brain is far from complete, in some cases these analogies are, at face value, only skin-deep.

Therefore, although it is tempting to classify MemComputing machines as 'brain-like', at this stage I do not think this classification goes too far beyond the superficial level of a few general features they share in common with the brain. Nonetheless, this analogy is quite intriguing and worth mentioning.

Neurons

Neurons are electrically excitable cells that transmit information, in the form of electrochemical signals, to other cells. They are composed of a *cell body* (or *soma*), an *axon* (a nerve fiber that transmits electrical polarization signals—*action potentials*—from the soma), and *dendrites* (which receive signals from other nerve cells).

Synapses

Connections between neurons are called *synapses*. See, e.g., (Sherwood, 2015) and (Kohonen, 1989) for more information.

I. Massively parallel architecture with combined information processing and storage

This is the very first (desirable) criterion I have put forward in Section 1.6 for MemComputing machines. Indeed, in this book I have shown that these machines have this feature, and may employ it to solve hard problems efficiently.

The animal brain seems to share this feature as well, or, at least, its representation as an artificial neural network (Kohonen, 1989). And although it is not fully clear how the brain actually processes and stores information, it is safe to say that the collection of its *neurons* and *synapses* (the *connectome*), viewed as a biological neural network, can perform both tasks simultaneously.

However, in the case of MemComputing, I also added the word 'scalable' as a desirable property. 'Scalability' in the context of MemComputing machines relates to their ability to solve complex problems efficiently (with polynomial resources), as the size of the problem increases. *Digital* MemComputing machines have indeed this extra property.

On the other hand, in the case of the animal brain, 'scalability' typically refers to the relation between number of neurons (or volume of the brain, or some other measure of 'size') to the characteristics of the living creatures. This is oftentimes called *allometry* (Huxley and Tessier, 1936), and therefore does not directly compare with the scalability of MemComputing machines.

II. Asynchronous computation

I have anticipated this property in Section 3.6, and it follows from the previous one. It means that all those units that are involved in the processing of information compute and exchange information *simultaneously*, without the need to wait for a specific period of time (such as a global clock period).

This property is a main feature of the animal brain (and most models of artificial neural networks) (Kohonen, 1989), and it is also a property of MemComputing machines, as I have amply discussed in this book.

III. Analog vs. digital computation

Although not incompatible with asynchronous computation, the brain seems to be more akin to an *analog* (and adaptive) machine rather than a digital one (see definition in Section 2.8). MemComputing machines can be defined as both analog and digital (or mixed)—see Chapter 5—but only the digital ones offer the scalability advantages required to solve efficiently the problems I have discussed in this book.

Of course, there may well be applications of *analog* MemComputing machines in problems where a digital representation may not be so straightforward or even necessary. Examples of these MemComputing applications are the *shortest-path* problem (see Section 4.3.5) as in (Pershin and Di Ventra, 2013), the *maze* problem (Pershin and Di Ventra, 2011b), or the *ant-colony optimization* problem (Pershin and Di Ventra, 2016).

The advantages of *asynchronous* (massively parallel) computation I already discussed, hold also for these types of applications, thus providing substantial benefits compared to traditional approaches.

IV. Short- and long-term memories

The animal brain, the human brain in particular, showcases both *long-term* and *short-term* memory (Sherwood, 2015). Short-term memory is thought to be a 'working memory' in the sense that allows us to accomplish certain tasks that can be forgotten within a relatively short time. As far as we know, it is mainly located in the *prefrontal cortex* (see Fig. 12.21).

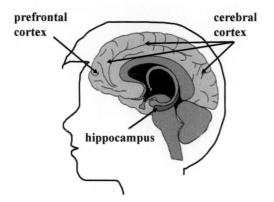

prefrontal cortex

cerebral cortex

hippocampus

Fig. 12.21 Sketch of the human brain. The *prefrontal cortex* makes up about one quarter of the whole *cerebral cortex*. Short-term memories are thought to be located in the prefrontal cortex. Long-term memories, instead, are created in the *hippocampus*, from which they are then distributed to the cerebral cortex (Sherwood, 2015).

In some sense, this is the type of memory (I also called 'short-term memory', see eqn (8.3) in Chapter 8) we have used all along in the practical realization of DMMs.

The long-term memory in the brain is instead responsible for allowing us to remember events far in the past.[59]

[59] And thus memorize skills learned.

The physical location in the brain where this memory is created—presumably from repeated short memories—seems to be the *hippocampus*, and from there, it is supposedly distributed to the *cerebral cortex* (Sherwood, 2015).

In the case of MemComputing machines, this is the type of memory I called 'long-term' in Chapter 8, eqn (8.4). In DMMs, this type of memory is indirectly 'reinforced' by the short one, and keeps track of the unsatisfiability of clauses between literals.

Finally, the relevant units in MemComputing, where the output of the computation is stored, are the ones where the output is read. For the *physical* DMMs I have introduced in Chapter 7, this can be easily accomplished when the system finds an equilibrium point of a search problem: all gates are in their satisfied configuration, and will persist in such a state, so long as the input voltages are on.

If the input voltages are switched off, the time it takes the output

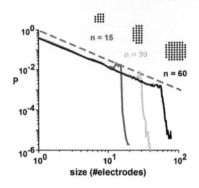

Fig. 12.22 Distribution of the size of neuronal avalanches in cortical networks deposited on an electrode grid with a number n of sensing electrodes. The distribution is a power law with an exponent close to $-3/2$ in the thermodynamic limit (dashed line). It also exhibits the typical 'bump' and decay due to finite-size scaling (cf. Fig. 12.11). Reprinted with permission from (Beggs and Plenz, 2003). Copyright 2003, *Society for Neuroscience*.

Criticality hypothesis
The results in Fig. 12.22, and many others that show similar trends (Muñoz, 2018), have been taken to support the *criticality hypothesis*, which states that the dynamics of the animal brain occur at the edge of a *continuous phase transition* separating different phases (Beggs and Plenz, 2003). However, whether this hypothesis is valid, or what these phases are, is still an open problem.

[60]Because scale-free distributions can emerge even *without* criticality (Touboul and Destexhe, 2017).

Neurogenesis
The process by which (the majority of) the animal brains generate new neurons is called *neurogenesis* (Kandel, 2013).

gates to return to their *thermodynamic* equilibrium will be dictated by the relaxation time of the system, which is again on the order of the RC time of the circuit (see Section 7.6). This time can be extremely short (even nanoseconds) for actual physical devices.

If the system does not have an equilibrium (fixed) point (e.g., the search problem has no solutions or it is of an optimization type), then the reading at the appropriate output terminals can be accomplished by measuring the voltages (or currents) at those terminals after a fixed amount of time dictated by the user.

V. Mechanisms of collective dynamics

An interesting physical property of the animal brain is the apparent *scale-free* behavior in the firing of neurons, even in the absence of external stimuli (Beggs and Plenz, 2003). This is an *emergent* property of the collection of neurons, and it has been demonstrated in several contexts, although much debate surrounds both its nature, and even its very existence (Muñoz, 2018).

For instance, in Fig. 12.22, I report an example taken from the literature (Beggs and Plenz, 2003), where cortical neurons are deposited on a grid of electrodes, and their firing is recorded.

The size of the number of neurons that fire collectively (the size of the *neuronal avalanches*) follows a power-law distribution, $S^{-\tau}$, with τ close to $3/2$.

This is similar to the critical Borel distribution I have derived in Section 12.5, using *mean-field theory*, for the size of the (voltage) avalanches in DMMs. In that Section, I also showed an explicit example of such a distribution when the machines solve random 3-SAT instances (see Fig. 12.11).

However, in the case of the brain, the reason for this particular power-law behavior of neuronal avalanches does not seem to be fully understood. In fact, it is not even clear if it is related to a critical state at all,[60] although several experiments, and various theoretical considerations point toward such a conclusion (Muñoz, 2018).

VI. Robustness against small imperfections and noise

Neurons in the brain both die off and get generated (*neurogenesis*) during the entire lifespan of the organism (Kandel, 2013). These mechanisms create changes in the architecture (physical topology) of the network of neurons.

Yet, the organism continues to function despite these changes (unless the damage to the brain is substantial). The brain is then *fault tolerant*.

It is not completely clear where this robustness originates from. It could be related to the fact that a network of elements with time non-locality (memory) is able to 're-route' information 'on the fly', even when parts of its architecture fail (e.g., the connectivity of some units is broken).

An example of this effect is demonstrated in (Pershin and Di Ventra, 2013), where this *self-organizing* mechanism in the presence of damage to the circuit, is explicitly shown in a network of resistive memories that solves the shortest-path problem (defined in Section 4.3.5).

In view of the fact that the animal brain seems to be robust against (relatively small) changes in the neuronal connectivity, it is not a stretch to think that it is also robust against small perturbations and noise (the latter, incidentally, being always present in physical systems).[61] Irrespective, taken all together, this type of robustness is *stronger* than the one of the DMMs I have discussed in this book.

In the latter case, a change of physical topology is *not* a minor perturbation: it would change *dramatically* the structure of the combinatorial optimization problem we are attempting to solve.[62]

Instead, as I have discussed in Section 12.7, DMMs are robust against *small* perturbations and noise that do not change the topological structure of the phase space. These considerations seem to suggest that the more specialized the physical topology (architecture) of the network is, the less robust it is to topological changes.

In fact, our brains are not specialized to solve specific problems. Rather, they can tackle an incredibly large variety of tasks, but are not particularly good at solving, e.g., the mathematical problems I have discussed in this book.[63]

VII. Information overhead

In Section 5.3.2, I have discussed a property, I called *information overhead*, that belongs to *all* MemComputing machines, not just the digital ones. In fact, such a property is *fundamental* to these machines.

It is related to the information we, the users, must provide about the *physical topology* of the network of memprocessors. As I have explained in Section 5.3.2, information overhead is a type of 'data compression' originating from the architecture of the network.

A similar concept has been discussed in the context of the brain. In fact, it seems that in brain networks, there is a high level of specific structures, in both the types of neurons and their network architecture, even at the microscopic level (Kohonen, 1989).

In other words, the brain physical topology is not completely random, as it is oftentimes assumed in artificial neural networks, and its structure may have important functional properties.

Fault tolerance
It is the ability of a system to accomplish a given task even if some of its elementary units fail. The brain has this property (at least to some degree of failure), while the DMMs designed for specific problems do not: if a self-organizing gate completely fails to work, the corresponding problem is *fundamentally* changed and cannot be solved any longer.

[61] This robustness would also pertain to *neurodynamics*, namely the firing of single neurons in the brain.

[62] For instance, even a single gate not working as it should, would result in the DMM not solving the original combinatorial optimization problem. Rather, the DMM would attempt to solve this 'newly defined' problem.

[63] It is also worth mentioning that experimental studies, employing neuro-imaging techniques, have shown structural differences in several brain regions in people with *autism spectrum disorder* (ASD) compared with individuals without ASD (van Rooij, 2018). This could also be the reason why children with ASD sometimes show superior mathematical skills compared to their non-autistic peers, while struggling in many other tasks (Iuculano *et al.*, 2014): their brain structure may be more topologically constrained.

VIII. Functional polymorphism

Recall from Section 5.3.3, that a universal MemComputing machine can, in principle, compute *different functions* without modifying the physical topology of the machine network, by simply applying the appropriate input signals. A practical example of this *functional polymorphism* is reported in (Traversa *et al.*, 2014). As I have also briefly mentioned in Section 5.3.3, another realization of such functional polymorphism in MemComputing may be accomplished with the aid of field programmable gate arrays (FPGAs), which offer user-specified variability in network topologies.

The biological brain seems to share this property to a very high degree: it can perform a wide range of tasks without changing the physical topology of its network, by simply responding to external stimuli (Kohonen, 1989).

With these analogies (and differences) between some characteristics of the brain and MemComputing I have now concluded the most 'technical' part of the book. In the next and final Chapter 13, I will offer some additional thoughts and ideas for future work. Of course, this is only a glimpse, guided by personal taste, of what can be explored in this field.

12.11   Chapter summary

— In this Chapter, I have shown that DMMs find the solution of a problem via a *composite instanton*, namely a succession of instantons/avalanches.

— The instantons allow the machine to rid itself of *logical defects*, similar to the elimination of *solitons/kinks* in non-linear media, to transition from one *perturbative vacuum* (ground state) to another.

— The number of instantonic steps, and the *physical* time to reach the solution (if it exists), scale at most *polynomially* with problem size.

— Instantons are the analogue of 'oracles' in non-deterministic Turing machines. However, unlike oracles, the instantons's 'choice' of correct path in phase space is dictated by well-defined physical laws of motion.

— The *distribution* of the *size* of the avalanches/instantons is *scale-free*, specifically a *critical Borel distribution*.

– This means that DMMs spontaneously *self-tune* to a *long-range ordered state*, and this state persists throughout the solution search. Namely, they showcase features similar to (although not necessarily the same as) *critical dynamics*.

– *Spontaneous self-tuning* means that no external parameters need to be varied for the machine to enter its *ordered state*. It will 'naturally' fall into it.

– Due to the topological nature of the solution search, DMMs are robust against noise and perturbations. This is the reason why even *simulations* of the ODEs of DMMs are *robust* against the unavoidable numerical errors.

– The solution search in the presence of noise can be modeled as a *directed percolation* of the state trajectory in phase space.

– MemComputing machines share some features that are observed also in *biological* neural networks.

Part V

Final thoughts

After the technical discussion of MemComputing machines, their applications and fundamental aspects, I am ready to conclude with some thoughts about future work.

I will offer suggestions regarding the applications of these machines, and directions of study to address more fundamental questions, which also affect the theory of computation in general.

These ideas are not exhaustive. In fact, I am confident that many more will arise over the years as new applications will be found, other physical concepts will be developed, and the design of these machines will improve.

(And I would not be surprised if a clearer exposition of this paradigm will emerge in the coming years; much better than the one I have attempted in this book.)

Therefore, I hope that this final Chapter will serve as a springboard for further developments and studies of this Physics-based MemComputing paradigm.

Epilogue and future work

13

He who sees things grow from the beginning will have the best view of them.
Aristotle (385−322 BC)

We are generally more effectually persuaded by reasons we have ourselves discovered than by those which have occurred to others.
Blaise Pascal (1623−1662)

Guessing before proving! Need I remind you that it is so that all important discoveries have been made?
Henri Poincaré (1854−1912)

13.1 Time non-locality as a computing tool

As I have amply discussed in this book, **MemComputing** is a portmanteau word that combines 'memory' and 'computing' together (Di Ventra and Pershin, 2013*b*). And by the word 'memory' I truly mean *time non-locality*, not necessarily 'storage'. In fact, I can now express this concept more clearly:

> **MemComputing employs the Physics of** *time non-locality* **to compute.**

This is a very important point! Although 'storage', sometimes called 'long-term memory' is a form of time non-locality,[1] the latter is truly a *dynamical* property of physical systems, not a static one. It refers to the fact that a physical system reacts to a given perturbation by 'keeping tabs' on how it has reacted at previous times.

As I have stressed in Sections 4.5 and 4.7, it is precisely this property that sets MemComputing apart from 'in-memory computing' (which is

[1] And, indeed, it is a necessary ingredient of *any* computing machine, including a MemComputing one (see criterion II of MemComputing in Section 1.6).

[2]And, of course, MemComputing is *fundamentally* different from Quantum Computing as well.

[3]It 'discovers' the *CW complex* (the topological manifold) of *critical points* and the *instantons* connecting them, required to find a solution (Chapters 11 and 12).

[4]We could call this *self-tuning* or *self-organization*.

[5]And if you recall the definition of Turing machines (Section 2.2), this 'history-dependent decision making on the fly' is *not* part of the Turing paradigm.

Scale-free correlations vs. criticality
The *critical state* of a continuous phase transition showcases *scale-free correlations* (onset of long-range order; cf. Section 4.3.2). The reverse is not true: scale-free correlations *do not* necessarily imply criticality.

Dynamical long-range ordered state
A state of a physical system out of thermodynamic equilibrium that showcases long-range order/correlations in its degrees of freedom (Section 4.3.2) may be called a *dynamical long-range ordered state*. It does not need to be the result of criticality of a continuous phase transition.

[6]In fact, recall from Section 4.3.4 that long-range order is *enough* to provide an advantage to the machine. Long-range order would correlate every degree of freedom with the whole system, even if the latter is not at criticality, namely at the critical point of a continuous phase transition.

[7]Unless we are incredibly lucky!

[8]Equivalently, we say that it *spontaneously* develops such order.

[9]Whose degrees of freedom are *strongly coupled*. For such systems perturbation theory does not work (cf. Section 2.7).

currently used in our modern computers), and makes it much more general than just 'computing with resistive memories'.[2]

Time non-locality is a physical property that allows a system to *adapt* to varying inputs, *learn* which dynamical paths are more advantageous, and, with this knowledge, take advantage of *feedback* to correct for mistaken 'decisions'. In simple terms, we may say that, due to time non-locality, a MemComputing machine *dynamically 'learns' the main (topological) structure* of the problem it is solving.[3]

All this, without 'external supervision', namely the system 'guides itself' toward the output.[4] This form of computation is what MemComputing is all about.[5]

13.2   MemComputing, long-range order, and all that

I have discussed in Section 2.6 that the main reason some combinatorial optimization problems are so difficult to solve with traditional algorithms is because they require *non-local* (*collective/simultaneous*) information about the truth value of several literals appearing in different clauses.

In fact, it is not a trivial task to generate such a non-local information with algorithms (see, e.g., Section 9.2.1 and (Pei and Di Ventra, 2021)). This is because non-local information on the truth value of literals requires *non-perturbative* changes of the variable assignments during the execution of an algorithm, namely changes that involve a large number of variables in the problem specification. Such non-perturbative changes are not so easy to guess from one iteration to the next.

This is the reason why *computing at criticality* (Section 4.3.4) is such an appealing alternative: it automatically puts the computing machine at a 'sweet spot', where it can take advantage of *long-range order*, which 'naturally' carries non-local information on the system.[6]

However, since the machine starts from an arbitrary state, which is *not* the solution of the problem to be solved,[7] this ordered state has to *emerge* during dynamics, namely the machine has to *self-tune* (has to be *attracted*) to a *dynamical long-range ordered state*, without supervision.[8]

As I have mentioned in Section 1.5, emergent properties are the hallmark of *complex systems*.[9] MemComputing machines are then *designed complex systems* (more on this in Section 13.3).

The particular dynamical ordered state of digital MemComputing machines (DMMs) is a *composite instanton* (Chapters 11 and 12), namely a succession of instantons (or *avalanches*) in the phase space. Instantons are *non-perturbative* phenomena: perturbation theory is inadequate to describe them. (In fact, they are the Euclidean analogue of quantum tunneling; Section 11.6.)

Instantons are also the ultimate *low-dimensional manifolds* of the phase space. Therefore, DMMs *break ergodicity*: they do not explore the whole phase space, just a tiny fraction of it. The dynamics of these machines are then highly *constrained*. In fact, they are *integrable* systems: they are *not* chaotic (cf. Section 3.10).[10]

Since DMMs can 'cut through' the (otherwise vast) phase space to reach the solution (equilibrium point), rather than wandering around it in a chaotic way, they can solve difficult problems efficiently. And if designed appropriately (see below), once the dynamics are initiated, they will be *driven* (directed) toward the equilibria (if they exist).

They have no other choice! Like the ball in the pinball machine of Example 1.3, or the water flowing down a hill in the Example 1.4.

Finally, instantons (and the critical points they connect) are of *topological* character. This means that the computation by DMMs is robust against noise (both physical and numerical) and perturbations; Section 12.7.

In Fig. 13.1, I summarize the main physical characteristics of the DMMs I have discussed in this book.

[10] The DMM's phase space is *foliated* with the instanton manifolds.

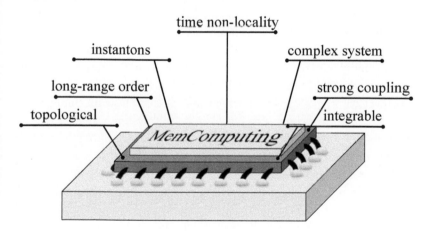

time non-locality

instantons

long-range order

complex system

topological

strong coupling

integrable

MemComputing

Fig. 13.1 Main characteristics of the physical systems representing the *digital* MemComputing machines I have discussed in this book.

Of course, some of these characteristics are also shared by other types of MemComputing machines, such as *analog* ones, namely those for which the input or output (or both) cannot be represented with finite means.[11]

For instance, time non-locality and strong coupling between degrees of freedom (hence *collective dynamics/intrinsic parallelism*) are features inherent to the MemComputing paradigm itself (cf. discussion in Chapter 5, and see (Traversa *et al.*, 2015) for an experimental demonstration of collective dynamics in *analog* MemComputing).

However, other features, such as *integrability* and *dynamical long-range order* are not necessarily universal in this paradigm of computation. The lack of some of these features may have substantial con-

[11] See Section 2.8 for the difference between digital and analog machines.

sequences on the *scalability* of these machines. This means that those lacking these features may not be ideal for the solution of the combinatorial optimization problems I have discussed in this book (Chapters 8, 9, and 10).

Nonetheless, as future work, it would be very interesting to study the *analog* version of these machines, and see what type of dynamical features they share with the digital ones. This research program is not just academic, since analog MemComputing machines can also find application in several important problems other than those I have discussed in this book (see Section 13.5 below).

13.3 How *not* to design a kludge

In addition to explaining the rationale behind MemComputing machines, and introducing their formal definition, I have suggested how to actually fabricate their *digital* version in *hardware* and simulate them in *software*.

The latter is definitely not a minor point. *Unlike* quantum computers, which *cannot* be simulated efficiently in software (hence, they need to be built in hardware to show any advantage), MemComputing machines can be realized as *non-quantum* dynamical systems.

And the ordinary differential equations (ODEs) of non-quantum systems *can* be efficiently simulated in software, running on our modern computers (see Chapters 8, 9, and 10).[12]

Irrespective, I have tried to convey the idea that these machines follow *well-defined design principles*. They are definitely *not* a kludge!

It is thanks to these design principles that DMMs support the *emergent phenomena* I have discussed in Chapters 11 and 12, such as *long-range order* and *critical branching*. As I have already mentioned, these phenomena are indeed key to their operation and ability to solve hard combinatorial optimization problems efficiently. A random assembly of the underlining components of these machines would not do it.

13.3.1 The hardware design

As for the hardware realization of DMMs, I have introduced the new concept of *self-organizing gates* (Traversa and Di Ventra, 2017; Di Ventra and Traversa, 2018*b*). This is a novel class of gates (whether Boolean or algebraic) that are *agnostic* to whether the logical signal is received by the traditional input or the traditional output (Chapter 7).

In other words, these gates have a *bi-directional logical topology*. As I have shown in Chapters 6 and 7, it is precisely because of the extra (memory) degrees of freedom that we can realize such gates.

These extra degrees of freedom transform the original problem into a dynamical system whose phase space contains only saddle points and

---

A kludge
A machine or a program that has been hastily improvised and poorly thought-out for the solution of a 'bug' (a problem) is sometimes called a *kludge*. In Computer Science the name is attributed to (Granholm, 1962).

---

[12]The efficient simulation of the ODEs of *analog* MemComputing machines is also possible, provided *integrability* of the flow vector field in eqns (3.1) is preserved. If not, chaotic dynamics would render the numerical simulations unstable and difficult to control.

the equilibrium points representing the problem solution.

Strength in variety

As I have already stressed in Chapter 7 the *design* of such gates is *not unique*. One could come up with *several ways* in which these gates are assembled, and use a variety of materials and devices, so long as the machine's properties are satisfied. This is truly a strength of this computing paradigm.

In fact, despite the focus of this book, electrical circuits may not be the only physical realizations of such gates. For instance, we could envision the use of *optical* devices to emulate such features, or even a combination of electrical and optical systems.[13] Of course, the electrical circuit realization—especially one employing only complementary metal-oxide-semiconductor (CMOS) (Section 3.3)—seems the most direct and easy route to follow. This is because CMOS is a well-established technology that can deliver sophisticated circuits with a high level of integration.[14]

For *industrial* applications, combinatorial optimization problems in the hundreds of thousands, or even millions of variables, are the norm, not the exception. This means that the self-organizing circuits of the DMMs solving those problems would require a comparable number of gates.

In turn, each one of these gates contains a certain number of circuit elements. Therefore, all in all, for industrial applications, the elementary devices (e.g., transistors) that would make up a DMM designed for such problems could easily number in the hundreds of millions, or even billions.

Beyond CMOS?

I do not see how, at the moment, this level of integration could be accomplished without CMOS, or, at least, with some components fabricated with this technology.[15]

Nonetheless, as materials and devices improve, and new technologies are developed, one could imagine that new ways in which these gates are built and integrated may emerge. A possible example could be the use of *spintronic* devices as suggested in (Finocchio *et al.*, 2020; Gypens *et al.*, 2021). Some of them are even compatible with CMOS, which would facilitate very large scale integration (Finocchio *et al.*, 2020).

A guiding principle

In fact, a strong *guiding principle* for the use of other types of devices and materials in building the gates and architectures of DMMs is that they have to be easy to *fabricate*[16] and *integrate*, so that one can scale

---

[13]Or even other, completely different, types of devices.

[14]Namely, it allows fitting a very large number of devices, e.g., transistors, in a single chip.

[15]Of course, as an aid to make DMMs's architectures.

Sprintronics
Electronics that takes advantage of the *spin* degree of freedom of electrons, in addition to their charge, is called *spintronics*, or 'spin electronics' (Bandyopadhyay and Cahay, 2008).

[16]Which means with low variability and high throughput.

their number, at least as easily as we do with our modern computers.

However, this guiding principle (of high-integration density) should not sacrifice *energy consumption*. In other words, the increased demand of cramming more gates into a single chip of a DMM should not come at an increased energy cost. This is where more research on materials and devices would be of great benefit.

General purpose DMMs

As I have shown in this book, given a combinatorial optimization problem, we can design a specific (although not necessarily unique) DMM to tackle such a task. And it is true that any NP problem can be mapped into a satisfiability (SAT) problem (Arora and Boaz, 2009). In fact, any combinatorial optimization problem can be expressed as a conjunctive normal form formula, with appropriate k-terminal OR gates (Section 2.4).

For each one of these problems, a DMM chip could be built, namely one would build *specialized* hardware for different applications. This seems to be a general trend in the computing industry anyway, so it would not represent an unusual direction for MemComputing as well.

However, it may be advantageous in some cases to transition from one problem formulation to another without relying on different chips. Namely, it would be ideal to have a single chip that can be *configured* by the user to solve a wide variety of problems, by simply changing the appropriate input signals (Di Ventra and Traversa, 2018b).

As I have mentioned in Section 5.3.3, such *functional polymorphism* can be implemented in hardware with field-programmable gate arrays (FPGAs), which allow the user to program the interconnects between gates.[17] This type of flexibility could come in handy for a variety of applications ranging from telecommunications to robotics to wearable devices.

13.3.2 The software design

Possible improvements in the software design of DMMs are many, and can proceed in parallel with the hardware design. In fact, the two do not need to be fully tied.

Software does not need to follow hardware

The reason is because the software simulation of the ODEs of DMMs brings with it a whole different set of (numerical) issues (see Chapter 8). These issues require different strategies than those employed in the hardware design. In fact, the hardware implementation of DMMs would solve problem instances on time scales dictated by the physical

---

Robotics
It is that field of science and technology that aims at developing machines (*robots*) that emulate and/or assist human actions. Many industrial plants employ plenty of robots.

Wearable electronics
Electronic devices that are worn by people (like clothing) to assist them in a variety of tasks are called *wearable electronics*. 'Smart watches' are prototypical examples.

[17]A practical example of functional polymorphism for DMMs has been proposed in (Traversa *et al.*, 2014).

devices employed. Given a problem size, these time scales may vary among instances, even by orders of magnitude (e.g., from microseconds to milliseconds or minutes). And yet the machine would still be useful for such a task.

In the numerical solution of the same instances, such a large variation of time scales is *not* desirable: it could easily make the difference between solving a problem instance in a few hours, versus several days, or even months![18] And note that this is not simply related to the *power-law scalability* of the numerical procedure.[19] Even if two numerical methods do not differ in terms of how they scale as a function of problem size, they can differ enormously in terms of actual CPU (or wall) *time* to reach a solution.[20]

Design strategies

So, which strategies should we use to speed up the numerical calculations?

For one, although I have provided a few examples of DMMs's equations of motion in Chapters 8, 9, and 10 for different types of combinatorial optimization problems, their *functional form* could still be improved. Of course, changes to these equations should still satisfy the mathematical properties I have discussed in Section 7.8. In particular, absence of chaos and, for search problems with solutions, absence of (quasi-)periodic orbits.[21]

In addition, the ODEs of DMMs have a few parameters that control the rate at which voltages and memory variables evolve in time. Those parameters have been kept fixed in the simulations I have shown in this book. But there is no reason for that to be the case.

Ideally, we would like those parameters to be *optimized* for *each* instance of the problem, so that considerable savings in simulation (whether CPU or wall) time may be achieved. Remember that the rates at which the dynamical variables change, determine how attractive and/or repulsive the manifolds at the critical points are (see, e.g., Section 11.4).

However, doing optimization of parameters for each instance, seems to defy the purpose. Instead, research should be done on an *a priori* relation between these rates and the instances of a given problem. If such a relation is found, the *time to solution* (TTS)—see Section 2.2.2—could be shortened considerably. How these parameters correlate to the various instances of a given problem, and among different problems, is then an open research direction worth pursuing.

13.3.3 Meta-MemComputing

We could even design a MemComputing machine as a 'probe', by solving an optimization problem where these parameters are the variables.

[18]Of course, if we put it into perspective, a few months is still better than the estimated age of Universe ($> 10^{10}$ years). Nevertheless, we should question the *usefulness* of finding solutions of a problem in months if done in software, when it can be done, say, in hours or even minutes in hardware.

[19]Although, the *coefficient* of the power law can make a large difference as well.

[20]This is because the *pre-factor* of the power law may be substantially different between two approaches with same scalability.

[21]Or, at the very least, if chaos and/or (quasi-)periodic orbits cannot be avoided, they should satisfy condition (iii) in Section 6.3.2.

Meta-learning

The word *meta* derives from a Greek word meaning 'transcend' or 'rise above'. When attached to 'learning', it typically means learning the learning process itself. In simple words, *meta-learning* means *learning to learn*. In the context of MemComputing, I use it to indicate a DMM that learns the features (parameters) of another DMM.

In view of what I have discussed in Chapter 9, this optimization problem could be efficiently tackled by a DMM. The best parameters found (*learned*) by this machine would then be used by the one that actually tries to solve the problem. In analogy with what I discussed in Chapter 10, I would call this a *MemComputing-assisted DMM* or, better, *meta-MemComputing*.

In the same vein, the search for an initial condition in the phase space that reduces the actual CPU (or wall) time of the computation could be again done by a 'probing DMM', whose results would aid the DMM that actually solves the problem. Finally, it may well be that better parameters, and/or initial conditions, may also affect the *power-law coefficient* in the simulations, not just its *pre-factor*.

Memory-informed restarts

Along similar lines, a better integration scheme (e.g., a better adaptive time-step procedure), may substantially improve the TTS (Bearden *et al.*, 2020).

In fact, from a numerical standpoint, the dynamics could be stopped after a set time, and *restarted*. Restarts were used, for instance, in the Machine Learning applications I have discussed in Chapter 10. This is because, for those applications, one needs to *sample* the low-energy state of the NNs, hence restarts allow collection of large statistics.

However, in the majority of cases I have shown in Chapter 10, the restarts have been performed by resetting the voltages and memory variables to random values (Manukian, Traversa and Di Ventra, 2019).

Instead, we could come up with several schemes that would initialize the voltages and the memory variables with specific rules learned in the previous run(s), rather than randomly. After all, the long-term memory in, e.g., eqn (8.4), collects a large amount of information on the previous trajectory of the system, and the state of satisfaction of the problem instance.

A clever *memory-informed* scheme that employs this information to restart at some time intervals could then be used to effectively reduce the TTS.

Writing the simulation code

Last, but not least, it is worth mentioning that most of the simulations I have presented in this book have been done with interpreted MATLAB.

It goes without saying that, while MATLAB is an incredibly useful tool for development purposes, it is definitely not ideal for optimizing a numerical code. Other programming languages should be used for that purpose. In fine, since we are integrating numerically differential equations, *parallelization* of a code to simulate them is doable.[22] This

---

[22] Unlike some of the algorithmic methods for combinatorial optimization problems that *cannot* be parallelized, apart from trivially running many instances on different processors.

could accelerate significantly the numerical performance, and push the simulations I have discussed in this book much further.

13.4   Real-time vs. offline computation

So, which one is better, hardware or software? That depends on the type of application.

If we need to solve some problem, whose solution is *not* needed in real time, a software simulation of DMMs should suffice, and the computation can be done *offline*. Examples of these applications abound.

For instance, checking the quality of a new computer operating system or a chip (which are difficult optimization problems), or finding the ground state of quantum Hamiltonians (Section 10.7), do not need to be done in real time.

In fact, in these cases, choosing the software route may be even more desirable. This is because software simulations can be tailored to specific needs much more easily than calculations done with specialized hardware. As a consequence, they are more easily deployable.

If, instead, the result of the computation needs to be used in a relatively short time,[23] then the hardware implementation becomes essential. The simulations of DMMs can still be used for testing, but they cannot replace the hardware for *real-time computing*.

> **Autonomous vehicles**
> A motorized vehicle (e.g., a car) that can move without much human intervention and supervision is called an *autonomous vehicle*. It requires continuous *sensing* of the vehicle's environment, and corresponding *adaptation*. Although partially realized with the second wave of Artificial Intelligence (AI), it is thought it can reach its full potential only with the third wave of AI (see Section 2.5).

[23]For instance, in the field of robotics or *autonomous vehicles*, the unsupervised training of NNs (Chapter 10) may need to be done 'on the fly', with varying input data.

13.5   More applications

I devoted the majority of this book to the *digital* version of MemComputing machines (DMMs). This is because they are the ones that are more easily scalable, in view of the fact that they map *finite* strings of symbols into *finite* strings of symbols (Chapter 6).

Their main application is then in the solution of combinatorial optimization problems (see definition of these problems in Section 2.5). Indeed, I have shown a few examples of how to use DMMs for this class of problems in Chapters 8, 9, and 10.

However, this is just a very small sample of instances that can be tackled with DMMs. Essentially, the majority of tasks in *logistics* (from transportation to energy delivery to packaging), encryption, data analytics, etc., belong to the class of problems for which DMMs can be designed.

For instance, in the case of encryption, it would be interesting to see how far the *simulations* of DMMs can be pushed to factor large numbers into primes (Section 9.1.2). It may turn out that some specific representations of the problem are more advantageous for this task than others (see also Section 13.8.4).

> **Logistics**
> The organization and execution of complex operations involving people, supplies, goods, etc., is called *logistics*. The majority of problems arising in this field are of the combinatorial optimization type, as those I have defined in Section 2.5.

Hashing function

A (cryptographic) function that maps data of arbitrary size to fixed-size values is called a *hashing* (or hash) *function* (Goldreich, 2001). It is an example of *one-way function*.

One-way function

We call a function, $f : \mathbb{Z}_2^n \to \mathbb{Z}_2^{n_o}$ (with n, n_o integers), *one-way* if it is 'easy' to evaluate (namely, in polynomial time) by an algorithm on every input, but computing its (pseudo-)inverse with a (randomized) algorithm (namely one that employs some degree of stochasticity) in polynomial time has negligible probability, when the size of the function domain increases (Goldreich, 2001). For instance, finding the primes of an integer, n, is an example of a one-way function: given the primes it is easy to find n. However, given n, it is not easy to find its primes, with known algorithms. Modern encryption protocols are based on these functions.

Along similar lines, it would be interesting to apply DMMs to *hashing functions* (Van Oorschot and Wiener, 1994), which are of importance to security and other protocols (Goldreich, 2001).

In addition, the applicability of DMMs to Machine Learning and Quantum Mechanics I have shown in Chapter 10, makes MemComputing relevant not just to Artificial Intelligence and Data Science, but also to Materials Science (e.g., in the study of physical properties of materials or design of new ones), Bioengineering (e.g., drug design), Biophysics (e.g., protein folding), and many other fields of study.

Finally, it is worth stressing that *analog* MemComputing machines may also find use in specific applications, as the ones I have briefly mentioned in Section 12.10. Indeed, it would be a mistake to think that only combinatorial optimization problems are of importance in modern science and technology (cf. discussion in Section 1.5).

In summary, the range of scientific and technological applications in which MemComputing (in both its digital and analog realization) can find use is quite vast. In fact, I expect that many more applications may be inspired by the opportunities offered by this paradigm of computation that are not, at the moment, particularly obvious (at least to me).

13.6   Quantum MemComputing?

By now, it should be abundantly clear that MemComputing is *radically* different from Quantum Computing. The former employs *non-quantum* features to compute, while the latter (at least for some applications) aims at leveraging phenomena, like *entanglement* (Section 4.3.1), that do not have a classical counterpart.

This difference reflects dramatically in the *mathematical* description of these two paradigms. MemComputing is described by ODEs whose state resides in a *phase space*, with dimension growing *linearly* with the number of degrees of freedom (see Chapter 3 and, in particular, Fig. 3.1).

On the other hand, the state of a quantum computer is a vector in a Hilbert space, whose dimension grows *exponentially* with the number of qubits (Fig. 3.1). This, by itself, precludes an efficient simulation of a quantum computer in software (cf. advantages of MemComputing over Quantum Computing in Section 3.3).

Of course, one could use some quantum phenomena in MemComputing as well. An example could be, again, the *spin* of the electron (see Section 13.3.1). Or we could employ particular devices where time non-locality is achieved via the coupling with a quantum system (Pfeiffer *et al.*, 2015).

All these suggestions are within the realm of possibilities for MemComputing. And some of these phenomena/devices may even provide advantages compared to traditional electronic components, e.g., reduced spatial footprint and/or energy consumption (Finocchio *et al.*, 2020).

Quantum MemComputing machines

However, none of these phenomena/devices change *fundamentally* the structure of MemComputing as I have introduced it in this book.[24]

Of interest, instead, would be the definition of a different *type* of machine, where Quantum Mechanics plays a prominent role. We could call it a *quantum MemComputing machine* (QMM).

We would then proceed along similar lines as the ones I used to define a universal MemComputing machine (UMM) in Chapter 5, by taking into account the possible states of a *quantum memprocessor*, the transition function(s) of a QMM, and finally determine how a QMM relates to a UMM, and consequently to a universal Turing machine.[25]

If this academic program is successful,[26] the next step would be to identify *actual* physical systems that could realize such a concept in practice (see also Section 13.7 below), and determine the type of advantages, if any, these machines would provide to both MemComputing and Quantum Computing.

Here, again, by 'advantages' I really mean with respect to *computing* difficult problems (cf. Section 3.3). If such machines do not solve the same range of problems as MemComputing does, and/or are too difficult to build, like quantum computers, then they are more of academic than of practical interest (see also discussion in Section 13.9).

13.7   Physical questions

Returning to the topic of MemComputing proper, it would be interesting to address several physical questions.

13.7.1   Instanton phase: Supersymmetry-protected topological order?

In Chapters 11 and 12, I have shown that a DMM goes through a succession of elementary *instantons* (a *composite instanton*) to reach the equilibria (solutions of a given problem). The size of these instantons/avalanches follows a critical Borel distribution, and the system *self-tunes* into such a state during the solution search (Section 12.5).

In fact, the 'rigid' state of the DMM's dynamics is reflected as well in the *scale-free* temporal and spatial correlations of the machine's voltages.[27] This behavior is further reinforced by the analogy of the state trajectory in phase space, in the presence of noise, with *bond directed percolation* in a high-dimensional lattice (Section 12.7.5).

[24] In other words, even though a proper description of the operation of some of the devices that make up MemComputing machines may require Quantum Mechanics, understanding how these devices work does not necessarily translate into a quantum description of how a MemComputing machine based on them operates.

[25] In addition, the properties of the Hilbert space, which is a Banach space, would most likely require a re-thinking of how to map the solutions of a given problem to the long-time dynamics of the machine (cf. Section 7.8).

[26] And I do not see why not, since we can *always* define *mathematical* quantities, whether they correspond to a physical reality or not. See von Neumann's quote at the beginning of Chapter 5.

[27] The DMM's dynamics are 'more rigid' during each instantonic transition, less so at the critical points.

[28]Recall from Section 12.9 that *topological order* refers to gapped (quantum) systems (in the thermodynamic limit) whose ground state has a degeneracy that cannot be lifted by any *local* perturbation, and depends only on the topology of the space. It gives rise to (long-range) quantum entanglement (Wen, 2013).

[29]Or is it another type of order altogether, that cannot be classified with either symmetry breaking or topology?

[30]See Section 1.3 for the difference between *syntactic* and *semantic* information.

[31]This is reminiscent of the dynamics of electrons in condensed-matter systems, with the electrons much faster (due to their smaller mass) than the atomic nuclei. Therefore, the electrons can be assumed to be always in their 'ground state', whenever the atomic nuclei move. One can then treat the positions of atomic nuclei as classical parameters (Messiah, 1958).

Gauge theory
A field theory in which the Lagrangian (cf. Section 11.5) is *invariant* with respect to some local transformations is called a *gauge theory*. The *gauge transformations* form a *symmetry (Lie) group* (or 'gauge group') of the theory. The field theory of Electromagnetism is the prototypical example of a gauge theory, since it is invariant under the scalar and vector potential gauge transformations.

Gauge fields and bosons
The fields associated with the generators of the gauge transformations are called *gauge fields*. (The scalar and vector potentials in the electromagnetic case.) All these concepts hold also for *quantum* gauge theories (Peskin and Schroeder, 1995). In this case, the *quanta* of the gauge fields are called *gauge bosons*.

All this points to the fact that DMMs self-tune into the state of some sort of *non-equilibrium phase transition*. If that is the case, is there an *order parameter* describing such a transition (cf. Section 12.7.5)? Or, as I have briefly argued in Section 12.9, the ordered state (in the thermodynamic limit) is more appropriately characterized by a (*supersymmetry-protected*) *topological order*,[28] for which the local order parameter (if it can be defined at all) is not gauge-invariant (cf. Section 13.7.3 below)?[29]

13.7.2   Information flow and low-energy effective field theory of DMMs

I have anticipated in Section 12.6.1 that the dynamics of DMMs can be viewed as a *quench* in which (syntactic) *information* is *transferred* from the fast degrees of freedom (voltages) to the slow degrees of freedom (memory variables).[30] Can we quantify this *information flow*? How does it affect the TTS?

This study also suggests that the fast degrees of freedom adjust almost 'instantaneously' to any change in the slow ones (cf. Section 7.4).[31] In turn, we could also think of the slow degrees of freedom as influenced by the fast ones. We can then 'trace out' the fast (high-frequency/high-energy) degrees of freedom, and attempt the derivation of a *low-energy effective field theory* for the memory dynamics.

Following this procedure, if the fast degrees of freedom are found to follow a particular distribution, their role is then to provide stochasticity to the memory variables, the latter following a *stochastic* (cf. eqn (12.6)) rather than a deterministic differential equation.

This line of research would then lead to a different low-energy effective theory of DMMs, that could reveal other aspects of their dynamics. In particular, it would improve our understanding of the memory variables's role, and could potentially suggest strategies for how to design DMMs with a minimal number of memory degrees of freedom.

13.7.3   Memory, gauge fields and asymptotic freedom

At this stage, I can make an interesting connection with another important physical theory: *Quantum Chromodynamics* (QCD). The latter is a Quantum Field Theory that describes the *strong force* between *quarks* as mediated by *gluons* to produce the observable *hadrons* (such as protons, neutrons, pions, etc.). Gluons are *gauge bosons*, namely *quanta* of the *gauge fields* of the theory, and *self-interact* (i.e., unlike photons, gluons interact with each other: they carry 'color charge').

I have already anticipated in Sections 2.6 and 11.10 that in QCD the coupling between quarks and gluons becomes stronger the lower the energy (hence the larger the distance between particles) (Peskin and Schroeder, 1995). This means that quarks (and gluons) cannot be observed as *free particles*.[32] Rather, we observe the mentioned hadronic *bound states*.[33]

This phenomenon is known as *asymptotic freedom*, because the particles' interaction strength becomes asymptotically weaker with increasing energy scale or, equivalently, with decreasing length scale (cf. discussion in Section 11.6 for the relation between energy and wavelengths).

A similar phenomenon occurs in DMMs: the memory variables *mediate*, and *participate* in (by 'self-interacting'[34]) the (strong) interactions between the fast degrees of freedom (the voltages in a circuit realization of DMMs) to produce a *composite instanton*. In addition, the interaction between the slow and fast degrees of freedom is *non-perturbative* at low-energies/long-wavelengths, similar to the non-perturbative interaction between quarks and gluons.[35]

On the other hand, at high energies, which correspond to high frequencies, hence short times (cf. Section 11.6), the memory dynamics are 'frozen', so that they have little dynamical influence on the voltages, and the voltage-memory system can be described by a 'free theory'.[36] We can then interpret the memory variables as some type of (*non-commuting*) 'gauge fields'[37], and the voltages as 'matter fields' of DMMs. The latter ones then satisfy some sort of 'asymptotic freedom', thus making the analogy with QCD very compelling.

If this analogy has any merit, what are the generators of the gauge transformations that correspond to these fields in the case of MemComputing? In other words, is the underlining theory of DMMs a *non-abelian* (Yang-Mills) theory, at least in some form?

We know for sure that the supersymmetric TFT I have discussed in Section 11.3 is *coordinate-free*, meaning that it is *invariant* with respect to any differentiable coordinate transformation (a *diffeomorphism*; cf. also Section 13.8.6) of the phase space on itself.[38] This invariance is a 'natural gauge' of the theory, but it may not be the only one.

It would then be very interesting to study the full consequences of this analogy as it could suggest interesting connections with other (possibly time non-local) physical theories.

'Mnemons' and 'voltons' as 'quantum particles' of DMMs

Following up on the previous analogy, I can push it further. In fact, this could lead to an interesting extension of MemComputing to the quantum domain (see Section 13.6), and may even help understand better the dynamics of these machines. Let me explain.

The equations of motion of DMMs I have introduced in this book can, in principle, be derived from a *Lagrangian* (cf. Section 11.3). Follow-

[32] Hence, the term (color) *confinement*.

[33] E.g., a pion (which is a *meson*) is a quark-antiquark bound state, while the proton and the neutron (*baryons*) are bound states of three quarks each.

Abelian and non-abeliean gauge theories

If the (Lie) group of gauge transformations of a gauge theory is *non-abelian* (i.e., there is at least a pair, (x, y), of elements of the group such that $x \star y \neq y \star x$, with $\star$ the binary operation of the group), then the theory is called a *non-abelian gauge theory* (e.g., a Yang-Mills theory). Otherwise, it is an *abelian gauge theory* (e.g., Electromagnetism).

---

[34] Hence, generating strong *non-linearities* in the dynamics.

[35] Note, in fact, that historically instantons have been first introduced in the context of *Yang-Mills* (non-abelian) gauge theories, of which QCD (with the coupling to fermionic fields describing quarks) is a prominent example (Peskin and Schroeder, 1995).

[36] In the sense that the memory variables, at short times, simply act as parameters for the voltage dynamics. The two types of degrees of freedom can then be considered 'decoupled' at short times.

[37] More precisely, as *Wilson lines* written in terms of gauge fields with 'color charge', because these fields can 'self-interact' (Peskin and Schroeder, 1995). The memory variables are *non-commuting* because the order in which they vary in time matters.

[38] Similar to the diffeomorphism invariance or *general covariance* of General Relativity (Misner, Thorne and Wheeler, 1973). In this sense, General Relativity is a *topological* theory: it is *independent* of the choice of the metric (cf. Section 1.6).

Memory quanta, namely quantum excitations of the memory degrees of freedom have already been discussed in (Cohen *et al.*, 2012).

Bose-Einstein condensation
A weakly interacting gas of bosons, cooled below a certain temperature, may transition into a *Bose-Einstein condensate*, in which a large fraction of the bosons occupies the lowest quantum state (Mahan, 2000).

Superfluidity
A fluid with *zero viscosity* is called a *superfluid* (Mahan, 2000). Although many superfluids emerge from the phenomenon of Bose-Einstein condensation, the two phenomena are not always related.

[40]In the sense that the 'unsolvable' side of that 'transition' is the noise-induced chaotic phase, and it occurs over a range of parameters, not just a single point.

ing techniques of Quantum Field Theory (Peskin and Schroeder, 1995), we can then *quantize* such a Lagrangian and obtain the *quantum-field* analogues of the voltage variables (following the tradition of Quantum Mechanics, we may call *voltons*) and memory variables (*mnemons*).[39]

Do these quantized fields correspond to actual physical quantities? And would this study suggest physically realizable Quantum MemComputing machines? What does the quantization tell us about the non-quantum (classical) dynamics? For instance, can we interpret the composite instanton as the *condensation* of these particles?

In other words, do the voltons condense into a *superfluid state*, due to the presence of the mnemons, or vice versa? (Indeed, this would be in line with the fact that DMMs are *strongly coupled systems*; see Section 11.10.)

These questions are particularly appealing from a fundamental point of view. However, they may also be relevant for the practical realization of MemComputing machines. In fact, would such a study tell us anything more about the underlying physical properties *any* machine needs to have to compute efficiently?

13.7.4 Condensation of (anti-)instantons and the solvable-unsolvable transition

Along similar ideas, it would be interesting to study the phenomenon of *dynamical breakdown* of topological supersymmetry due to the *condensation* of instantons and (noise-induced) anti-instantons, I mentioned in Section 11.4.1.

Although not yet proved, this phenomenon may be related to the 'solvable-unsolvable' transition I discussed in Section 8.6.1. If that is the case, then such a 'transition' would lead to a *phase* with a finite width.[40]

This study could then reveal the maximum noise strength DMMs can support before *noise-induced chaos* emerges (Ovchinnikov, 2016). Let me stress again that such chaotic behavior is *not* due to the non-integrability of the flow vector field in eqns (3.1): it would be a totally different *phase* of the (stochastic) dynamical system.

13.7.5 MemComputing and the Renormalization Group

I have discussed in this book that the difficulty of certain combinatorial optimization problems originates from the fact that variables appear in 'far-away' clauses: a logical assignment of a variable in a clause affects another clause, containing the same variable, with the two clauses some 'distance' away from each other (cf. discussion in Section 2.6). This concept of 'clause distance' is justified by the representation of such problems as actual *physical* circuits, like the ones that realize DMMs in hardware.

A DMM designed to solve such problems then starts from a random initial condition of the literal values at each terminal. At this initial stage of the dynamics, the DMM only exploits the *local* (*short-range*) features of the underlying Boolean (or algebraic) circuit. In fact, this is typically the only information we have of the problem we are attempting to solve.

The DMM then proceeds by employing specific trajectories (instantons) in the phase space. These trajectories showcase *collective*, or *long-range* behavior. Once the DMM enters the instantonic phase of the dynamics, it is able to 'extract' the *long-range* features of the problem it is attempting to solve. And indeed, these features not only render a problem really difficult to solve, they are the most challenging ones to discover.

Recall that short wavelengths correspond to high energies—*ultraviolet* (UV) limit—while long wavelengths to low energies—*infrared* (IR) limit (Section 11.6).

The solution of a given problem by a DMM then proceeds from the UV to IR energy scales. Or, equivalently, a DMM solves the problem by 'exploring' short length scales first, and then moving on to longer and longer length scales (Fig. 13.2).

It is as if a DMM performs a *coarse-graining procedure* 'on the fly'. Intuitively, this is like varying the magnifying power of a microscope (we use to image a sample) toward larger and larger features, so that short length-scale details of the sample become less and less visible.

As I have mentioned in Section 12.9, during dynamics the (bosonic) degrees of freedom (voltages and memory variables) decrease their 'entropy' (increase their 'knowledge' of the problem to solve) at the expense of fermionic degrees of freedom, whose entropy increases.

Along the way, the DMM 'gets rid' of the *irrelevant* interactions (coupling constants) between variables, while keeping only the *relevant* ones in the IR limit. In other words, there is some sort of 'universality', whereby the *majority* of coupling constants, which are present in the UV limit, are *not* important in the IR limit, and only a few survive, those relevant to the solution of the problem.

The procedure I have just described is known in (Quantum) Field Theory and Statistical Mechanics as the *Renormalization Group* (RG): a theoretical framework that allows the study of a physical system at different length (energy) scales—from short to long length scales (Goldenfeld, 1992).

The RG has found wide applicability in Physics. For instance, it can be used to explain the reason behind the *universality classes* of

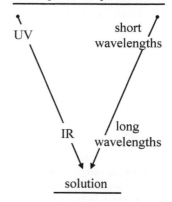

DMM flow

initial problem specification

UV — short wavelengths

IR — long wavelengths

solution

Fig. 13.2 The 'RG flow' of a DMM solving a problem: from high energies (UV limit) to low energies (IR limit). Or, equivalently, from short distances to long distances.

The Renormalization Group
The theoretical framework that integrates out *high-energy* degrees of freedom of the action of a physical system, leaving only an *effective action* containing the (relevant) *low-energy* degrees of freedom (hence, changing the energy scale of the theory) is called the *Renormalization Group* (RG) (Goldenfeld, 1992). In other words, the RG allows us to understand how a given theory changes as we probe it in the infrared limit. We can then determine how physical properties of the system change (or 'run') by varying the energy scale. For instance, for the interaction couplings (g_i) of the system, this is done by defining their *β-functions* ($\beta_i \equiv \partial g_i / \partial \ln \Lambda$, with Λ the energy scale), from which we can obtain the equations of the *RG flow* of the couplings. The RG methodology has found wide applicability in High-Energy Physics, critical phenomena, and Condensed Matter Physics in general.

continuous phase transitions, namely that physical systems differing only by microscopic details, but sharing the same symmetries, dimensionality, and range of interactions, should have the same *critical exponents* (cf. Section 4.3.4).

The DMMs I have discussed in this book seem to perform this *RG flow* from the UV to the IR limit 'naturally', namely without any external intervention (Fig. 13.2).[41] It is then not far-fetched to expect that we may make the previous statements even more formal by applying the RG methodology to MemComputing machines.[42]

If so, would the RG tell us more about the physical operation of DMMs? What about the problems they solve?[43]

Can we employ it to show that the 'solvable-unsolvable' transition (Section 8.6.1) in the presence of noise, whether physical or numerical, belongs to the same universality class as *directed percolation*?

And/or, can the RG analysis demonstrate that the unsolvable phase is a different phase of the system (as I mentioned previously)?

Finally, could this RG study show explicitly that DMMs are described by an asymptotically free theory?[44] Would it help in the design of such machines?

13.7.6   Topological computation and its physical vacua

In Section 1.6, I mentioned that Topological Quantum Computing is based on the idea of using the *topological* features (and quantum entanglement; Section 4.3.1) of the ground states of some *strongly correlated* quantum systems to compute unencumbered by noise and perturbations.

More precisely, such states may provide computation that is *topologically protected* against *decoherence* (Freedman *et al.*, 2002). In Chapters 11 and 12 I have shown that MemComputing—in particular, its digital version—is another example of *topological computing*.

The topological features employed by digital MemComputing machines are *critical points* and trajectories connecting them, *instantons—* forming a *CW complex* (Section 12.8).

The notable upshot of all this is that these features pertain to *non-quantum* dynamical systems. Therefore, (digital) MemComputing has all the advantages of topological computing, without the practical disadvantages of quantum computers (cf. Section 3.3).

Irrespective, in both quantum and non-quantum dynamical systems, topological computation seems to require the state of some *strongly coupled* systems. However, the reverse is not necessarily true: not all strongly coupled systems are good candidates for topological computation.

In fact, it seems that the *physical vacua* for such a computation must

[41] In fact, it is thanks to this type of 'RG-flow' behavior that we could simplify the dynamics of actual, hardware DMMs, and represent them with much simpler equations of motion showcasing similar physical characteristics (cf. discussion in Section 8.1).

[42] This methodology is directly tied to, and can help in the development of the *low-energy effective field theories* of DMMs I have discussed in Section 13.7.2.

[43] The solutions of these problems would be (stable/attractive) fixed points of the RG flow. Therefore, are the computational complexity classes just a manifestation of universality classes of physical systems designed to solve such problems?

[44] Like in QCD, this would reflect in a *negative β*-function of the appropriate running coupling(s), so that the theory is 'free' at high energies, and strongly coupled at low energies (Peskin and Schroeder, 1995). This also means that the vacuum of the theory at zero coupling (between memory and voltages/fast degrees of freedom) is *unstable*, and a fundamentally different ground state must emerge when the two types of degrees of freedom are coupled, *even weakly*. This new vacuum would correspond to the *composite instanton* I discussed in Section 12.1.

all be described by Topological Field Theories, of the *Schwarz*-type—in particular, the *Chern-Simons* subclass—in the quantum case; of the *Witten*-type in the non-quantum case (Section 11.3). It would then be interesting to explore this connection, and see if it leads to a *unifying picture* of topological computing.

13.8   Mathematical questions

After these physical questions, let me now turn to several mathematical ones that I think should be addressed.

13.8.1   Chaos and periodic orbits

For one, the demonstration that the equations of motion I have used in this book do not support chaos and (quasi-)periodic orbits, should in principle be repeated every time we modify their *functional form*.

In the case of chaos, absence of *topological transitivity* can be checked relatively easily, based on the existence of fixed points and the compact nature of the phase space (cf. Section 3.9).[45]

As for (quasi-)periodic orbits, instead, it can be quite a cumbersome exercise to determine their absence (Bearden, Pei and Di Ventra, 2020).

Is there then a general *mathematical recipe* (or 'principle') that allows us to change these equations while preserving the desired properties? If the answer is positive, we would have a very powerful tool to choose particular ODEs instead of others.

And this choice would also influence the type of *physical* devices that make up a given DMM. Namely, it would guide the choice of which devices are more conducive to showcase the necessary dynamical features when assembled in the appropriate architectures.

13.8.2   Lyapunov function for DMMs

Regarding the dissipative character of DMMs (Section 3.8), since the long-time stability (robustness against perturbations) of the equilibria (solutions of a given problem) is of fundamental importance, can we construct a general *Lyapunov function* for these dynamical systems? If so, would this function aid in the *design* of DMMs?

13.8.3   Discrete maps

The ODEs integrated numerically define *discrete maps* (Section 8.4). And, as I have explained in Section 12.7.3, the maps that originate from continuous-time ODEs may have a *different* phase space than the original dynamics.

The two types of Topological Field Theories (TFTs)

There are essentially two types of TFTs: *Schwarz*-type and *Witten*-type. *Chern—Simons* theory is the best known example of a Schwarz-type TFT, and has found application in the description of some strongly correlated electron systems, and the corresponding concept of Topological Quantum Computing (Freedman *et al.*, 2002). The Witten-type TFT has been, so far, mostly used in Mathematics, e.g., in *Morse theory* (Milnor, 1963). Section 11.3 describes a Witten-type TFT applied to *non-quantum* dynamical systems (Ovchinnikov, 2016). See (Birmingham *et al.*, 1991) for a thorough discussion of these two types of TFTs.

[45] Another way of proving absence of chaos is by noting that the global *topological supersymmetry* of the DMM's dynamics is 'protected' (*restored*) by instantons, namely it is not broken; see Section 12.9.

Lyapunov function
Given an ODE, a *scalar function* may be defined that characterizes the stability of the equilibria of such an equation. Such a function is called *Lyapunov function* (Hale, 2010). Unfortunately, there is no straightforward prescription to construct such functions for general ODEs.

322 Epilogue and future work

In particular, some integration schemes can introduce additional (*ghost*) critical points in the phase space, with a consequent degradation of the dynamics (cf. Section 8.6).

The mathematical field of research that studies the critical points of discrete maps derived from continuous-time ODEs is relatively young (Stuart, 1994). It is, nonetheless, very relevant for the development of DMMs. Further advancements in this field could address important questions.

For instance, given some ODEs of DMMs, is there a general way of determining how many ghost critical points are introduced as a function of the integration time step?

What 'type' are these critical points? Are they additional local minima, or saddle points?[46] How does the functional form of the ODEs of DMMs affect these ghost critical points?

These questions are mostly relevant for the *software* simulation of DMMs, but could even help improve their *hardware* design. For instance, if we knew that specific functional forms of ODEs may easily introduce ghost local minima, we could steer clear from them, and focus our efforts on different types of physical systems described by different ODEs.

13.8.4 Problem representation and phase-space engineering

As I have already mentioned, it is known that any problem in the NP class can be mapped (with polynomial resources) into any NP-complete problem (Arora and Boaz, 2009).

For instance, we could transform these problems into a 3-SAT (with exactly three literals per clause; Section 2.4). Or we could transform them into a SAT with a mixed number of clauses, with two and three literals, or any other combination of clauses with k literals.

According to Computational Complexity Theory, all of them should be solvable with comparable difficulty.[47] However, this does not necessarily translate into the same efficiency with which DMMs would solve all these problems.

Consider, for example, the Ising spin glass I have discussed in Section 9.2.2. I have shown in that Section that it can be 'naturally' formulated as a QUBO problem. However, I may transform it into a weighted MAX-SAT, or an Integer Linear Programming problem (Section 9.2.5).

Are all these mappings always *advantageous*, as far as the TTS is concerned? Not necessarily.

Recall that DMMs do *not* solve combinatorial optimization problems algorithmically. The given problem is first mapped into a dynamical system, and then one looks for the steady states of such a system. The topology of the phase space is therefore of fundamental importance (cf.

[46] If the ghost critical points were only saddle points, they would not have any detrimental effect on the dynamics. They would just change the instantonic trajectory to reach the equilibria.

[47] In the following sense: if an NP-complete problem is solved in polynomial time, another NP problem could be mapped into it, and solved also in polynomial time, possibly with a *different* degree of the polynomial (Arora and Boaz, 2009).

Chapters 11 and 12).

If the mapping from one representation of a problem to another introduces a substantial number of local minima, other than the minima corresponding to its solution,[48] then one needs to modify the *functional form* of the equations derived from this mapping, before they can be used efficiently to find the steady states.

In other words, the mapping can *fundamentally* change the topological properties of the phase space of the resulting DMM, which could, in turn, affect the TTS. Therefore, particular care needs to be put in choosing the type of representation of a combinatorial optimization problem.

Research that would uncover the relation between a problem representation and the corresponding DMM's phase space properties would then be very beneficial. In other words, a more rigorous way of doing *phase-space engineering* would be desirable.

13.8.5 Duality between Computational Complexity Theory and Algebraic Topology

In Section 12.3, I have shown that, given a combinatorial problem, the number of instantonic steps to solution (if it exists) can be easily counted. In fact, in Section 12.8, I have argued that the *composite instanton* is a finite (hence compact) *CW complex* of the phase space.

In this way, I have transformed the task of finding the number of steps to solution of the original problem into the counting of topological objects (*instantons* in a topological space) necessary to reach it. And in order to do so, I have re-written the original combinatorial problem in the language of a *Cohomological Field Theory*.

Namely, I have transformed the time-to-solution problem into an exercise in *Algebraic Topology* (Hatcher, 2002). This seems to suggest a interesting *duality* between *Computational Complexity Theory* and Algebraic Topology.

As in other branches of Physics and Mathematics, this duality could lead to many fruitful developments. For instance, in Section 11.4, I have discussed that instantons can be *parametrized* according to their *moduli*, which encode their *collective* nature.

The resulting space, the *moduli space*, is then the space of these parameters and classifies the family of instantons in DMMs.

Moduli spaces appear in a wide variety of mathematical classifications of objects (see, e.g., (Hori *et al.*, 2000) for an application to *String Theory*). It would then be interesting to see if there exists a classification of different computational problems according to the moduli spaces that emerge from their solution using DMMs. This classification may tie more closely the different problems in computational complexity with the dynamical systems that can solve them efficiently.

[48]Even if no additional local minima are introduced, a different problem representation may generate saddle points that have unstable directions with quasi-flat curvatures. If that is the case, the theoretical scalability may remain polynomial but with possibly a large degree. For all practical purposes, then such a large-degree polynomial 'resembles' an exponential.

Duality in Physics and Mathematics

The concept of *duality* has been extremely fruitful in both Physics and Mathematics (see, e.g., Poincaré duality in Section 11.9). In its most practical aspect, duality means that one can solve some problems that are especially difficult in one theoretical framework by means of ideas pertaining to, and calculations done in another. However, it could reveal a much more profound relation between different aspects of a theory, like the 'wave-particle duality' of Quantum Mechanics (Messiah, 1958).

Moduli space

A space of *parameters* that characterizes (classifies) a family of mathematical structures (e.g., instantons) is called a *moduli space*. Every instance of that mathematical structure is represented by a point in that space. The *geometry* of the moduli space reflects how these structures vary under small perturbations (Hartshorne, 2010).

String Theory
The body of mathematical statements that purports to describe particles and their interactions as vibrations (modes) and dynamics of one-dimensional extended objects (strings) is called *String Theory* (Hori *et al.*, 2000). Although this mathematical program has given us many tools (e.g., the supersymmetric Topological Field Theory of Section 11.3 is built upon some of these tools), it has yet to make an experimentally verifiable prediction.

[49]They are Turing-complete, but not necessarily equivalent (Section 5.4).

[50]Here, I am talking about the *mathematical* Turing machine, not its *approximate* physical representation (cf. Section 2.8).

[51]They may be *suggestive* of the validity of a thesis. And, in fact, we may even use the *duck test* which states that 'if it looks like a duck, swims like a duck, and quacks like a duck, then it probably *is* a duck'—attributed to J. W. Riley (1849–1916). However, in no way all this can be used as a rigorous mathematical proof.

Diffeomorphism
An invertible function that maps a differentiable manifold to another, in such a way that both the function and its inverse are smooth (differentiable), is called a *diffeomorphism.*

[52]In fact, as I already mentioned in Section 13.7, (supersymmetric) TFT is *coordinate-free*, as it can be seen from, e.g., the fact that it can be written in the exterior algebra language; its action does not depend on any metric (Section 11.3). However, it is worth mentioning that *any* theory that describes physical phenomena should be independent of the choice of coordinates.

13.8.6   Is NP = P?

Let me now turn to a question I have repeatedly said cannot be answered with the results I have presented so far. This is the famous question of whether NP = P or not (Section 2.3.2).

It could be quite easy to dismiss—in fact, bypass altogether—this mathematical question on the grounds that it pertains *only* to Turing machines. This is indeed true. And since MemComputing machines are *not* Turing machines,[49] we could simply say that question is *irrelevant* in the MemComputing paradigm of computation.

However, I have also shown *simulations* of the ODEs of DMMs in Chapters 8, 9, and 10. In that case, I have transformed a *continuous-time* dynamics into a *discrete map*, which can be integrated with any explicit (or implicit) numerical method.

In summary, a numerical simulation of DMMs *is* an algorithm. And an algorithm can be executed on a Turing machine.[50] However, *numerical* results cannot be used as *proof* of the validity of a mathematical statement![51]

In fact, even the power-law trends I have shown in Chapters 8, 9, and 10 cannot completely rule out the possibility that the same fits may be just the beginning (albeit of very-large size problems) of an exponential function with an *extremely small* coefficient.

Of course, even if this were the case, the *practical* usefulness of the simulations of DMMs would still hold, since they can be pushed to problem sizes that *cannot* be easily reached by traditional algorithms.

However, as suggestive as those trends are, they are definitely *not* a proof that NP = P!

Theorems for continuous and discrete dynamics
A rigorous proof of the famous conjecture would then require extra work, and, quite possibly, new mathematical tools.

In fact, the supersymmetric Topological Field Theory I have briefly introduced in Chapter 11, and which I have used to show that the *physical* (continuous-time) DMMs can solve hard problems in polynomial time (Chapter 12), does not necessarily hold for all types of *discrete* maps.

Such a theory relies on the fact that physical (continuous-time) dynamics generate evolution operators that are *diffeomorphisms* of the phase space on itself (Ovchinnikov, 2016).[52] The discrete maps we obtain from them may not be.

Therefore, discrete maps may have different dynamical behavior than the continuous-time dynamics they originate from.

For instance, as I mentioned in Section 11.3, the discrete map may break supersymmetry *explicitly*, hence may support chaotic behavior (Ovchinnikov and Di Ventra, 2019). Although we have not observed such a behavior in all numerical cases we have considered so far, a rigorous proof of this result is still lacking.

In addition, the *stability* of the discrete map in terms of the time step, Δt, in any given integration scheme, appears to scale *polynomially* with problem size (as I have shown in Chapter 8).

However, these are *empirical* results. They are, again, *not* rigorous mathematical proofs! Therefore, much more theoretical work is needed to address these specific issues.[53]

Consensus in Science? What's that?

To this, I can only add that, being pragmatic about what computing machines should be useful for (see Section 13.9 below), I am personally agnostic about the resolution of the NP vs. P question, either way.

However, I find it risible and, quite frankly, utterly *unscientific* when I hear statements of the type: 'Since *no one* has ever demonstrated it, then NP *cannot* be equal to P', or 'The majority of scientists *believe* that NP $\neq$ P, therefore it *must* be true'.

There are several serious issues with these sentences. For one, the word 'belief' should not even be used in a scientific discussion, since Science does not rely on dogmas.

In addition, *Science is not a democracy*: it does not, and should not, proceed by consensus. (For a relatively brief and simple discussion of the *Scientific Method* see, e.g., (Di Ventra, 2018); cover in Fig. 13.3.)

Moreover, the fact that a conjecture has not been proved yet—no matter how many brilliant scientists have worked on it—should not preclude or discourage scientific inquiry into new research directions.

Going against the prevailing paradigm, and changing the starting point of view in *any* field, always brings benefits to Science (Kuhn, 1962).

If anything, it is bound to inject into a mature and established field, new and 'unorthodox' ideas that may eventually tickle the imagination of its practitioners.

Violation of the extended Church–Turing hypothesis?

Even if the NP vs. P question has not been settled yet, the results I have discussed in this book indicate that we should be able to fabricate *physical* DMMs capable of solving combinatorial optimization problems *efficiently*, namely with resources that grow polynomially with input size.

In Section 2.2.2, I have introduced a couple of hypotheses (not theorems!) named after the mathematicians Church (Fig. 13.4) and Turing (Fig 2.2). The 'original' hypothesis only states that a function on the natural numbers can be calculated by an algorithmic procedure, *if* ('sufficient' condition) and *only if* ('necessary' condition) it is computable by

[53]For instance, the duality between Computational Complexity Theory and Algebraic Topology, I mentioned in the previous Section 13.8.5, seems a particularly good direction to follow for the development of novel theoretical tools to settle many of these questions.

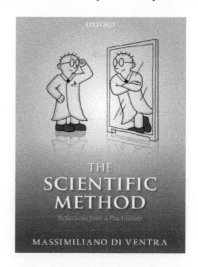

Fig. 13.3 Cover of the author's book *The Scientific Method: Reflections from a Practitioner* (Oxford University Press, 2018).

Fig. 13.4 Alonzo Church (1903–1995). Sketch by the author.

[54]Namely, those that do not require infinite precision.

[55]The latter ones being only a mathematical construct, and 'non-reasonable' on the grounds that they cannot be realized in practice, not even approximately.

Will we replace our modern computers soon?
To answer this question, let's not forget that our modern computers (in all their hardware variants) have been, and *still are* extremely useful in an extraordinarily wide range of applications. Therefore, I do not see their dominance as main computational tools waning any time soon. It is much more likely, and quite frankly much more realistic, that any new computing paradigm (whether quantum or not) leading to practical solutions of some difficult computational problems will play the role (at least in the beginning) of *co-processor*, or *co-computer*, of our traditional computers. This means that such a paradigm will not simply replace the existing one. Rather, it will likely *enhance* the functions, and *extend* the reach, of the one we are currently using.

a Turing machine.

In Section 5.4, I have shown that MemComputing machines are Turing-*complete*. However, they have not been shown to be Turing-*equivalent*. Therefore, at the moment, the 'necessary' part of this hypothesis is still unverified.

On the other hand, in regard to the 'extended' Church-Turing hypothesis, MemComputing—in particular its *digital* version—provides a stronger statement.

In fact, the extended hypothesis posits that *all* 'reasonable' models of computation[54] yield the same class of problems that can be computed in a time that grows *polynomially* with input size.

The DMMs *are* 'reasonable' models of computation. In fact, we have seen in Chapter 5, that DMMs (and UMMs in general) have the same computational power of *nondeterministic* Turing machines.[55]

This means that the class of problems DMMs can solve in polynomial time is *not the same* as the class of problems a *deterministic* Turing machine can solve. This seems to suggest that the 'extended' Church-Turing hypothesis may not be valid.

13.9   On the usefulness of a computing machine

Let me conclude this book with a comment that I think is very relevant, not only for MemComputing, but also for any type of computing paradigm of current interest, and even those that may be eventually conceived in the future.

Whether the mathematical NP vs. P question is ultimately answered in the affirmative, and/or the Church-Turing hypotheses are found to be invalid—which are certainly important mathematical questions worth answering—we should not forget the *practical utility* of computing.

As I have already mentioned in the Prologue of this book (Section 1.1), the question of 'usefulness' in *any* technology should be addressed, first and foremost, in terms of its *final goal*. In the case of computing, the end goal is precisely *to compute*!

This means solving problems that are challenging, and cannot be easily tackled, within a reasonable time, with traditional approaches, or any other type of machine.

A computing machine that does *not* satisfy this basic requirement of usefulness, may still be of academic interest, but it does not rise to the level of practical worth and applicability, irrespective of whether it employs quantum phenomena or not.

Taking into account this important criterion, we can conclude that for many problems of interest to both academia and industry—as those I have discussed in this book—MemComputing has *already* shown considerable advantages compared to other computing paradigms of current interest (such as Quantum Computing), as well as traditional algorithmic approaches.

In this respect, MemComputing has already demonstrated its practical utility.[56]

The mind that opens up to a new idea never returns to its original size.

The more I learn, the more I realize how much I don't know.

The important thing is to never stop questioning. Curiosity has its own reason for existing.

Albert Einstein (1879—1955)

[56] *Apri la mente a quel ch'io ti paleso e fermalvi entro; ché non fa scienza, senza lo ritenere, aver inteso.*
(Open your mind to what I shall disclose, and hold it fast within you; he who hears, but does not hold what he has heard, learns nothing.)

Dante Alighieri (1265—1321)
La Divina Commedia: Paradiso

About the Author

Massimiliano Di Ventra is Professor of Physics at the University of California, San Diego. His research interests range from condensed matter theory to unconventional computing. He co-edited the textbook *Introduction to Nanoscale Science and Technology* (Springer, 2004) for undergraduate students; he is single author of the graduate-level textbook *Electrical Transport in Nanoscale Systems* (Cambridge University Press, 2008), and of the trade book *The Scientific Method: Reflections from a Practitioner* (Oxford University Press, 2018). He has published more than 200 papers in refereed journals and delivered more than 300 invited talks worldwide. He is a fellow of the American Physical Society, the Institute of Physics, and a foreign member of Academia Europaea. In 2018 he was named Highly Cited Researcher by Clarivate Analytics and he is the recipient of the 2020 Feynman Prize for theory in Nanotechnology.

Discovery is a child's privilege. I mean the small child, the child who is not afraid to be wrong, to look silly, to not be serious, and to act differently from everyone else. He is also not afraid that the things he is interested in are in bad taste or turn out to be different from his expectations, from what they should be, or rather he is not afraid of what they actually are. He ignores the silent and flawless consensus that is part of the air we breathe—the consensus of all the people who are, or are reputed to be, reasonable.
Alexander Grothendieck
(1928–2014)

New ideas pass through three periods: 1) It can't be done. 2) It probably can be done, but it's not worth doing. 3) I knew it was a good idea all along!
Arthur C. Clarke (1917–2008)

Bibliography

Arfken, G. B., Weber, H. J., and Harris, F. E. (1989). *Mathematical Methods for Physicists: A Comprehensive Guide*. Springer-Verlag, New York. [Cited on pages 11, 38, 39, 40, 41, 46, 49, 100, 101, 106, 135, 153, 273, and 285.]

Arnold, V. I. (1989). *Mathematical Methods of Classical Mechanics*. Springer-Verlag, New York. [Cited on pages 11, 40, 53, 54, 239, and 240.]

Arora, S. and Boaz, B. (2009). *Computational Complexity: A Modern Approach*. Cambridge University Press, Cambridge, UK. [Cited on pages 18, 20, 22, 23, 24, 25, 26, 27, 28, 29, 30, 92, 93, 112, 140, 151, 163, 170, 310, and 322.]

Athreya, K. B. and Ney, P. E. (2004). *Branching Processes*. Dover Publications, New York. [Cited on page 272.]

Aulbach, B. and Kieninger, B. (2001). On three definitions of chaos. *Nonlinear Dynamics and Systems Theory*, **1**, 23–37. [Cited on page 51.]

Bandyopadhyay, S. and Cahay, M. (2008). *Introduction to Spintronics*. CRC Press, Boca Raton, Florida. [Cited on page 309.]

Banks, J., Brooks, J., Cairns, G., Davis, G., and Stacey, P. (1992). On Devaney's definition of chaos. *The American Mathematical Monthly*, **99**, 332. [Cited on page 51.]

Bearden, S. R. B., Manukian, H., Traversa, F. L., and Di Ventra, M. (2018). Instantons in self-organizing logic gates. *Physical Review Applied*, **9**, 034029. [Cited on pages 122, 123, 125, 161, 172, 226, 235, 260, 266, 277, 281, and 282.]

Bearden, S. R. B., Pei, Y. R., and Di Ventra, M. (2020). Efficient solution of Boolean satisfiability problems with digital MemComputing. *Scientific Reports*, **10**, 19741. [Cited on pages 140, 144, 145, 146, 153, 155, 161, 162, 163, 164, 312, and 321.]

Bearden, S. R. B., Sheldon, F., and Di Ventra, M. (2019). Critical branching processes in digital MemComputing machines. *Europhysics Letters*, **127**(3), 30005. [Cited on pages 161, 271, 272, and 274.]

Beggs, J. M. and Plenz, D. (2003). Neuronal avalanches in neocortical circuits. *Journal of Neuroscience*, **23**, 11167–11177. [Cited on page 300.]

Bennett, C. H. (1973). Logical reversibility of computation. *IBM Journal of Research and Development*, **17**, 525. [Cited on page 112.]

Birmingham, D., Blau, M., Rakowski, M., and Thompson, G. (1991). Topological field theory. *Physics Reports*, **209**, 129. [Cited on pages 229, 231, 238, 253, 265, and 321.]

Blum, C. (2005). Ant colony optimization: Introduction and recent trends. *Physics of Life Reviews*, **2**(4), 353–373. [Cited on pages 68 and 298.]

Bonani, F., Cappelluti, F., Guerrieri, S. D., and Traversa, F. L. (2014). *Wiley Encyclopedia of Electrical and Electronics Engineering*. John Wiley and Sons, Inc., New York. [Cited on pages 78 and 79.]

Caravelli, F., Hamma, A., and Di Ventra, M. (2015). Scale-free networks as an epiphenomenon of memory. *Europhysics Letters*, **109**, 28006. [Cited on page 68.]

Carleo, G. and Troyer, M. (2017). Solving the quantum many-body problem with artificial neural networks. *Science*, **355**, 602. [Cited on pages 216 and 219.]

Chua, L. O. (1971). Memristor - the missing circuit element. *IEEE Transactions of Circuit Theory*, **18**, 507–519. [Cited on page 75.]

Chua, L. O. and Kang, S. M. (1976). Memristive devices and systems. *Proceedings of IEEE*, **64**, 209–223. [Cited on page 70.]

Cohen, G.Z., Pershin, Y. V., and Di Ventra, M. (2012). Lagrange formalism of memory circuit elements: Classical and quantum formulations. *Physical Review B*, **85**, 165428. [Cited on page 318.]

Coleman, S. (1977). *Aspects of Symmetry, Chapter 7*. Cambridge University Press, Cambridge, UK. [Cited on pages 31, 234, 235, 239, 242, 244, and 255.]

Cover, T. M. and Thomas, J. A. (2006). *Elements of Information Theory* (2nd edn). Series in Telecommunications and Signal Processing. John Wiley and Sons, Inc., New York. [Cited on pages 8, 9, and 90.]

CPlex (2018). IBM ILOG CPlex Optimizer, New York. *www-01.ibm.com/software/integration/optimization/cplex-optimizer*. [Cited on page 172.]

Dauphin, Y. N., Pascanu, R., Gulcehre, C., Cho, K., Ganguli, S., and Bengio, Y. (2014). Identifying and attacking the saddle point problem in high-dimensional non-convex optimization. *Advances in Neural Information Processing Systems (NIPS 2014)*, 1–9. [Cited on page 226.]

Deutsch, D. (1997). *The Fabric of Reality*. Penguin Books, London. [Cited on page 4.]

Devaney, R. L. (1992). *A First Course in Chaotic Dynamical Systems: Theory and Experiment*. Addison-Wesley, Boston. [Cited on page 51.]

Di Ventra, M. (2008). *Electrical Transport in Nanoscale Systems*. Cambridge University Press, Cambridge, UK. [Cited on pages 50, 60, 71, 276, and 278.]

Di Ventra, M. (2018). *The Scientific Method: Reflections from a Practitioner*. Oxford University Press, Oxford. [Cited on pages 3, 18, 31, and 325.]

Di Ventra, M. and Ovchinnikov, I. V. (2019). Digital MemComputing: from logic to dynamics to topology. *Annals of Physics*, **409**, 167935. [Cited on pages 31, 134, 229, 237, 240, 243, 249, 252, 260, 263, 264, and 271.]

Di Ventra, M. and Pershin, Y. V. (2013a). On the physical properties of memristive, memcapacitive and meminductive systems. *Nanotechnology*, **24**, 255201. [Cited on pages 61, 70, 73, 75, and 115.]

Di Ventra, M. and Pershin, Y. V. (2013b). The parallel approach. *Nature Physics*, **9**, 200. [Cited on pages 11, 12, 76, and 305.]

Di Ventra, M., Pershin, Y. V., and Chua, L. O. (2009). Circuit elements with memory: Memristors, memcapacitors, and meminductors. *Proceedings of the IEEE*, **97**, 1717–1724. [Cited on pages 73, 74, 75, and 76.]

Di Ventra, M. and Traversa, F. L. (2017a). Absence of chaos in digital memcomputing machines with solutions. *Physics Letters A*, **381**, 3255. [Cited on pages 43, 49, 52, 104, 107, 131, and 133.]

Di Ventra, M. and Traversa, F. L. (2017b). Absence of periodic orbits in digital memcomputing machines with solutions. *Chaos: An Interdisciplinary Journal of Nonlinear Science*, **27**, 101101. [Cited on pages 49, 104, 107, and 133.]

Di Ventra, M. and Traversa, F. L. (2018a). MemComputing: Leveraging memory and physics to compute efficiently. *Journal of Applied Physics*, **123**, 180901. [Cited on pages 28, 45, 63, 111, 127, 151, 161, 265, and 269.]

Di Ventra, M. and Traversa, F. L. (2018b). Self-organizing logic gates and circuits and complex problem solving with self-organizing logic circuits, US patent No. 9,911,080. [Cited on pages 97, 100, 119, 120, 308, and 310.]

Di Ventra, M., Traversa, F. L., and Ovchinnikov, I. V. (2017). Topological field theory and computing with instantons. *Annalen der Physik (Berlin)*, **529**, 1700123. [Cited on pages 31, 227, 243, 245, 247, and 252.]

DiVincenzo, D. P. (2000). The physical implementation of quantum computation. *Fortschritte der Physik*, **48**, 771. [Cited on page 12.]

Dyakonov, M. (2019). The case against quantum computers. *IEEE Spectrum*, **56**, 24. [Cited on page 39.]

Ercsey-Ravasz, M. and Toroczkai, Z. (2011). Optimization hardness as transient chaos in an analog approach to constraint satisfaction. *Nature Physics*, **7**(12), 966. [Cited on pages 33 and 53.]

Finocchio, G., Di Ventra, M., Camsari, K. Y., Everschor-Sitte, K., Khalili-Amiri, P., and Zeng, Z. (2020). The promise of spintronics for unconventional computing. *Journal of Magnetism and Magnetic Materials*, **48**, 167506. [Cited on pages 309 and 315.]

Fomenko, A. T. (1994). *Visual Geometry and Topology*. Springer-Verlag, New York. [Cited on pages 14, 40, 46, 47, 53, 148, 229, 233, 249, 250, 278, 279, and 293.]

Freedman, M. H., Kitaev, A., Larsen, M. J., and Wang, Z. (2002). Topological quantum computation. *The Bulletin of the London Mathematical Society*, **40**, 31. [Cited on pages 14, 296, 320, and 321.]

Frenkel, E., Losev, A., and Nekrasov, N. (2007). Notes on instantons in topological field theory and beyond. *Nuclear Physics B*, **171**, 215. [Cited on page 260.]

Gamerman, D. and Lopes, H. F. (2006). *Markov Chain Monte Carlo: Stochastic Simulation for Bayesian Inference*. Chapman & Hall/CRC, Boca Raton, Florida. [Cited on pages 167 and 190.]

Garey, M. R. and Johnson, D. S. (1990). *Computers and Intractability; A Guide to the Theory of NP-Completeness*. W. H. Freeman & Co., New York. [Cited on pages 18, 22, 23, 24, 25, 27, 96, 176, 225, and 268.]

Gilmore, R. (1998). Topological analysis of chaotic dynamical systems. *Reviews of Modern Physics*, **70**, 1455. [Cited on pages 43, 52, and 54.]

Göckeler, M. and Schücker, T. (1989). *Differential Geometry, Gauge Theories, and Gravity*. Cambridge University Press, Cambridge, UK. [Cited on pages 229, 230, 231, 232, 251, and 293.]

Goldenfeld, N. (1992). *Lectures on Phase Tranitions and the Renormalization Group* (3rd edn). Perseus Publishing, New York. [Cited on pages 50, 63, 64, 65, 140, 166, 183, 244, 246, 289, and 319.]

Goldreich, O. (2001). *Foundations of Cryptography*. Cambridge University Press, Cambridge, UK. [Cited on pages 109, 111, 131, 165, and 314.]

Goldstein, H. (1965). *Classical Mechanics*. Addison-Wesley, Boston. [Cited on pages 9, 11, 40, 41, 48, 53, 54, 55, 167, 233, and 241.]

Goodfellow, I., Bengio, Y., and Courville, A. (2016). *Deep Learning*. Volume 1. MIT Press, Cambridge, MA. [Cited on pages 29, 181, 182, 183, 184, 185, 186, 187, 188, 191, 192, 193, 194, 207, 208, 209, 211, 212, and 214.]

Gottschalk, W. E. and Hedlund, G. A. (1955). *Topological Dynamics*. American Mathematical Society, Providence, Rhode Island. [Cited on page 52.]

Granholm, J. W. (1962). How to design a kludge. *Datamation*. [Cited on page 308.]

Grassberger, P. (1982). On phase transitions in Schlögl's second model. *Zeitschrift für Physik B Condensed Matter*, **47**(4), 365–374. [Cited on page 288.]

Gypens, P., Leliaert, J., Di Ventra, M., Van Waeyenberge, B., and Pinna, D. (2021). Nanomagnetic self-organizing logic gates. *Physical Review Applied*, **16**, 024055. [Cited on page 309.]

Hale, J. K. (2010). *Asymptotic Behavior of Dissipative Systems* (2nd edn). Volume 25, Mathematical Surveys and Monographs. American Mathematical Society, Providence, Rhode Island. [Cited on pages 102, 103, 104, 134, 135, 260, 264, 276, and 321.]

Hamming, R. W. (1997). *The Art of Doing Science and Engineering: Learning to Learn*. Taylor & Francis, Amsterdam. [Cited on pages 68 and 87.]

Hamze, F., Jacob, D. C., Ochoa, A. J., Perera, D., Wang, W., and Katzgraber, H. G. (2018). From near to eternity: spin-glass planting, tiling puzzles, and constraint-satisfaction problems. *Physical Review E*, **97**(4), 043303. [Cited on pages 168 and 169.]

Hardy, G. H. and Wright, E. M. (1979). *An Introduction to the Theory of Numbers*. Oxford University Press, Oxford. [Cited on page 110.]

Hartmann, A. K. and Rieger, H. (2004). *New Optimization Algorithms in Physics*. John Wiley and Sons, Inc., New York. [Cited on pages 155 and 162.]

Hartshorne, R. (2010). *Deformation Theory*. Springer-Verlag, New York. [Cited on page 323.]

Håstad, J. (2001). Some optimal inapproximability results. *Journal of the ACM*, **48**(4), 798–859. [Cited on pages 174 and 175.]

Hatcher, A. (2002). *Algebraic Topology*. Cambridge University Press, Cambridge, UK. [Cited on pages 264, 290, 295, 296, and 323.]

Haykin, S. (1998). *Neural Networks: A Comprehensive Foundation* (2nd edn). Prentice Hall PTR. [Cited on pages 85 and 181.]

Henkel, M., Hinrichsen, H., and Lübeck, S. (2008). *Non-Equilibrium Phase Transitions*. Springer-Verlag, New York. [Cited on pages 13, 244, 283, 284, 285, 288, 289, and 290.]

Hennessy, J. L. and Patterson, D. A. (2006). *Computer Architecture, Fourth Edition: A Quantitative Approach*. Morgan Kaufmann Publishers, Inc., Burlington, MA. [Cited on pages 8, 12, 13, 34, 42, 45, 59, 69, 80, 86, 87, 92, 161, and 174.]

Hilborn, R. C. (2001). *Chaos and Non-Linear Dynamics*. Oxford University Press, Oxford. [Cited on pages 50, 51, 52, 104, 237, and 282.]

Holland, J. H. (1998). *Emergence: From Chaos to Order*. Perseus Books, New York. [Cited on page 12.]

Hori, K., Katz, S., Klemm, A. Pandharipande, R., Thomas, R., Vafa, C., Vakil, R., and Zaslow, E. (2000). *Mirror Symmetry*. Clay Mathematics Institute, Cambridge, MA. [Cited on pages 240, 243, 253, 254, 271, 323, and 324.]

Horowitz, P. (2015). *The Art of Electronics.* Cambridge University Press, Cambridge, UK. [Cited on pages 34, 69, 78, 79, 120, 121, 122, 128, and 181.]

Houdayer, J. and Hartmann, A. K. (2004). Low-temperature behavior of two-dimensional Gaussian Ising spin glasses. *Physical Review B*, **70**(1), 014418. [Cited on page 171.]

Hradil, Z. (1997). Quantum-state estimation. *Physical Review A*, **55**(3), R1561. [Cited on pages 218 and 219.]

Huxley, J. S. and Tessier, G. (1936). Terminology of relative growth. *Nature*, **268**, 780–781. [Cited on page 298.]

Ielmini, D. and Wong, H.-S. P. (2018). In-memory computing with resistive switching devices. *Nature Electronics*, **1**, 333. [Cited on page 80.]

Iuculano, T., Rosenberg-Lee, M., Supekar, K., Lynch, C.J., Khouzam, A., Phillips, J., Uddin, L.Q., and Menon, V. (2014). Brain organization underlying superior mathematical abilities in children with autism. *Biological Phsychiatry*, **75**, 223. [Cited on page 301.]

Kandel, E. R. ed. (2013). *Principles of Neural Science.* McGraw-Hill. [Cited on page 300.]

Kim, J., Pershin, Y. V., Yin, M., Datta, T., and Di Ventra, M (2020). An experimental proof that resistance-switching memory cells are not memristors. *Advanced Electronic Materials*, **2000010**, 1–6. [Cited on pages 70 and 75.]

Kohonen, T. (1989). *Self-organization and Associative Memory* (3rd edn). Springer-Verlag, New York. [Cited on pages 59, 90, 92, 297, 298, 301, and 302.]

Koopman, B. O. (1931). Hamiltonian systems and transformations in Hilbert space. *Proceedings of the National Academy of Sciences USA*, **17**, 315. [Cited on page 229.]

Kubo, R. (1957). Statistical-mechanical theory of irreversible processes. I. General theory and simple applications to magnetic and conduction problems. *J. Phys. Soc. Japan*, **12**(6), 570–586. [Cited on pages 60, 61, 70, 71, 73, 265, and 296.]

Kuhn, T. S. (1962). *The Structure of Scientific Revolutions.* University of Chicago Press. [Cited on pages 3 and 325.]

Lanczos, C. (1986). *Variational Principles of Mechanics.* Dover Publications, New York. [Cited on pages 239 and 240.]

Landau, L. D. and Lifshitz, E. M. (1959). *Fluid Mechanics.* Pergamon Press, Oxford. [Cited on page 246.]

Landau, L. D. and Lifshitz, E. M. (1975). *The Classical Theory of Fields.* Butterworth-Heinemann, Oxford. [Cited on pages 40 and 105.]

Landauer, R. (1991). Information is physical. *Physics Today*, **44**, 23. [Cited on pages 4 and 112.]

Langton, C. G. (1990). Computation at the edge of chaos: phase transitions and emergent computation. *Physica D*, **42**, 12–37. [Cited on pages 13, 31, 66, and 271.]

Le Roux, N. and Bengio, Y. (2008). Representational power of restricted Boltzmann machines and deep belief networks. *Neural Computation*, **20**, 1631–1649. [Cited on pages 182, 187, and 216.]

Mahan, G. D. (2000). *Many-Particle Physics*. Springer-Verlag, New York. [Cited on page 318.]

Manukian, H., Bearden, S. R. B., Pei, Y. R., and Di Ventra, M. (2020). Mode-assisted unsupervised learning of restricted Boltzmann machines. *Communications Physics*, **3**, 105. [Cited on pages 161, 197, 199, 200, 201, 202, 203, 204, 205, 206, and 213.]

Manukian, H. and Di Ventra, M. (2021). Mode-assisted joint training of deep Boltzmann machines. *Scientific Reports*, **11**, 19000. [Cited on pages 210, 211, and 212.]

Manukian, H., Traversa, F. L., and Di Ventra, M. (2019). Accelerating deep learning with memcomputing. *Neural Networks*, **110**, 1. [Cited on pages 161, 196, 213, 215, and 312.]

Melchior, J., Fischer, A., and Wiskott, L. (2016). How to center deep Boltzmann machines. *The Journal of Machine Learning Research*, **17**(1), 3387. [Cited on page 212.]

Messiah, A. (1958). *Quantum Mechanics*. Dover Publications, New York. [Cited on pages 9, 28, 38, 39, 49, 61, 62, 104, 181, 216, 218, 234, 242, 250, 252, 272, 279, 316, and 323.]

Mezard, M and Montanari, A (2009). *Information, Physics, and Computation*. Oxford University Press, Oxford. [Cited on pages 141, 162, and 163.]

Milnor, J. (1963). *Morse Theory*. Princeton University Press, Princeton, New Jersey. [Cited on pages 259, 290, and 321.]

Misner, C. W., Thorne, K. S., and Wheeler, J. A. (1973). *Gravitation*. W. H. Freeman and Company, New York. [Cited on page 317.]

Mostafazadeh, A. (2002). Pseudo-Hermiticity *versus* PT-symmetry III: Equivalence of pseudo-Hermiticity and the presence of anti-linear symmetries. *J. Math. Phys.*, **43**, 3944–3951. [Cited on page 230.]

Muñoz, M. A. (2018). Colloquium: Criticality and dynamical scaling in living systems. *Reviews of Modern Physics*, **90**, 031001. [Cited on pages 66 and 300.]

Nielsen, M. A. and Chuang, I. L. (2010). *Quantum Computation and Quantum Information* (10th Anniversary edn). Cambridge University Press, Cambridge, UK. [Cited on pages 38, 61, 62, 94, 105, 215, and 217.]

Ott, E. (1993). *Chaos in Dynamical Systems*. Cambridge University Press, Cambridge, UK. [Cited on pages 52, 55, and 282.]

Ovchinnikov, I. V. (2016). Introduction to supersymmetric theory of stochastics. *Entropy*, **18**, 108. [Cited on pages 228, 229, 232, 239, 271, 296, 318, 321, and 324.]

Ovchinnikov, I. V. and Di Ventra, M. (2019). Chaos as a symmetry-breaking phenomenon. *Modern Physics Letters B*, **33**, 1950287. [Cited on pages 51, 52, 233, 294, and 324.]

Pei, Y. R., Traversa, F. L., and Di Ventra, M. (2019). On the universality of MemComputing machines. *IEEE Transactions on Neural Networks and Learning Systems*, **30**(6), 1610. [Cited on page 93.]

Pei, Y. R. and Di Ventra, M. (2021). Non-equilibrium criticality and efficient exploration of glassy landscapes with memory dynamics. *arXiv:2102.04557*. [Cited on pages 161, 168, 169, and 306.]

Pei, Y. R., Manukian, H., and Di Ventra, M. (2020). Generating weighted MAX-2-SAT instances of tunable difficulty with frustrated loops. *Journal of Machine Learning Research*, **21**, 1. [Cited on pages 145 and 203.]

Peotta, S. and Di Ventra, M. (2014). Superconducting memristors. *Physical Review Applied*, **2**, 034011. [Cited on page 75.]

Perko, L. (2001). *Differential Equations and Dynamical Systems* (3nd edn). Volume 7. Springer Science & Business Media, New York. [Cited on pages 37, 41, 46, 50, and 51.]

Pershin, Y. V. and Di Ventra, M. (2008). Spin memristive systems: Spin memory effects in semiconductor spintronics. *Physical Review B*, **78**, 113309. [Cited on page 38.]

Pershin, Y. V. and Di Ventra, M. (2010). Experimental demonstration of associative memory with memristive neural networks. *Neural Networks*, **23**(7), 881–886. [Cited on page 76.]

Pershin, Y. V. and Di Ventra, M. (2011a). Memory effects in complex materials and nanoscale systems. *Advances in Physics*, **60**, 145–227. [Cited on pages vii, 67, 73, 74, 76, 115, 122, and 164.]

Pershin, Y. V. and Di Ventra, M. (2011b). Solving mazes with memristors: A massively parallel approach. *Physical Review E*, **84**, 046703. [Cited on page 298.]

Pershin, Y. V. and Di Ventra, M. (2013). Self-organization and solution of shortest-path optimization problems with memristive networks. *Physical Review E*, **88**, 013305. [Cited on pages 298 and 301.]

Pershin, Y. V. and Di Ventra, M. (2016). MemComputing implementation of ant colony optimization. *Neural Processing Letters*, **44**, pages265. [Cited on page 298.]

Pershin, Y. V., La Fontaine, S., and Di Ventra, M. (2009). Memristive model of amoeba learning. *Physical Review E*, **80**, 021926. [Cited on page 59.]

Peskin, M. E. and Schroeder, D. V. (1995). *An Introduction to Quantum Field Theory*. Westview Press, Boulder. [Cited on pages 31, 52, 67, 229, 231, 234, 236, 239, 241, 244, 256, 316, 317, 318, and 320.]

Pettini, M. (2007). *Geometry and Topology in Hamiltonian Systems and Statistical Mechanics*. Springer-Verlag, New York. [Cited on page 289.]

Pfeiffer, P., Egusquiza, I. L., Di Ventra, M., Sanz, M., and Solano, E. (2015). Quantum memristors. *Scientific Reports*, **6**, 29507. [Cited on page 314.]

Plattner, H. and Zeier, A. (2011). *In-Memory Data Management: Technology and Applications*. Springer-Verlag, New York. [Cited on pages 79 and 80.]

Polyakov, A. M. (1977). Quark confinement and topology of gauge theories. *Nuclear Physics B*, **120**, 429. [Cited on page 293.]

Poole, D., Mackworth, A., and Goebel, R. (1998). *Computational Intelligence: A Logical Approach*. Oxford University Press, Oxford. [Cited on page 28.]

Press, W. H., Teukolsky, S. A., Vetterling, W. T., and Flannery, B. P. (2007). *Numerical Recipes: The Art of Scientific Computing*. Cambridge University Press, Cambridge, UK. [Cited on pages 41, 150, 152, and 276.]

Pruessner, G. (2012). *Self-Organised Criticality: Theory, Models and Characterisation*. Cambridge University Press, Cambridge, UK. [Cited on pages 228, 274, and 275.]

Rajaraman, R. (1982). *Solitons and Instantons: An Introduction to Solitons and Instantons in Quantum Field Theory*. North Holland. [Cited on pages 229, 234, 235, 237, 238, 241, 262, and 263.]

Rautenberg, W. (2009). *A Concise Introduction to Mathematical Logic*. Springer-Verlag, New York. [Cited on pages 25 and 109.]

Saigusa, T., Tero, A., Nakagaki, T., and Kuramoto, Y. (2008). Amoebae anticipate periodic events. *Physical Review Letters*, **100**, 018101. [Cited on page 59.]

Salakhutdinov, R. and Hinton, G. (2009). Deep Boltzmann machines. In *Artificial Intelligence and Statistics*, pp. 448–455. [Cited on pages 208 and 212.]

Salakhutdinov, R. and Hinton, G. (2012). An efficient learning procedure for deep Boltzmann machines. *Neural Computation*, **24**, 1967–2006. [Cited on page 207.]

Sauer, T. (2018). *Numerical Analysis*. Pearson, New York. [Cited on pages 147, 148, 150, 152, 153, 201, 224, 269, 278, and 280.]

Schwarz, A. S. (1993). *Quantum Field Theory and Topology*. Springer-Verlag, New York. [Cited on pages 229, 237, 240, 262, 292, and 293.]

Sheldon, F., Cicotti, P., Traversa, F. L., and Di Ventra, M. (2020). Stress-testing MemComputing on hard combinatorial optimization problems. *IEEE Transactions on Neural Networks and Learning Systems*, **31**, 2222. [Cited on pages 175, 176, and 177.]

Sheldon, F., Traversa, F. L., and Di Ventra, M. (2019). Taming a nonconvex landscape with long-range order: MemComputing the Ising spin glass. *Physical Review E*, **100**, 053311. [Cited on pages 161, 170, 171, 172, 247, 263, and 269.]

Sherwood, L. (2015). *Human Physiology: From Cells to Systems*. Cengage Learning, Boston. [Cited on pages 59, 297, and 299.]

Shor, P. W. (1994). Algorithms for quantum computation: Discrete logarithms and factoring. In *Foundations of Computer Science, 1994 Proceedings., 35th Annual Symposium on*, pp. 124–134. [Cited on page 43.]

Shor, P. W. (1997). Polynomial-time algorithms for prime factorization and discrete logarithms on a quantum computer. *SIAM Journal on Computing*, **26**(5), 1484–1509. [Cited on pages 11, 38, 61, 130, and 131.]

Slipko, V. A., Pershin, Y. V., and Di Ventra, M. (2013). Changing the state of a memristive system with white noise. *Physical Review E*, **87**, 042103. [Cited on page 126.]

Stuart, A. M. (1994). Numerical analysis of dynamical systems. *Acta Numerica*, **3**(1), 467. [Cited on pages 148, 156, 278, 279, and 322.]

Tanner, J. C. (1961). A derivation of the Borel distribution. *Biometrika*, **48**(1-2), 222–224. [Cited on page 273.]

Thompson, N.C., Greenewald, K., Lee, K., and Manso, G.F. (2021). Deep learning diminishing's returns. *IEEE Spectrum*. [Cited on page 186.]

Toffoli, T. and Margolus, N. (1987). *Cellular Automata Machines: A New Environment for Modeling*. MIT Press, Cambridge, MA. [Cited on page 66.]

Touboul, J. and Destexhe, A. (2017). Power-law statistics and universal scaling in the absence of criticality. *Physical Review E*, **95**, 012413. [Cited on pages 66 and 300.]

Traversa, F. L. (2019). Aircraft loading optimization: MemComputing the 5th Airbus problem. *arXiv:1903.08189*. [Cited on pages 178 and 179.]

Traversa, F. L., Bonani, F., Pershin, Y. V., and Di Ventra, M. (2014). Dynamic computing random access memory. *Nanotechnology*, **25**, 285201. [Cited on pages 76, 80, 92, 302, and 310.]

Traversa, F. L., Cicotti, P., Sheldon, F., and Di Ventra, M. (2018). Evidence of exponential speed-up in the solution of hard optimization problems. *Complexity*, **2018**, 7982851. [Cited on pages 161, 174, and 175.]

Traversa, F. L. and Di Ventra, M. (2015). Universal MemComputing machines. *IEEE Transactions on Neural Networks and Learning Systems*, **26**(11), 2702. [Cited on pages 77, 85, 87, 91, 93, 94, and 297.]

Traversa, F. L. and Di Ventra, M. (2017). Polynomial-time solution of prime factorization and NP-complete problems with digital MemComputing machines. *Chaos: An Interdisciplinary Journal of Nonlinear Science*, **27**, 023107. [Cited on pages 28, 76, 90, 97, 98, 100, 101, 104, 107, 113, 114, 119, 120, 121, 122, 125, 127, 128, 130, 131, 133, 134, 136, 151, 152, 161, 164, 226, 227, 265, 267, and 308.]

Traversa, F. L. and Di Ventra, M. (2018). MemComputing integer linear programming. *arXiv:1808.09999*. [Cited on pages 113, 118, 127, 161, 176, and 178.]

Traversa, F. L., Ramella, C., Bonani, F., and Di Ventra, M. (2015). MemComputing NP-complete problems in polynomial time using polynomial resources and collective states. *Science Advances*, **1**(6), e1500031. [Cited on pages 86 and 307.]

Trefethen, L. N. and Bau III, D. (1997). *Numerical Linear Algebra*. SIAM. [Cited on pages 280 and 281.]

Turing, A. M. (1936). On computational numbers, with an application to the entscheidungsproblem. *Proceedings of the London Mathematical Society*, **42**, 230–265. [Cited on pages 3 and 17.]

Turing, A. M. (2004). *The Essential Turing: Seminal Writings in Computing, Logic, Philosophy, Artificial Intelligence, and Artificial Life, Plus The Secrets of Enigma*. Oxford University Press, Oxford. [Cited on pages 18, 25, and 93.]

van Kampen, N.G. (1992). *Stochastic Processes in Physics and Chemistry*. Elsevier, Amsterdam. [Cited on pages 185, 190, 226, 230, 273, 274, 276, and 277.]

Van Oorschot, P. C. and Wiener, M. J. (1994). Parallel collision search with application to hash functions and discrete logarithms. In *Proceedings of the 2nd ACM Conference on Computer and Communications Security*, pp. 210–218. [Cited on page 314.]

van Rooij, D. *et al.* (2018). Cortical and subcortical brain morphometry differences between patients with autism spectrum disorder and healthy individuals across the lifespan: Results from the ENIGMA

ASD working group. *American Journal of Psychiatry*, **175**, 359. [Cited on page 301.]

von Neumann, J. (1932). Zur operatorenmethode in der klassischen mechanik. *Annals of Mathematics*, **33**, 587. [Cited on page 229.]

von Neumann, J. (1993). First draft of a report on the EDVAC. *Annals of the History of Computing, IEEE*, **15**(4), 27–75. [Cited on page 86.]

Wegner, F. (2016). *Supermathematics and its Application to Statistical Physics*. Springer-Verlag, New York. [Cited on pages 250, 251, and 254.]

Wen, X.-G. (2013). Topological order. From long-range entangled quantum matter to a unified origin of light and electrons. *International Scholarly Research Notices*, **2013**, 198710. [Cited on pages 63, 65, 295, and 316.]

White, S. R. (1992). Density matrix formulation for quantum renormalization groups. *Physical Review Letters*, **69**, 2863. [Cited on page 220.]

Witten, E. (1988). Topological quantum field theory. *Communications in Mathematical Physics*, **117**, 353–386. [Cited on page 229.]

Yang, Z., Tackett, A., and Di Ventra, M. (2002). Variational and nonvariational principles in quantum transport calculations. *Physical Review B*, **66**, 041405(R). [Cited on page 216.]

Zhang, Y.-H. and Di Ventra, M. (2021). Directed percolation and numerical stability of simulations of digital MemComputing machines. *Chaos: An Interdisciplinary Journal of Nonlinear Science*, **31**, 063127. [Cited on pages 140, 155, 156, 157, 158, 161, 284, 285, 286, and 288.]

Index

FREE online study and revision support available at
www.oup.com/lawrevision

Take your learning further with:

- Multiple-choice questions with instant feedback
- Interactive glossaries and flashcards of key cases
- Tips, tricks and audio advice
- Annotated outline answers
- Diagnostic tests show you where to concentrate
- Extra questions, key facts checklists, and topic overviews

unique features

student-focused online support

New to this edition

- Fully updated in light of recent key case law by the International Court of Justice, such as *Gambia v Myanmar*, Advisory Opinion on *Chagos Archipelago, Palestine v USA (Relocation of US Embassy)*, and *Immunities and Criminal Proceedings (Equatorial Guinea v France)*.
- Includes international developments such as self-determination debates and the contractualisation of international law.
- A brand new chapter on international criminal law.

Acknowledgements

The authors would like to thank the editorial and commissioning team at Oxford University Press, who worked tirelessly and with meticulous dedication to see this work to fruition, as well as Dr Ioannis Kalpouzos and Dr Solon Solomon for commenting on certain chapters. Ilias Bantekas is responsible for chapters 4–7 and 11–13 and Efthymios Papastavridis for the rest.